Application of Information Theory to Computer Vision and Image Processing

Application of Information Theory to Computer Vision and Image Processing

Editors

Wendy Flores-Fuentes
Oleg Sergiyenko
Julio Cesar Rodríguez-Quiñonez
Jesús Elías Miranda-Vega

MDPI

Basel • Beijing • Wuhan • Barcelona • Belgrade • Novi Sad • Cluj • Manchester

Editors

Wendy Flores-Fuentes
Facultad de Ingeniería
Universidad Autónoma de Baja California
Mexicali, Baja California
Mexico

Oleg Sergiyenko
Instituto de Ingeniería
Universidad Autónoma de Baja California
Mexicali, Baja California
Mexico

Julio Cesar Rodríguez-Quiñonez
Facultad de Ingeniería
Universidad Autónoma de Baja California
Mexicali, Baja California
Mexico

Jesús Elías Miranda-Vega
ITM de Mexicali
Tecnológico Nacional de México
Mexicali, Baja California
Mexico

Editorial Office

MDPI
St. Alban-Anlage 66
4052 Basel, Switzerland

This is a reprint of articles from the Special Issue published online in the open access journal *Entropy* (ISSN 1099-4300) (available at: www.mdpi.com/journal/entropy/special_issues/MWI13854O7).

For citation purposes, cite each article independently as indicated on the article page online and as indicated below:

Lastname, A.A.; Lastname, B.B. Article Title. *Journal Name* **Year**, *Volume Number*, Page Range.

ISBN 978-3-7258-0292-0 (Hbk)
ISBN 978-3-7258-0291-3 (PDF)
doi.org/10.3390/books978-3-7258-0291-3

Contents

About the Editors

Wendy Flores-Fuentes

Wendy Flores-Fuentes received her bachelor's degree in electronic engineering from the Autonomous University of Baja California in 2001; her master's degree in engineering from the Technological Institute of Mexicali in 2006; and her Ph.D. degree in science, applied physics, with emphasis on optoelectronic scanning systems for SHM, from the Autonomous University of Baja California in June 2014. She is currently the author of 33 journal articles in Elsevier, IEEE, Emerald and Springer; 18 book chapters and 8 books in Springer, Intech, IGI Global Lambert Academic and Springer; and 43 proceedings articles in IEEE ISIE 2014–2020, IECON 2014, 2018, 2019, the World Congress on Engineering and Computer Science (IAENG 2013), IEEE Section Mexico IEEE ROCC2011, and the VII International Conference on Industrial Engineering ARGOS 2014. Recently, she organized and participated as the Chair of a Special Session on "Machine Vision, Control and Navigation" at IEEE ISIE 2015, 2016, 2017, 2019, 2020, and 2021 and IECON 2018 and 2019. She holds one patent in Mexico and one patent in Ukraine. She has been a reviewer of several articles in Taylor and Francis, IEEE, Elsevier, and EEMJ. Currently, she is a full-time professor at Universidad Autónoma de Baja California, in the Faculty of Engineering. She was incorporated into CONACYT National Research System in 2015. She received the award for "Best session presentation" at WSECS2013 in San-Francisco, USA, and she received, as a coauthor, the award for "Outstanding Paper in the 2017 Emerald Literati Network Awards for Excellence". Her interests include optoelectronics, robotics, artificial intelligence, measurement systems, and machine vision systems.

Oleg Sergiyenko

Oleg Sergiyenko received his B.S. and M.S. degrees at the Kharkiv National University of Automobiles and Highways, Kharkiv, Ukraine, in 1991 and 1993, respectively. He received his Ph.D. degree at the Kharkiv National Polytechnic University, with a specialization in "Tools and methods of non-destructive control" in 1997. He received his DSc. (habit.) degree at the Kharkiv National University of Radioelectronics in 2018. He is the author of 1 book, was the editor of 8 books (in Springer, IGI Global, etc.), has written 31 book chapters and over 150 papers indexed in Scopus, and holds 3 patents (from Ukraine and Mexico). From 1994 to the present, he has presented his research at several international congresses such as IEEE, ICROS, SICE, and IMEKO in the USA, England, Japan, Italy, Austria, Ukraine, Canada, Portugal, Brazil, and Mexico. At many of these congresses, he was a Session Chair. He received the "Best presentation award" at the IEEE conferences IECON2014 in Dallas, USA; IECON2016 in Florence, Italy; and ISIE2019 in Vancouver, Canada. He also received the 2017 Outstanding Paper Emarald Literati Award for his article published in *Industrial Robot: an International Journal.*

In December 2004, Dr. Sergiyenko was invited to join the Engineering Institute of Baja California Autonomous University as a researcher, but he is now the current head of the Applied Physics Department at the Engineering Institute of Baja California Autonomous University, Mexico, and the director of several master's and doctoral theses. He is a full member (Academician) of the Academy of Applied Radioelectronics of Belorussia, Ukraine, and Russia. His current research interests include automated metrology, machine vision systems, fast electrical measurements, control systems, robot navigation, and 3D laser scanners.

Julio Cesar Rodríguez-Quiñonez

Julio Cesar Rodríguez-Quiñonez received his B.S. degree at CETYS, Mexico, in 2007 and his Ph.D. degree from Baja California Autonomous University, México, in 2013. He is currently a full-time researcher–professor in the Engineering Faculty of the Autonomous University of Baja California and a member of the National Research System Level 1. Since 2016, he has been a senior member of IEEE. He is involved in the development of optical scanning prototypes in the Applied Physics Department and is a research leader in the development of a new stereo vision system prototype. He has been the thesis director of two doctorate degree students and three master's degree students. He holds two patents for a dynamic triangulation method; has been the editor of four books and a guest editor of the *IEEE Sensors Journal, the International Journal of Advanced Robotic Systems,* and the *Journal of Sensors*; has written over 70 papers and 8 book chapters; has been a reviewer for the *IEEE Sensors Journal, Optics and Lasers in Engineering, IEEE Transaction on Mechatronics and Neural Computing and Applications of Springer*; and has participated as a reviewer and Session Chair for IEEE ISIE conferences in 2014 (Turkey), 2015 (Brazil), 2016 (USA), 2017 (UK), 2019 (Canada), IECON 2018 (USA), IECON 2019 (Portugal), ISIE 2020 (Netherlands), and ISIE 2021 (Kyoto). His current research interests include automated metrology, stereo vision systems, control systems, robot navigation, and 3D laser scanners.

Jesús Elías Miranda-Vega

Jesús Elías Miranda-Vega received his B.S. degree in electrical and electronic engineering from Tecnológico Nacional de México/IT de Los Mochis, in 2007; his master's degree in electronic engineering from Tecnológico Nacional de México/IT de Mexicali, in 2014; and his Ph.D. degree in science and applied physics from the Autonomous University of Baja California, in December 2019, with Honorable Mention (Cum Laude). He has written five book chapters, and several journal and conference proceeding papers. He has also been the co-director thesis of a master's degree student. He was incorporated into the CONACYT National Research System in 2021. His research interests include machine vision, machine learning, data signal processing, optoelectronics theory and devices, and their applications.

entropy

Editorial

Application of Information Theory to Computer Vision and Image Processing

Wendy Flores-Fuentes [1,*], Oleg Sergiyenko [2], Julio C. Rodríguez-Quiñonez [1] and Jesús E. Miranda-Vega [3]

1 Facultad de Ingeniería, Universidad Autónoma de Baja California, Mexicali 21100, Mexico; julio.rodriguez81@uabc.edu.mx
2 Instituto de Ingeniería, Universidad Autónoma de Baja California, Mexicali 21100, Mexico; srgnk@uabc.edu.mx
3 Tecnológico Nacional de México, IT de Mexicali, Mexicali 21376, Mexico; elias.miranda@itmexicali.edu.mx
* Correspondence: flores.wendy@uabc.edu.mx

1. Introduction

Our perception of the world is the product of the human visual system's complex optical and physical process. When we open our eyes, light stimuli enter our pupils, which are the gateway to our visual experience.

These incoming rays of light then pass through the various structures of the eye, such as the cornea and lens, which help the light to focus onto the retina. The retina, located at the back of the eye, is a crucial component in the process of perceiving the world. It is composed of specialized cells called photoreceptors, namely rods and cones. Rods are responsible for vision in low-light conditions and help us perceive shades of gray, while cones enable us to see colors and function best in bright light.

As light reaches the retina, the photoreceptors initiate a remarkable transformation. They convert the incoming light into electrochemical signals that can be transmitted to the brain through the optic nerve. This process involves the absorption of light by pigments in the photoreceptor cells, triggering a cascade of chemical reactions that generate electrical impulses.

The transmitted electrical signals, laden with visual information, travel along the optic nerve to the visual cortex in the brain. Here, the incoming data undergo a complex process that allows us to organize, interpret, and analyze the information received. The brain seamlessly integrates this visual input with other sensory cues, such as auditory and tactile information, to create a coherent and multi-dimensional perceived reality. It is important to note that perception is not a direct replication of the external world but rather a constructed representation based on the available sensory input. Factors like individual differences in perception, attention, and previous experiences can shape how we interpret and make sense of the visual information received.

The process underlying humans' perception of the world involves intricate interplay between the eye's optical components, the retina's photoreceptors, and the brain's complex neural networks. Together, they transform light into meaningful visual experiences, allowing us to navigate and interact with the world around us.

In a similar way to the intricate optical and physical processes of human vision, machine vision serves as the "eyes" of cybernetic systems. Machine vision refers to technology that enables machines to process and interpret visual information, much like how human eyes perceive and understand their surroundings, facilitating the coexistence of the virtual and real world in our daily lives. Cybernetic systems are involved in multiple disciplines, and they address the emerging challenges of managing the information provided from the virtual and physical world to offer solutions that adhere to human needs and demands [1]. Machine vision, as a part of cybernetic systems, is vital for enabling these systems to

Citation: Flores-Fuentes, W.; Sergiyenko, O.; Rodríguez-Quiñonez, J.C.; Miranda-Vega, J.E. Application of Information Theory to Computer Vision and Image Processing. *Entropy* **2024**, *26*, 114. https://doi.org/10.3390/e26020114

Received: 13 January 2024
Accepted: 19 January 2024
Published: 26 January 2024

navigate and interact within both virtual and real-world environments in diverse applications, including in smart cities, factories, and homes, via monitoring, analyzing, and controlling machinery, devices, and objects based on end-to-end data collected by smart sensors connected to the internet and a cloud network [2].

Machine vision systems are based on technologies that strive for seamless integration into our lives, are driven by creativity and a global perspective, are enabled by the power of the intelligent interconnectivity of several surrounding environments related to an application [3], and are continuously evolving due to ongoing research and technological innovations, including improvements in efficiency, accuracy, and the development of novel information theories for computer vision and image processing models [4–6] and applications like those based on collaborative multi-agent approaches applied mainly in swarm robotics [7].

This remarkable collaboration between agents and the fusion of their information has been made possible through the advancement of sensor technologies and sophisticated systems that acquire and process vast amounts of information through the Internet of Things [8–10]. Machine vision relies on a harmonious amalgamation of optoelectronics devices, sensors, cameras, and technical vision systems. These components work together to capture visual data, which form the foundation for subsequent analysis and interpretation. In this era of big data, the main technological challenges are related to handling high-throughput tasks that are both complex and efficient, which requires the development of new materials, new operational principles, and new designs to fulfil the requirements. These developments require the mimicking of the relationship between the structures and functions found in the human visual system, demonstrating significant potential for efficiently processing optical information while consuming minimal power [11].

The field of machine vision encompasses a diverse range of technologies and methodologies, including artificial intelligence algorithms like deep learning algorithms and neural networks for recognizing [12] and classifying objects in images or videos [13], enhancing image quality and reducing noise in images [14], and 3D vision and depth sensing [15]. These algorithms are robust and adaptable, and they are used in embedded systems [16], robust control mechanisms [17], inertial navigation systems, robotics, interconnectivity, big data applications, and cloud computing applications [18]. These elements are at the core of machine vision advancements, enabling cyber–physical systems to collaborate with humans in both their real and virtual environments and activities [19].

Sensors play a pivotal role in machine vision, acting as the first point of contact for acquiring data from the environment. These carefully designed and calibrated sensors are capable of detecting and measuring various physical properties, such as light, temperature, pressure, and motion. The acquired data are then processed through sophisticated algorithms and computer vision techniques, which extract meaningful information and patterns from the raw sensory input [20].

Artificial intelligence (AI) algorithms, a driving force behind machine vision, allow systems to understand, interpret, and make decisions based on the captured data. These algorithms leverage deep learning, neural networks, and pattern recognition to discern objects, recognize faces, analyze scenes, and even predict future events. The integration of AI algorithms empowers machine vision systems to adapt and learn from their interactions with the environment, continuously improving their performance and enhancing their ability to assist humans in diverse tasks [21].

Embedded systems and robust control mechanisms ensure the seamless integration and synchronization of various components within machine vision systems. These systems coordinate the operation of sensors, cameras, actuators, and other peripherals, ensuring precise data acquisition and processing. By tightly controlling the system's behavior, machine vision can deliver accurate and reliable results, even in challenging and dynamic environments.

Interconnectivity, big data, and cloud computing further augment the capabilities of machine vision systems. The ability to connect to the internet and share data allows for real-time collaboration, remote monitoring, and the analysis of visual information.

With the integration of cloud computing, machine vision systems can access vast computing resources and leverage sophisticated algorithms for complex tasks such as object recognition, scene understanding, and predictive analytics. This interconnected ecosystem facilitates seamless communication between cyber–physical systems, enabling humans to simultaneously interact with the virtual and real worlds [22].

2. An Overview of Published Articles

This Special Issue collates articles on information theory, measurement methods, data processing tools, and techniques for the design of machine vision systems and the instrumentation used in machine vision systems via the application of computer vision and image processing. Short summaries for each of the articles included within this Special Issue are provided below.

In the article by Garcia-Gonzalez et al. (contribution 1), a novel signal processing method is proposed for a technical vision system in order to deal with random fluctuations in electrical voltages during data acquisition, specifically the acquisition of an optoelectrical signal. An information theory-based method centering around the use of Shannon Entropy for extracting the features of optical patterns is presented to deal with the random processes presented in the acquisition of the signal. It is implemented in structural health monitoring to augment the accuracy of optoelectronic signal classifiers for a metrology subsystem of the technical vision system in order to enhance the system's spatial coordinate measurement performance under real operation conditions in noisy electrical and optical environments, as well as to better estimate structural displacement and for an improved estimation of its health. In this study, five different machine learning (ML) techniques were used to classify the optical patterns captured. Linear predictive coding (LPC) and the autocorrelation function (ACC) were used for the extraction of optical patterns. The Shannon entropy segmentation (SH) method was used to extract relevant information from optical patterns, and the model's performance was shown to be improved. The results reveal that segmentation with Shannon entropy achieved over 95.33% accuracy. Without Shannon entropy, the worst accuracy was 33.33%.

Wei et al. (contribution 2) propose a low-illumination image enhancement method based on structural and detail layer images to improve an image's brightness while effectively maintaining the texture and details of the image, guaranteeing a high-quality image. A network called the SRetinex-Net model was designed and subsequently divided into two parts: a decomposition module and an enhancement module. The decomposition module mainly adopts the SU-Net structure, which is an unsupervised network that decomposes the input image into a structural layer image and detail layer image. The enhancement module mainly adopts the SDE-Net structure, which is divided into two branches: the SDE-S branch and the SDE-D branch. The SDE-S branch mainly enhances and adjusts the brightness of the structural layer image through Ehnet and Adnet to prevent insufficient or excessive enhancements of the brightness of the image. The SDE-D branch was denoised and enhanced with textural details through the use of a denoising module. The results of numerous experiments show that the proposed structure has a more significant impact on the brightness and detail preservation of restored images.

Stasenko et al. (contribution 3) present a promising approach for food quality control during the postharvest stage that leverages the power of Generative Adversarial Network (GAN) and Convolutional Neural Network (CNN) techniques to use synthesized and segmented Visible Near-infrared (VNIR) imaging data ("400–1100 nm") collected under various environmental conditions (temperature and humidity) for early postharvest decay and fungal zone predictions, as well as for assessing the quality of stored food. Synthesized images were obtained via the pairing of Visible (V) "380–700 nm" images and Near-infrared (NIR) "780–2500 nm" images. By achieving accurate predictions and segmenting the decay and fungal zones, this approach offers significant advantages over traditional methods. NIR imagery provides detailed information about the diseased areas in stored fruits, which is why the hyperspectral cameras containing thousands of bands are used for food quality monitoring at postharvest stages. However, hyperspectral devices are expensive and are not suitable for use among farmers and sellers. Future research directions may include further comparisons with existing methodologies, exploring its applicability to different crops and storage conditions, and evaluating scalability for larger and more diverse datasets. The authors concluded that by harnessing deep learning (DL) and computer vision (CV) techniques in precision agriculture, significant strides forward in reducing food losses and ensuring a sustainable and secure food supply chain can be made.

Haipeng et al. (contribution 4) asserted that infrared and visible image fusion methods can be used to address the challenges of low-light scenes. This paper addresses the challenges of weak textural details, low-contrast infrared targets, and poor visual perception in existing deep learning fusion algorithms for low-light visible images to generate high-quality fused images under the conditions for such scenes. The authors propose a novel fusion method that exploits the characteristics of infrared and visible images to generate high-quality fused images under such conditions. The methodology followed consisted of the design of a Multi-Scale Edge Gradient Module (MEGB), which extracts texture information from both infrared and visible images. Additionally, they employed the Salient Dense Residual Module (SRDB) to extract salient features through pre-training with salient loss. The saliency map obtained from the SRDB was incorporated into the overall network training process. To fuse global and local information, the authors proposed the Spatial Bias Module (SBM). Extensive comparison experiments with existing methods were conducted to validate the effectiveness of the proposed approach in describing target features and global scenes. The results of the ablation experiments demonstrate the efficacy of the proposed modules. Furthermore, the authors evaluated the method's facilitation for high-level vision tasks, specifically semantic segmentation in diverse low-light scene images. The proposed method was evaluated qualitatively and quantitatively on three datasets: TNO, MSRS, and M3FD. The authors compared their method with seven other fusion algorithms to demonstrate its superiority. The evaluation metrics used include Standard Deviation (SD), Visual Information Fidelity (VIF), Average Gradient (AG), Difference Correlation Sum (DCS), Entropy (EN), and Structure Fidelity (SF). However, the authors acknowledge that their method has limitations, including its inability to remove the overexposure effect caused by strong light interference. The results of the comprehensive evaluation and comparison experiments validate the proposed method's superiority over existing algorithms.

Yichun et al. (contribution 5) aimed to reconstruct high-frequency details in the images of a scene by applying the thermal infrared image super-resolution method. They proposed an improved thermal infrared image super-resolution reconstruction method to solve the problem of poor image quality caused by the imaging mechanisms related to imaging sensors, such as motion blur, optical blur, and electronic noise, which lead to degradation in the quality of infrared images. The proposed method is based on multimodal sensor fusion; as inputs, it uses low-resolution (LR) versions of infrared images, visible light images as the reference images, and high-resolution (HR) versions of infrared images to obtain a super-resolution (SR) image. Primary feature encoding, super-resolution reconstruction, and high-frequency detail fusion subnetworks were also included in this study. The network incorporates hierarchical dilated distillation modules and a cross-attention transformation module to extract and transmit image features effectively. A hybrid loss function was introduced to guide the network in extracting salient features from both thermal infrared and reference images while maintaining accurate thermal information. Additionally, a learning strategy is proposed to ensure high-quality super-resolution reconstruction performance, even in the absence of reference images.

The identification of text clusters under the sparsity of feature points derived from characters was achieved by Huei-Yung Lin and Chin-Yu Hsu in contribution 6. The proposed method was applied to invoices and banknotes for text region detection. The proposed approach involves the distillation of local image features combined with clustering analysis to identify meaningful regions of interest. This approach incorporates application-specific reference images for feature learning and extraction, enabling the identification of text clusters even in the presence of sparse character features. The method involves calculating clusters with high feature density and iteratively expanding the regions of interest for complete text coverage (feature extraction, clustering analysis, and region selection), enabling the detection of text clusters despite sparse feature points in real-world applications (adaptability to various application scenarios, including regions with different orientations, size changes, or perspective distortions), as it can achieve fast detection using limited computational resources. Unlike deep neural network approaches, it does not require extensive model training or high computational power, making it easily implementable with hardware-oriented acceleration. Additionally, a multi-stage algorithm with a robust receptor descriptor is presented for character recognition. The technique offers fast region detection and can be implemented with hardware acceleration. However, one limitation of the proposed approach is that its detection capability is limited to man-made structures. The authors state that their future work will center around investigating structural patterns in natural scenes, specifically for agriculture applications.

In contribution 7, Zheng, Siming, Mingyu Zhu, and Mingliang Chen propose a method called the hybrid multi-dimensional attention U-Net (HMDAU-Net) for reconstructing hyperspectral images from a single-shot 2D measurement in the context of spectral snapshot compressive imaging (SCI). The traditional methods for capturing spatial–spectral information involve scanning-based techniques, while SCI utilizes compressive sensing to capture 3D spatial–spectral data efficiently in a single measurement. However, the reconstruction process of retrieving the 3D cube from the 2D measurement is a challenging problem. The HMDAU-Net addresses this challenge by integrating 3D and 2D convolutions in an encoder–decoder structure, striking a balance between computational cost and performance. The network incorporates attention gates to highlight important features and suppress noise from skip connections. The authors observe that, for SCI reconstruction tasks, the depth of the backbone network (e.g., U-Net) is not as crucial as its width (number of kernels in each layer) in achieving good results. This observation is attributed to the difference in tasks between image reconstruction and image classification. Additionally, the attention gate is employed to extract essential correlations in the spectral data cube and improve the reconstruction performance of the network. Furthermore, the authors suggest that the HMDAU-Net could potentially be applied in tasks related to other domains, such as medical imaging, image compression, temporal compressive coherent diffraction imaging, and video compressive sensing.

As described by Pang, Xiyu, Yilong Yin, and Yanli Zheng in contribution 8, vehicle re-identification across multiple cameras is one of the main problems of intelligent transportation systems (ITSs) due to the small differences in appearance between vehicles of the same model and the significant changes in appearance that arise when viewing from different viewpoints. In this study, a model called multi-receptive field soft attention part learning (MRF-SAPL) was established by learning semantically diverse vehicle part-level features under different receptive fields through multiple local branches. In this model, soft attention is used to adaptively locate the positions of the vehicle parts on the final feature map, ensuring alignment and maintaining internal semantics. In particular, the soft-attention part learning module (SAPL) in this model does not require any part-related labels and can adaptively learn to localize the locations of the parts on the feature map to suppress severe spatial misalignments in vehicle Re-ID. A new loss function is proposed to obtain parts with different semantic patterns by penalizing overlapping regions. The main contributions of MRF-SAPL are flexible part-level feature learning, adaptive part localization using soft attention, and the use of multiple local branches with different receptive fields. The authors show that the model outperforms previous methods on vehicle re-identification datasets, demonstrating its effectiveness in learning fine-grained local features at multiple semantic levels to effectively distinguish different vehicles with similar appearances.

Junqing et al. (contribution 9) introduced an encryption scheme designed specifically for high-pixel-density images for ensuring the security of data transmission. The proposed scheme leverages the quantum random walk algorithm in combination with the long short-term memory (LSTM) model to address the efficiency- and statistical property-based challenges of generating large-scale pseudorandom matrices. The LSTM was divided into columns and utilized for training purposes. However, due to the random nature of the input matrix, effective training of the LSTM was not possible. To overcome this, the output matrix was predicted to possess a high level of randomness. This LSTM prediction matrix, matching the size of the key matrix, was generated based on the pixel density of the encrypted image, effectively facilitating image encryption. In terms of statistical performance, the proposed encryption scheme demonstrates an average information entropy of 7.9992, an average number of pixels changed rate (NPCR) of 99.6231%, an average uniform average change intensity (UACI) of 33.6029%, and an average correlation of 0.0032. Additionally, various noise simulation tests were conducted to evaluate the scheme's robustness against common noise and attack interference in real-world applications. This approach harnesses the nearly infinite key space provided by the quantum random walk algorithm while addressing its low generation efficiency. Furthermore, the permutation and obfuscation processes in the proposed scheme make use of the key space of the quantum random walk, avoiding limitations related to the key space in a specific process.

Lei et al. (contribution 10) propose a novel method named NMYOLO for detecting infusion containers using the You Only Look Once version 4 (YOLOv4) approach to support medical staff in complex clinical environment by alleviating the pressure they face. The proposed method introduces several improvements to enhance the detection of infusion containers. First, a coordinate attention module was added after establishing YOLOv4 as the backbone to improve the model's perception of direction and location of information. Next, the spatial pyramid pooling (SPP) module was replaced with the cross-stage partial spatial pyramid pooling (CSP-SPP) module, allowing for the reuse of input information features. Additionally, an adaptively spatial feature fusion (ASFF) module was added after the path aggregation network (PANet) to facilitate the fusion of feature maps at different scales. The method also utilizes the EIoU (Enhanced Intersection over Union) as a loss function to address the anchor frame aspect ratio problem, resulting in more stable and accurate detection. The experimental results reported in this article demonstrate the advantages of the proposed method in terms of recall, timeliness, and mean average precision (mAP). Although the proposed NMYOLO method achieved the desired detection performance, it has the drawback of reduced frame rate compared to YOLOv4. The authors suggest possible future improvements, such as using a lightweight backbone or removing the non-essential convolution modules to reduce the model's parameters. They also mention the possibility of replacing modules or modifying the architecture to reduce the model's size while maintaining its detection accuracy.

Shengping et al. (contribution 11) discuss the limitations of the Magnetic Flux Leakage (MFL) visualization technique used in the surface defect inspection of ferromagnetic materials when detecting complex defects, particularly cracks, and the loss of information during unidirectional magnetization. To address this problem, they propose a novel image registration method for MFL visualization that aligns images captured under different magnetization orientations. The method utilizes mutual information and Particle Swarm Optimization (PSO) to optimize the registration process. In this study, the design of a new registration method for MFL images under different magnetization orientations was achieved, a solenoid model was utilized in MFL image registration, and higher accuracy compared to traditional methods was demonstrated through comparative experiments, suggesting that the proposed method has the potential to enhance crack detection in MFL testing.

Jian et al. (contribution 12) introduce a one-stage scale enhancement pyramid network (SEPNet) to address the challenges of object detection in large-scale images captured by unmanned aerial vehicles (UAVs), particularly when detecting small objects with significant scale variation. The proposed SEPNet consists of two core modules: the context enhancement module (CEM) and the feature alignment module (FAM). The CEM module produces more salient context information by combining multi-scale atrous convolution and multi-branch grouped convolution to model global relationships and enhance object feature representation at different scales. It prevents the flow of features with lost spatial information into the feature pyramid network (FPN). The FAM module learns the transformation offsets of pixels to preserve aggregate feature space translation invariance, addressing feature inconsistency issues in the FPN. It also adaptively adjusts the location of sampling points in the convolutional kernel to preserve feature consistency and alleviate information conflict caused by the fusion of adjacent features. This module ensures that small objects are not drowned in feature conflicts. Additionally, this paper introduces channel attention to refine pre-aggregated features, allowing the network to focus on the target area rather than the background. Looking ahead, the authors of this paper suggest that designing lightweight structures for deployment on embedded devices could be a valuable topic to explore in future research. This implies a focus on optimizing the model's efficiency without compromising its performance.

In conclusion, the application of information theory to computer vision and image processing represents a convergence of advanced technologies that bridge the gap between the virtual and real world. Through the integration of optoelectronic devices, sensors, artificial intelligence algorithms, embedded systems, robust control mechanisms, interconnectivity, big data, and cloud computing, machine vision empowers cyber–physical systems to collaborate with humans in their daily activities. As this field continues to evolve, we can anticipate a future where machine vision seamlessly integrates into our lives, unlocking new possibilities and transforming the way we perceive, interact with, and navigate both the physical and digital realms. The Guest Editors hope that after exploring the articles published in this Special Issue, entitled "**Application of Information Theory to Computer Vision and Image Processing**" (https://www.mdpi.com/journal/entropy/special_issues/MWI13854O7)—from the Information Theory, Probability and Statistics section of the ***Entropy*** **journal**—readers can take inspiration for their future research and publications.

Funding: This research received no external funding.

Institutional Review Board Statement: Not applicable.

Acknowledgments: The Guest Editors of this Special Issue acknowledge the authors and reviewers that contributed to the successful organization of the Special Issue. We would like to express our appreciation to the Universidad Autónoma de Baja California, to the Tecnológico Nacional de México/IT de Mexicali, and to the journal *Entropy* and MDPI for their constant and valuable support during this project.

Conflicts of Interest: The authors declare no conflicts of interest.

List of Contributions

1. Garcia-González, W.; Flores-Fuentes, W.; Sergiyenko, O.; Rodríguez-Quiñonez, J.C.; Miranda-Vega, J.E.; Hernán-dez-Balbuena, D. Shannon Entropy Used for Feature Extractions of Optical Patterns in the Context of Structural Health Monitoring. *Entropy* **2023**, *25*, 1207. https://doi.org/10.3390/e25081207.
2. Ge, W.; Zhang, L.; Zhan, W.; Wang, J.; Zhu, D.; Hong, Y. A Low-Illumination Enhancement Method Based on Structural Layer and Detail Layer. *Entropy* **2023**, *25*, 1201. https://doi.org/10.3390/e25081201.
3. Stasenko, N.; Shukhratov, I.; Savinov, M.; Shadrin, D.; Somov, A. Deep Learning in Precision Agriculture: Artificially Generated VNIR Images Segmentation for Early Postharvest Decay Prediction in Apples. *Entropy* **2023**, *25*, 987. https://doi.org/10.3390/e25070987.

4. Liu, H.; Ma, M.; Wang, M.; Chen, Z.; Zhao, Y. SCFusion: Infrared and Visible Fusion Based on Salient Compensation. *Entropy* **2023**, *25*, 985. https://doi.org/10.3390/e25070985.
5. Jiang, Y.; Liu, Y.; Zhan, W.; Zhu, D. Improved Thermal Infrared Image Super-Resolution Reconstruction Method Base on Multimodal Sensor Fusion. *Entropy* **2023**, *25*, 914. https://doi.org/10.3390/e25060914.
6. Lin, H.Y.; Hsu, C.Y. Structured Cluster Detection from Local Feature Learning for Text Region Extraction. *Entropy* **2023**, *25*, 658. https://doi.org/10.3390/e25040658.
7. Zheng, S.; Zhu, M.; Chen, M. Hybrid Multi-Dimensional Attention U-Net for Hyperspectral Snapshot Compressive Imaging Reconstruction. *Entropy* **2023**, *25*, 649. https://doi.org/10.3390/e25040649.
8. Pang, X.; Yin, Y.; Zheng, Y. Multi-receptive field soft attention part learning for vehicle re-identification. *Entropy* **2023**, *25*, 594. https://doi.org/10.3390/e25040594.
9. Liang, J.; Song, Z.; Sun, Z.; Lv, M.; Ma, H. Coupling Quantum Random Walks with Long-and Short-Term Memory for High Pixel Image Encryption Schemes. *Entropy* **2023**, *25*, 353. https://doi.org/10.3390/e25020353.
10. Ju, L.; Zou, X.; Zhang, X.; Xiong, X.; Liu, X.; Zhou, L. An Infusion Containers Detection Method Based on YOLOv4 with Enhanced Image Feature Fusion. *Entropy* **2023**, *25*, 275. https://doi.org/10.3390/e25020275.
11. Li, S.; Zhang, J.; Liu, G.; Chen, N.; Tian, L.; Bai, L.; Chen, C. Image Registration for Visualizing Magnetic Flux Leakage Testing under Different Orientations of Magnetization. *Entropy* **2023**, *25*, 167. https://doi.org/10.3390/e25010167.
12. Sun, J.; Gao, H.; Wang, X.; Yu, J. Scale Enhancement Pyramid Network for Small Object Detection from UAV Images. *Entropy* **2022**, *24*, 1699. https://doi.org/10.3390/e24111699.

References

1. Yang, B.; Serrano, J.V.; Launer, M.A.; Wang, L.; Rabiei, K. A comprehensive and systematic study on the cybernetics management systems. *Syst. Pract. Action Res.* **2023**, *36*, 479–504. [CrossRef]
2. Mudhivarthi, B.R.; Shah, P.; Sekhar, R.; Murugesan, D.; Bhole, K. Cybernetic Technologies in Industry 4.0. In Proceedings of the 2023 14th International Conference on Computing Communication and Networking Technologies (ICCCNT), Delhi, India, 6–8 July 2023; pp. 1–6.
3. Sergiyenko, O.; Flores-Fuentes, W.; Mercorelli, P. (Eds.) *Machine Vision and Navigation*; Springer: Berlin/Heidelberg, Germany, 2020; pp. 5–30.
4. Fusiello, A. *Computer Vision: Three-Dimensional Reconstruction Techniques*; Springer Nature: Cham, Switzerland, 2024.
5. Yuille, A. An information theory perspective on computational vision. *Front. Electr. Electron. Eng. China* **2010**, *5*, 329–346. [CrossRef]
6. Ruiz, F.E.; Pérez, P.S.; Bonev, B.I. *Information Theory in Computer Vision and Pattern Recognition*; Springer Science & Business Media: Berlin, Germany, 2009.
7. Podpora, M.; Kawala-Sterniuk, A.; Kovalchuk, V.; Bialic, G.; Piekielny, P. A distributed cognitive approach in cybernetic modelling of human vision in a robotic swarm. *Bio-Algorithms Med-Syst.* **2020**, *16*, 20200025. [CrossRef]
8. Han, H.; Tang, J.; Jing, Z. Wireless sensor network routing optimization based on improved ant colony algorithm in the Internet of Things. *Heliyon* **2023**, *10*, e23577. [CrossRef] [PubMed]
9. Hallyburton, R.S.; Zelter, N.; Hunt, D.; Angell, K.; Pajic, M. A Modular Platform For Collaborative, Distributed Sensor Fusion. In Proceedings of the ACM/IEEE 14th International Conference on Cyber-Physical Systems (with CPS-IoT Week 2023), San Antonio, TX, USA, 9–12 May 2023; pp. 268–269.
10. Souli, N.; Kolios, P.; Ellinas, G. Online Distributed Relative Positioning Utilizing Multiple Cooperative Autonomous Agents. *J. Intell. Robot. Syst.* **2023**, *109*, 87. [CrossRef]
11. Chen, W.; Liu, G. Intelligent Optoelectronic Devices for Next-Generation Artificial Machine Vision. *Adv. Electron. Mater.* **2022**, *8*, 2200668. [CrossRef]
12. Sirimewan, D.; Bazli, M.; Raman, S.; Mohandes, S.R.; Kineber, A.F.; Arashpour, M. Deep learning-based models for environmental management: Recognizing construction, renovation, and demolition waste in-the-wild. *J. Environ. Manag.* **2024**, *351*, 119908. [CrossRef] [PubMed]
13. Lee, J.W.; Kang, H.S. Three-Stage Deep Learning Framework for Video Surveillance. *Appl. Sci.* **2024**, *14*, 408. [CrossRef]
14. Liang, Z.; Wei, H.; Liu, G.; Cheng, M.; Gao, J.; Li, S.; Tian, X. Leveraging GAN-based CBCT-to-CT translation models for enhanced image quality and accurate photon and proton dose calculation in adaptive radiotherapy. *J. Radiat. Res. Appl. Sci.* **2024**, *17*, 100809. [CrossRef]

15. Clemente, C.; Chambel, G.; Silva, D.C.; Montes, A.M.; Pinto, J.F.; Silva, H.P.D. Feasibility of 3D Body Tracking from Monocular 2D Video Feeds in Musculoskeletal Telerehabilitation. *Sensors* **2023**, *24*, 206. [CrossRef] [PubMed]
16. Meribout, M.; Baobaid, A.; Khaoua, M.O.; Tiwari, V.K.; Pena, J.P. State of art IoT and Edge embedded systems for real-time machine vision applications. *IEEE Access* **2022**, *10*, 58287–58301. [CrossRef]
17. Kitchatr, S.; Sirimangkalalo, A.; Chaichaowarat, R. Visual Servo Control for Ball-on-Plate Balancing: Effect of PID Controller Gain on Tracking Performance. In Proceedings of the 2023 IEEE International Conference on Robotics and Biomimetics (ROBIO), Koh Samui, Thailand, 4–9 December 2023; pp. 1–6.
18. Malik JJ, S.; Saxena, G.D.; Mukkapati, N.; Chacko, S.; Thirumoorthy, P.; Dilip, R. A Review on Augmented Reality Application in Industrial 4.0. *NeuroQuantology* **2023**, *21*, 278.
19. Wang, P.; Yang, L.T.; Li, J.; Chen, J.; Hu, S. Data fusion in cyber-physical-social systems: State-of-the-art and perspectives. *Inf. Fusion* **2019**, *51*, 42–57. [CrossRef]
20. Sergiyenko, O.; Flores-Fuentes, W.; Mercorelli, P.; Rodriguez-Quinonez, J.C.; Kawabe, T. Guest editorial special issue on sensors in machine vision of automated systems. *IEEE Sens. J.* **2021**, *21*, 11242–11243. [CrossRef]
21. Real-Moreno, O.; Rodríguez-Quiñonez, J.C.; Sergiyenko, O.; Flores-Fuentes, W.; Mercorelli, P.; Ramírez-Hernández, L.R. Obtaining object information from stereo vision system for autonomous vehicles. In Proceedings of the 2021 IEEE 30th International Symposium on Industrial Electronics (ISIE), Kyoto, Japan, 20–23 June 2021.
22. Nikishina, L.B. Industry 4.0: History of emergence, development, prospects of transformation into Industry 5.0. *E3S Web Conf.* **2023**, *458*, 06023. [CrossRef]

entropy

MDPI

Article

Shannon Entropy Used for Feature Extractions of Optical Patterns in the Context of Structural Health Monitoring

Wendy Garcia-González [1], Wendy Flores-Fuentes [1], Oleg Sergiyenko [2], Julio C. Rodríguez-Quiñonez [1], Jesús E. Miranda-Vega [3,*] and Daniel Hernández-Balbuena [1]

1 Engineering Faculty, Universidad Autónoma de Baja California, Mexicali 21280, BC, Mexico; wendy.garcia26@uabc.edu.mx (W.G.-G.); flores.wendy@uabc.edu.mx (W.F.-F.); julio.rodriguez81@uabc.edu.mx (J.C.R.-Q.); dhernan@uabc.edu.mx (D.H.-B.)

2 Engineering Institute, Universidad Autónoma de Baja California, Mexicali 21100, BC, Mexico; srgnk@uabc.edu.mx

3 Department of Computer Systems, Tecnológico Nacional de México, IT de Mexicali, Mexicali 21376, BC, Mexico

* Correspondence: elias.miranda@itmexicali.edu.mx

Abstract: A novelty signal processing method is proposed for a technical vision system (TVS). During data acquisition of an optoelectrical signal, part of this is random electrical fluctuation of voltages. Information theory (IT) is a well-known field that deals with random processes. A method based on using of the Shannon Entropy for feature extractions of optical patterns is presented. IT is implemented in structural health monitoring (SHM) to augment the accuracy of optoelectronic signal classifiers for a metrology subsystem of the TVS. To enhance the TVS spatial coordinate measurement performance at real operation conditions with electrical and optical noisy environments to estimate structural displacement better and evaluate its health for a better estimation of structural displacement and the evaluation of its health. Five different machine learning (ML) techniques are used in this work to classify optical patterns captured with the TVS. Linear predictive coding (LPC) and Autocorrelation function (ACC) are for extraction of optical patterns. The Shannon entropy segmentation (SH) method extracts relevant information from optical patterns, and the model's performance can be improved. The results reveal that segmentation with Shannon's entropy can achieve over 95.33%. Without Shannon's entropy, the worst accuracy was 33.33%.

Keywords: machine learning; data augmentation; sensor data processing; technical vision system; optical patterns; random process; entropy

Citation: Garcia-González, W.; Flores-Fuentes, W.; Sergiyenko, O.; Rodríguez-Quiñonez, J.C.; Miranda-Vega, J.E.; Hernández-Balbuena, D. Shannon Entropy Used for Feature Extractions of Optical Patterns in the Context of Structural Health Monitoring. *Entropy* **2023**, *25*, 1207. https://doi.org/10.3390/e25081207

Academic Editor: Renaldas Urniezius

Received: 2 July 2023
Revised: 7 August 2023
Accepted: 9 August 2023
Published: 14 August 2023

1. Introduction

Modern society requires infrastructure to perform indispensable activities such as transportation, communication, power grid, and water supply systems. These urban infrastructures (UI) are necessary to sustain a city's economy. The rapid urban growth allows testing the strengths of the civil infrastructure (CI). Traffic loads and natural hazards are factors that can cause deterioration of the UI. To pursue sustainable development goals, it is necessary to consider the monitoring and control of current infrastructure. Urban sensing deals with collecting relevant information about the urban environment to develop early warning systems to make sustainable urban systems through technology. Different types of data sets of large amounts of information can be gathered, like air quality [1], traffic patterns [2], and CI [3], just to mention a few examples.

The most common variables studied in CI that need adequate maintenance are related to energy, transportation, and building. The datasets taken from CI are urban area, population density, energy utilization by each consumer, etc. All this information is collected and used to understand complex problems that are hard to solve, like the challenge of aging infrastructure. To face this challenge, a TVS is proposed to capture optical signals to create datasets from particular behaviors of the laser beam reflected from a CI.

Optimizing maintenance activities of a key UI is an important task that should be adopted to ensure good performance under challenging conditions. Sustaining key infrastructures requires constant monitoring for the safety of citizens. These elements must follow strict safety requirements to avoid stopping the economy. Adequate maintenance of these infrastructures can save lives. For these reasons, preserving civil and industrial infrastructures through programmed maintenance is important. The field of structural health monitoring (SHM) is a convenient and organized way to address the current challenges. The methodologies addressed by SHM are designed to develop technologies for monitoring and analyzing data to prevent damage to infrastructure. SHM can help us to evaluate the risks and manage assets better.

Nowadays, many engineers have implemented technologies to solve problems related to damage detection. The current technology improved diagnostic accuracy by identifying problems to make decisions objectively. On the other hand, sensor information is one of the most important elements in quantifying risk. This allows for defining strategies to minimize the likelihood of critical damage. To collect the information from structures there are different technologies based on materials such as fiber Bragg grating (FBG) for strain sensor applications in road [4], piezoelectric nanofiber membranes sensor based on PAN/$BaTiO_3$ (polyacrylonitrile and flexible barium titanate) [5], Carbon nanotubes (CNTs) [6]. FBG is a technology based on optical fiber that reflects certain wavelengths and transmits others. This material also be used for sensing applications. PAN/BaTiO3 is a novel nanomaterial proposed for electromechanical conversion in SHM tasks due to its strong piezoelectricity capacity. A CNT is a tiny hollow tube made of cylindrical molecules of carbon that is widely used in many fields of science due to its electromechanical and thermal properties. In the field of SHM, the CNT is used for measuring the strain, stress, load, temperature, displacement, and pressure.

Recent literature based on numerical and experimental models to address classification problems can be solved using supervised and unsupervised machine learning techniques. These can be described as follows. The following authors [7] proposed a methodology based on the acceleration and shear time histories evaluated on the rails. The work is treated as a binary classification. The methodology proposed could automatically distinguish a defective wheel from a healthy one. The development of an easy-to-implement, low-cost monitoring system is a relevant contribution. The continuous wavelet transform (CWT) model was used as a feature extractor from acquired responses.

A general ML framework to deal with the railway wheel flats identification can be consulted in [8]. They deal with damage identification based on the acceleration measurements on the rails. A numerical approach was performed to evaluate whether the number of sensors used to detect and classify wheel flats. An autoregressive (AR) model was performed as a feature extractor to take meaningful information from measurements.

The following research [9] studies the different vibration-based damage detection methods, such as fundamental modal examination, local diagnostic method, non-probabilistic methodology, and the time series method.

A Singular spectrum analysis (SSA) is a nonparametric method for analyzing time series. This tool can enhance the sensitivity of the acceleration signals. SSA uses time history data obtained from each sensor separately, and the singular value decomposition (SVD) is performed on the Hankel matrix formed [10]. The work [11] was focused on detecting and identifying damage in a structure in an online framework. They proposed a methodology for real-time based on recursive singular spectrum analysis (RSSA). According to the findings, RSSA facilitates the monitoring of structural systems and real-time data processing through acceleration data using single and multiple sensors. The exact damage instant can be identified by extracting damage-sensitive features from measurements.

The authors [12] give a broader discussion of first-order perturbation (FOP) techniques that solve SHM problems in online real-time structural damage detection for vibrating systems. The following authors performed a novel framework by applying Recursive Principal Component Analysis (RPCA) in conjunction with Time Varying Auto-Regressive Modeling (TVAR) for an online damage detection method for real time processes [13].

A literature review of next-generation smart sensing technology in SHM, such as smartphones, unmanned aerial vehicles (UAVs), cameras, and robotic sensors, are used in acquiring and analyzing the vibration data [14]. A LiDAR (Light Detection and Ranging) device is an instrument that has significant potential for damage detection based on laser scanning providing geometric information about the structures [15].

Although a Light Detection and Ranging (LiDAR) system is highly precise and reliable, the cost of its implementation for SHM tasks can be expensive in the case of a 64-beam model that can cost around $75,000 (USD) [16]. Despite their high cost, LiDAR is mainly used for perception and localization tasks at most high level [17]. The advantage of these systems is that the performance of the system can be determined by using non-destructive techniques (NDT).

This work is focused on a TVS system for displacement measurements. The current TVS has the patent number MX2014000647, which uses a dynamical triangulation method to get angular position and 3D coordinates from objects or surfaces. This system can also perform the same tasks as cameras and LiDAR for SHM tasks. But with advantages like high accuracy, low computational cost, and low volume of data requirement for measurements.

A laser source obtains the geometrical coordinates of a surface under study. A photosensor detects the laser beam reflected. However, in a real operation, interferences of other radiation sources can affect the information collected with a TVS. For example, sunlight is the main interference that should be filtered. For this reason, the reflected laser beam is mixed up with undesired signals.

A novelty signal processing method is proposed for a technical vision system (TVS). A method based on the use of the Shannon entropy for feature extractions of optical patterns in the context of SHM to augment the accuracy of optoelectronic signal classifiers implemented in the metrology subsystem of the TVS. To enhance the TVS spatial coordinate measurement performance at real operation conditions with electrical and optical noisy environments to estimate structural displacement better and evaluate its health.

The following research faces the same problem in reconstructing the real returning signal shape, and its problem is exacerbated by the presence of strong solar background illumination [18]. Using optical filters and higher-power lasers would be a solution. However, these increase the cost of manufacturing a TVS and increase larger usage risks and augment energy consumption. An alternative solution is to apply Artificial intelligence (AI) to detect what signal corresponds to the laser beam. ML can solve the interference issue as a recognition pattern problem. To enhance the accuracy of ML models, Shannon's entropy is proposed to remove parts that contain random signals and isolate them from the optical patterns.

In this work the following novelty signal processing method is proposed to enhance the TVS accuracy.

A method based on the use of the Shannon entropy for feature extractions of optical patterns in the context of SHM to augment the accuracy of optoelectrical signal classifications implemented in the metrology subsystem of the TVS. To enhance the TVS spatial coordinate measurement performance at real operation conditions with electrical and optical noisy environments to estimate structural displacement better and evaluate its health.

Relevant procedures in the method are:

1. Using a phototransistor with black daylight filter as a photosensor of a TVS to reduce the influence of solar radiation as much as possible.
2. Calibrating the TVS with a turned-off laser and obtaining raw signals (Class 1)
3. Calibrating the TVS with a turned-on laser and obtaining raw signals (Class 2).
4. Creating a Class 3 with data augmentation to create robust ML models.
5. Comparing the performance of five different ML models with LPC and ACC.

6. Comparing the performance of five different ML models with LPC and ACC and Shannon's entropy as a segmentation process.

This work aims to find the configuration that enhances the performance of ML models to discriminate against sunlight interference. One of the main goals is to implement a pipeline that can recognize the reflected laser beam pattern. For that reason, this research compares the accuracy of five different ML techniques with LPC, ACC, and Shannon's entropy. The following classifiers such as Naïve Bayes (NB), support vector machines (SVM), linear discriminant analysis (LDA), K-Nearest Neighbors (KNN), and neural network (NN), were used. Data augmentation was implemented to enhance the accuracy of these classifiers.

This paper is organized as follows. Section 2 gives details about the problem statement. Section 3 describes the operational principle of a TVS and the latest improvements. Section 4 provides a brief overview of the feature extractions used. Section 5 summarizes the ML methods used in this work. Section 6 presents the proposed ML pipeline to solve the problem of interference. Section 7 discusses the results and highlights of the experiments carried out in this work. Finally, some conclusions and recommendations from the experiments are shared in Section 8.

2. Problem Statement

Recent studies were conducted outdoors and compared with experimentation under indoor (Laboratory) conditions, from which the results showed that undesired signals affected the performance of the TVS. This was primarily due to the conditions of intense radiation [19–21]. Consequently, a laser beam cannot be captured by a TVS system. The solar radiation spectrum shows that infrared light is reflected more than ultraviolet (UV) or visible light due to its longer wavelength. This is important to consider because several devices can work with these wavelengths, such as phototransistors (PT) and photodiodes (PD).

PT is more sensitive to light than PD due to high gain. Another advantage of PT over PD is that it can be obtained at low-cost. The PT used in this work minimizes outside interference thanks to its daylight filter. Although PT was chosen, TVS is still detecting low interference outdoors. The interference can be discriminated against using ML models to address this issue. Particular optical patterns only can appear in three different scenarios. The first scenario corresponds to when TVS is turned off. The second scenario appears when TVS is turned on. Finally, the third scenario represents a saturation of a signal captured.

Figure 1, shows raw signals detected with the PT. Figure 1a corresponds to the background or possible interferences found outdoors; at that moment, TVS is turned off. This optical pattern has a peak voltage of 2.5 Volts, labeled class 1. The reason is to create an ML model that can discriminate between the interferences and laser scanning of the TVS system. Figure 1b shows a particular pattern at the moment the TVS is turned on. Thanks to laser power, this signal can be captured by the PT. This signal has a peak voltage of 4.5 Volts and is labeled class 2. Note that the peak of voltage of class 1 and class 2 is different. Figure 1c is class 3 created by a data augmentation stage (synthetic signal). This signal represents a random signal created by external factors.

The three classes contain low voltages, captured when radiation is not detected. These are a part of the optical patterns that can be regarded as a problem because there is no relevant information. Random voltage variations are redundant information that needs to be addressed.

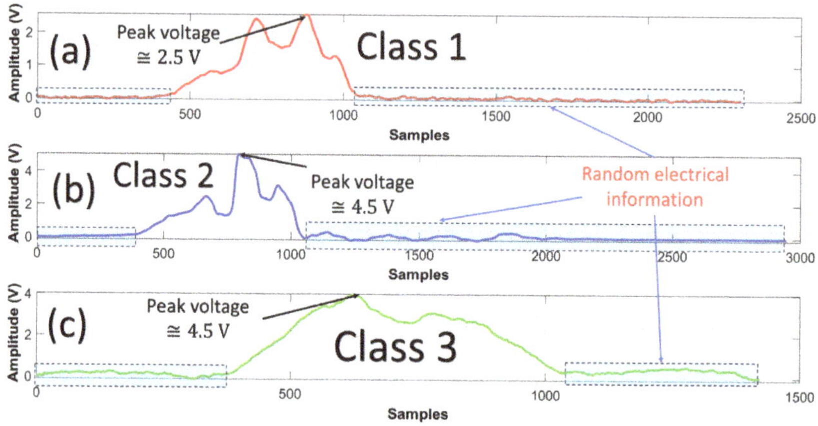

Figure 1. Raw signals mixed up with interference are used for pattern recognition. (**a**) The solid red line represents TVS is turned off. (**b**) Blue color shows the signal captured when TVS is turned on. (**c**) The solid green line signal represents an unknown pattern (this signal was created as a synthetic pattern). Note that an important part of the signal is random electrical information.

3. Operational Principle of TVS

In this section, a brief overview of evolution and operational principles of TVS are given.

TVS system is a device that can solve real-time tasks to measure three-dimensional (3D) coordinates. These tasks are needed in many contexts of SHM, such as displacement measurements or surface estimation. This system has two main parts to realize depth measurements. The first component of a TVS is the positioning laser (PL) that uses an active laser in conjunction with mechanical elements, such as a step/servo motor and gears, whereby the space of interest can be radiated. The second component is the scanning aperture (SA), which contains photosensors to receive the radiation reflected from objects under study.

The distance between PL and SA is known, and can be identified as a which is illustrated in Figure 2. Angular position of the PL can be controlled by a step motor or servo motor. PL is an angle known by the user that corresponds to $C_{i,j}$. The angular position of SA is measured by knowing the peak time of the Gaussian signal and period of the DC motor with a speed constant, this is denoted by $B_{i,j}$.

Figure 2 explains how to determine the angular position of SA when a Gaussian signal appears. This signal has the shape of a normal distribution bell (Gaussian). A rotational mirror at an angle of 45° reflects the radiation of an object to the photosensor placed on SA. As a consequence, the Gaussian shape of a signal is formed. The capacitance of a photosensor and signal processing can smooth the Gaussian signal.

Since TVS is scanning at a constant angular velocity, the time elapsed between the start of the first pulse and the second pulse can be used to estimate the angular position. For instance, the angular position β of the PT can be calculated as follows with Equation (1).

$$\beta = 2\pi \frac{t_\alpha}{T_{2\pi}} \quad (1)$$

where the time t_α is defined as the interval between the signal m_1 and the position of the energy center is m_2.

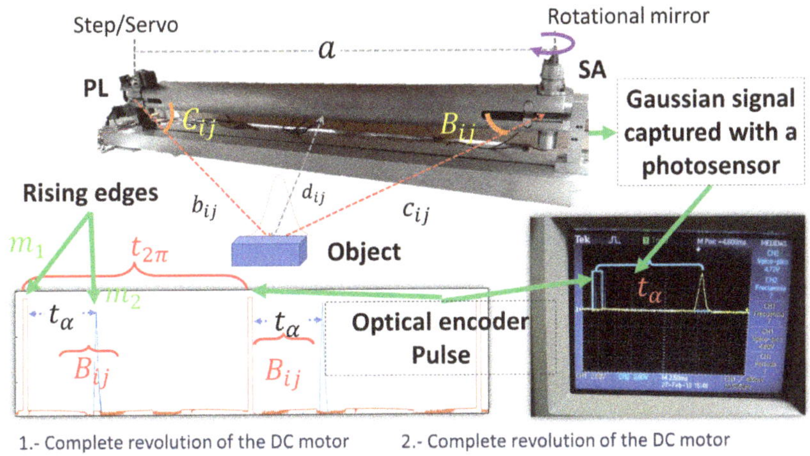

Figure 2. Graphical aid to visualize the main parts of TVS in operation. $B_{i,j}$ is the angular position according to a frame reference. The oscilloscope shows the pulse and a laser beam (Gaussian Signal) without interference.

The local maximum of the Gaussian signal is related to the energetic center of the radiation reflected on the surface studied. A complete revolution of a motor is the period of time used to know the angular position of a local maximum of a Gaussian signal, and this is called $B_{i,j}$. Opto-interrupters or Hall sensors are usually used to calculate the pulses per revolution of a DC motor. In this paper, ITR8102 (Everlight Electronics, New Taipei City, Taiwan) was implemented to know the position of the motor on SA. This opto-interrupter sends pulses for every revolution of the rotational mirror. For each revolution of a DC motor, a Gaussian shape will appear during the scanning, as illustrated in Figure 2.

Knowing the scanning frequency makes it possible to calculate $B_{i,j}$. With this information, the object position is estimated at two different times. If $B_{i,j}$ moves, $d_{i,j}$ can be determined in real-time. Note that distance $d_{i,j}$ corresponds to depth. According to sine theorems and the values of the angles $B_{i,j}$ and $C_{i,j}$ depth information $d_{i,j}$ is estimated with Equation (2).

$$d_{i,j} = a\frac{sin(B_{i,j})sin(C_{i,j})}{sin[180^{\circ} - (B_{i,j} + C_{i,j})]} \quad (2)$$

Figure 3, details relevant information about different TVS prototypes.

To extend the information of a TVS, the following work [22] shows typical laser scanner constructions and their constraints.

The following researchers [23–25] worked with the first version of the TVS prototype number one. They used the TVS for remote sensing and obstacle detection in an unknown environment. This prototype presented simplicity, versatility, and economic accessibility to realize 3D coordinates measurements. An inconvenience of this prototype is that it could only scan in a discontinuous way. In other words, point clouds give shape to the object studied. The next work [26] involved the substitution of the previous (prototype No. 2) and changed the stepper-motor by servo-motors to achieve a continuous laser scan (newly developed prototype No. 3).

A complete mathematical apparatus for processing digital information inside the system and for determining the distances and angle measurements in the system proposed is developed [27].

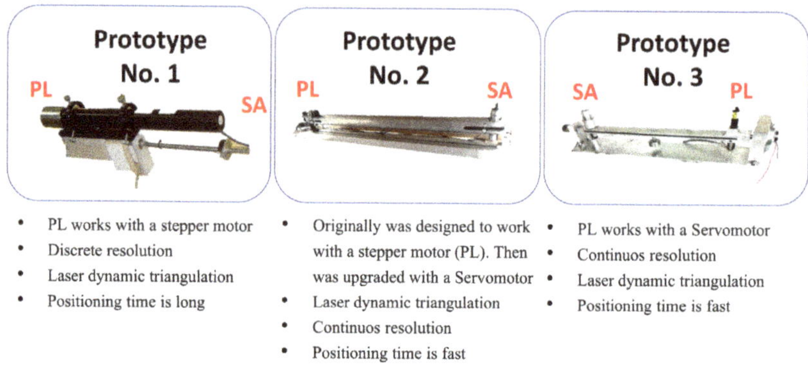

Figure 3. Comparison of different TVS systems developed for measuring 3D coordinates.

4. Feature Extraction Methods

Feature extractors are mathematical algorithms that recover relevant information (attributes) from a phenomenon like our raw signals captured with a PT. This process is known as feature engineering, and the main purpose is to use representative data with less information. These features enhance the ML models, and redundant data are minimized. There are several methods, such as Autoregressive (AR) Modelling, Linear Predictive Coding (LPC), Autocorrelation coefficients (ACC), Mel Frequency Cepstral Coefficients (MFCC), Fast Fourier transform (FFT), Hilbert transform, just to mention a few. Trends and a zoomed-in perspective of feature extraction methodologies can be consulted in [28].

This work implements ACC and LPC as feature extractors. However, Shannon's entropy is used to segment only the optical pattern from the electrical signal and remove the rest of the signal.

The description of these techniques are detailed as follows.

4.1. Autocorrelation Function

The autocorrelation function (ACF) vector can be used for extract the features (ACC) of the TVS system by measuring the correlation between y_t and y_{t+k} where $x = 0, \ldots, k$ and y_t is a stochastic process.

The correlation for lag k can be estimated by applying Equation (3). For more details, see the following work [29].

$$r_k = \frac{c_k}{c_0} \tag{3}$$

where c_0 represents the sample variance of the time series and c_k can be estimated by Equation (4).

$$c_k = \frac{1}{T} \sum_{t=1}^{T-k} (y_t - \bar{y})(y_{t+k} - \bar{y}) \tag{4}$$

Figure 4 shows the feature space of ACC as a feature extractor from a raw signal.

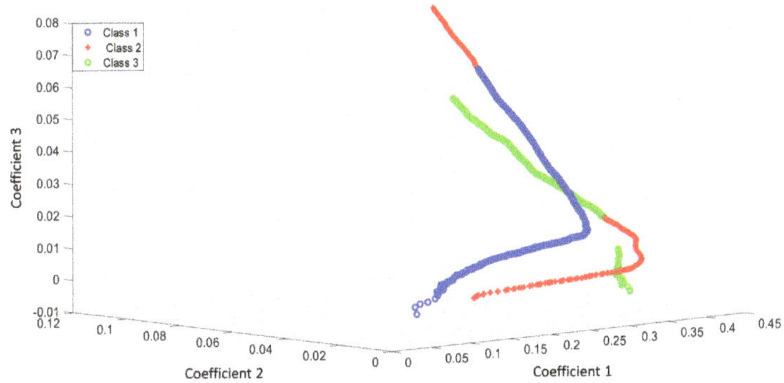

Figure 4. Feature space of ACC as a feature extractor of a raw signal.

Figure 5 shows the feature space of ACC with the Shannon entropy. Note that the segmentation with the Shannon entropy separates each class in comparison with Figure 4. The coefficients 1,2,3 represent the first three features of ACC or LPC. For this study, we extracted 11 features for each optical pattern. These figures differences rely on the useful features extracted with the Shannon entropy as a segmentation process. The ideal case is when the feature extraction process can separate all classes.

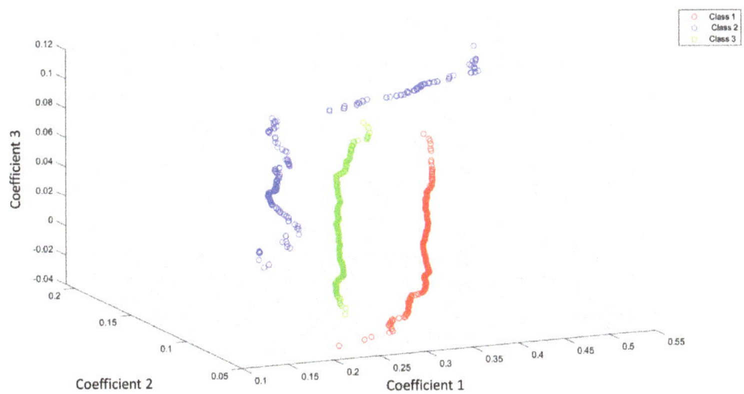

Figure 5. Feature space of ACC and the Shannon entropy as a segmentation process.

4.2. Linear Predictor Coefficients

The procedure to calculate the LPC coefficients can be described as follows. First, Equation (5) was applied to calculate the FFT of a desired signal to compute the autocorrelation vector.

$$x_k = \sum_{n=1}^{N} x_n e^{-j2\pi(k-1)(n-1)/N} \quad (5)$$

After obtaining FFT X_k, the inverse discrete Fourier transform of the absolute value of X_k value is taken and squared to compute the autocorrelation vector R by Equation (6).

$$X_j = \frac{1}{n} \sum_{k=1}^{n} Y_k e^{2i\pi(j-1)(k-1)/n} \quad (6)$$

A scaling is applied to the output and the bias of the autocorrelation is estimated $B = R./m$, where m represents the number of the length of the vector or signal segment under study.

The Hermitian Toeplitz system of equations is built as follows:

$$\begin{bmatrix} B(1) & B(2)^* & \dots & B(n)^* \\ B(2) & B(1) & \dots & B(n-1))^* \\ \vdots & \ddots & \ddots & \vdots \\ B(n) & \dots & B(2)) & B(1)) \end{bmatrix} \begin{bmatrix} A(2) \\ A(3) \\ \vdots \\ A(n+1) \end{bmatrix} = \begin{bmatrix} -B(2) \\ B(3) \\ \vdots \\ -B(n+1) \end{bmatrix} \quad (7)$$

This system of equations Equation (7) can be solved by Levinson-Durbin, recursion and the real coefficients A for the predictor are taken.

Figure 6 illustrates the feature space of LPC as a feature extractor from a raw signal. Figure 7 shows how the classes were separated with Shannon's entropy segmentation process.

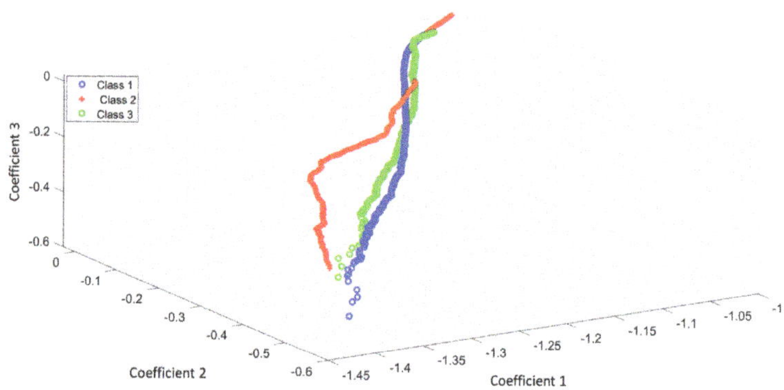

Figure 6. Feature space of LPC as a feature extractor of a raw signal.

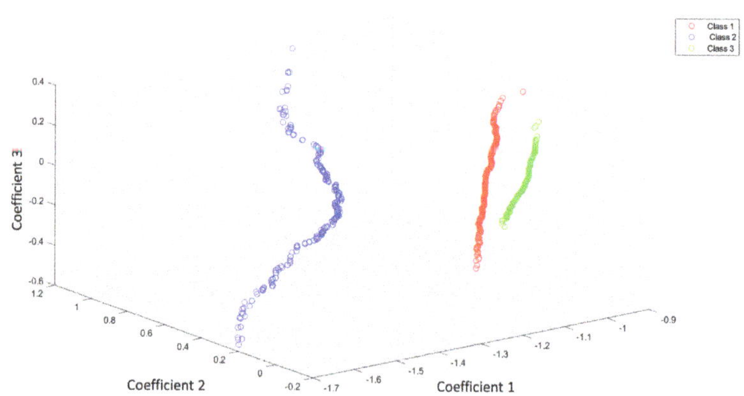

Figure 7. Feature space of LPC and the Shannon entropy as a segmentation process.

4.3. Entropy as an Optical Feature

Entropy is a broad concept that measures the disorder in a random system. According to Shannon, a non-linear measurement in dynamic signals, measures the average information contents associated with the data randomness encountered in a signal or event [30]. The relevance of SHM is that SE is a useful tool to show changes in the measured signals associated with the structure condition.

Given a source of random events from the discrete set of possible events $a_1, a_2, \ldots, a_n$, with associated probability distribution $P(a_1), P(a_2), \ldots, P(a_n)$, the average information per source output can be called as the entropy of the source, see Equation (8)

$$H = -\sum_{n=1}^{N} P(a_n) log P(a_n) \quad (8)$$

H may alternatively be understood as a measure of unpredictability of information content [31].

In the case of this work, Shannon's entropy is used as a segmentation process. As mentioned before, part of the optical patterns are random electrical signals that increase the size of the dataset. It measures the entropy of a signal divided into frames with 80 samples of window length and with 30 samples of overlap.

5. Machine Learning Classifiers

The power of ML techniques is based on the algorithm chosen to solve a complex problem. The following ML classifiers, such as NB, SVM, LDA, KNN, and NN, were selected to discriminate between the reflected laser beam and sunlight or other radiation sources. These techniques are described below.

5.1. Naives Bayes

NB is a probabilistic classifier that works well as most distributions of related features follow probabilistic nature [32]. This classifier assumes that the properties of the features on a given class are independent of the values of the other features. By knowing our class labels and training data set as $T = (x_1, y_1), \ldots (x_N, y_N)$. Each instance from the data set is represented by n-dimensional feature vector, $X = x_1, x_2, \ldots, x_n$. Each label is represented by $y \in \{1, \ldots, K\}$, where K is the number of a class. In this work there are three classes, C_1, C_2 and C_3.

According to [33], given a sample X, NB will predict that X belongs to the class having the highest a posteriori probability, which is conditioned on X. X is predicted to belong to the class C_i if and only if.

$$P(C_i \mid X) > P(C_j \mid X) \; for \; 1 \leq j \leq m, \; j \neq i \quad (9)$$

Based on Bayes' theorem,

$$P(C_i \mid X) = \frac{P(X \mid C_i)P(C_i)}{P(X)} \quad (10)$$

where $P(C_i \mid X)$ is the class prior probability and posteriori probability. $P(X \mid C_i)$ is the likelihood, which is the probability of the predictor given class.

5.2. Support Vector Machines

A SVM classifier starts with a construction of a decision function $f(x, \omega) = sign(h(x, \omega))$ with outputs $\{\pm 1\}$, where $h(x, \omega))$ is a separating hyperplane that can be expressed as follows:

$$h(x, \omega) = \langle \omega + \phi(x) \rangle + b \quad (11)$$

where ω defines a direction perpendicular to the hyperplane.

Although SVM was initially designed for binary classification, several methods have been proposed to create a multiclass classifier. One-versus-rest (1VR) and one-versus-one (1V1) are representative ensemble schemes for discrimination for more than two categories. This approach's main issue is constructing a good Error Correcting Output Codes (ECOC) matrix [34]. This work is based on 1V1 that fits $K(K-1)/2$ by individual binary classifiers SVM models.

One of the main problems of applying multi-class classification is usually solved by a decomposing and reconstruction procedure when two-class decision machines are implied [35].

5.3. Linear Discriminant Analysis

An extension of Fisher's linear discriminant for n classes can be represented as an intra-class matrix according to [36].

$$\hat{\Sigma}_w = S_1 + \cdots + S_n = \sum_{i=1}^{n} \sum_{x \in c_i} (x - \bar{x}_i)(x - \bar{x}_i)' \tag{12}$$

where $\bar{x}$ is the mean value and $\bar{x}_i$ corresponds to the mean for each class.

The inter-class matrix can be represented as follows:

$$\hat{\Sigma}_b = \sum_{i=1}^{n} m_i (\bar{x}_i - \bar{x})(\bar{x}_i - \bar{x})' \tag{13}$$

where m_i represents the number of training samples for each class.

5.4. K-Nearest Neighbors

KNN is one of the most popular algorithms to be used as a multi-class ML technique. This technique is a non-probabilistic classifier as well as SVM. It is well known as lazy learning because it does not carry out a training phase [37]. The latest trends and applications of KNN in Big Data [38] are used in the context of smart cities [39]. This algorithm compares the k nearest neighbors to be used as a decision rule to classify a new distance as belonging to a class. Furthermore, KNN can be applied for regression problems. KNN algorithm is a distance-based classifier, and the main functions employed by this algorithm are Euclidean, Mahalanobis, Hamming, and Citiblock. The Euclidean distance is used in this work. This metric can be expressed as follows.

$$d(x, y) = \sum_{i=1}^{N} \sqrt{x_i^2 - y_i^2} \tag{14}$$

where, N is the total number of samples.

5.5. Neural Network

Learning and memory are complex processes that AI tries to imitate or understand. An artificial NN is a computational tool inspired by human brain behavior to perform these tasks. Recent development in neural networks profoundly showed incredible results in object classification, pattern recognition, and natural language processing [40].

A NN can be classified into two categories such as feed-forward (FNN) and feed-backward or recurrent (RNN), according to their interconnection between the neuron layers.

Convolutional neural networks (CNN) are another type of NN well-known by machine vision designers. The NN classifier used in this work is based on FNN.

Weight matrix and activation functions are important parts of designing a NN. The activation function transforms an input signal into an output signal. The functions such as Rectified Linear Unit (RELU), SoftMax, Binary Step Function, Linear, Sigmoid, Tanh, Exponential Linear Unit, and Swish are commonly used.

A RELU function has the main goal of establishing a threshold operation for input, and this can be expressed as follows:

$$f(x) = \begin{cases} 0\ if\ x < 0 \\ x\ if\ x \geq 0 \end{cases} \tag{15}$$

The SoftMax activation function regularly is applied to the final fully connected layer. This function can be implemented as follows:

$$softmax(z_i) = \frac{exp(z_i)}{\sum_{j=1}^{K} exp(z_i)} \tag{16}$$

where z are the real values of the output layer and K is the number of classes. SoftMax functions convert these values into probabilities (classification scores).

The parameters of activation functions applied in this work are to create a NN classifier using two fully connected layers, each with three outputs. The RELU activation function was implemented for every fully connected layer of the NN model. As an output layer of the NN classifier, a SoftMax activation function was used.

The following authors [41], implemented an FNN application to evaluate sustainable urban environmental quality to deal with air pollution. Ref. [42] reviews on the basic theories and recent algorithms for optimizing NN.

6. Proposed Classification Schemes

This section gives an overview of a practical implementation of ML with different techniques. Figure 8, shows the complete procedure for classifying the laser beam. This procedure starts with inputting of raw signal captured with a phototransistor (peak wavelength 940 nm). A data augmentation process was used to enhance the performance of the ML models. In that stage, time scale modification (TSM) is applied on the input raw signals by applying different speedup factors. In the literature, this is known as alpha. Then, a white Gaussian noise is added to a raw signal. Finally, the signal is filtered a mean filter by using 14 coefficients. Different signals were created with this procedure to make more robust predictions. The normalization process is realized to work with uniform data with values between 0 and 1. The main goal of the segmentation process is to divide the input signal into several windows, and each segment should also be meaningful. Particular features of each window are evident thanks to this process. As shown in Figure 8, a Hamming window of a 240-point is created to convolve with each segment. The following process shows that ACC or LPC are estimated and stored in a feature matrix. This matrix is split into two data sets. Training and test data sets are used in the learning process. Finally, the classification process gives the accuracy of the feature vectors tested. Figure 9, shows the procedure of how the optical patterns were collected. Each frame has an entropy value representing the optical pattern's randomness level. Smaller values of total entropy are removed. Note that in Figure 9 the size of the feature matrix with Shannon's entropy is reduced in comparison with Figure 8. Appendix A shows the pseudocode of two proposed classification schemes, as described previously.

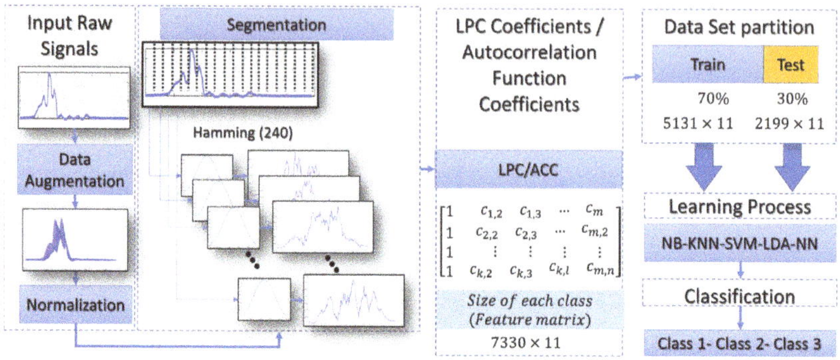

Figure 8. Flow chart of the reflected laser beam classification without Shannon's entropy for removing frames.

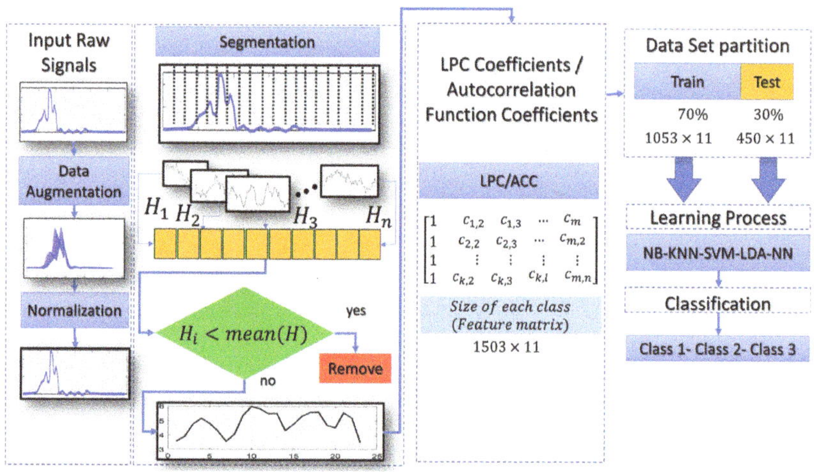

Figure 9. Flow chart of the laser beam classification based on Shannon's entropy segmentation.

7. Results and Discussion

This section explains how the results of this study were obtained to classify the optical patterns into class 1, class 2 and class 3 (see Figure 1 as reference for each kind of class).

7.1. Results without Entropy Segmentation

Table 1 shows the percentages of correctly and incorrectly classified instances for each true class carried out in this experiment. True positives of class 1, class 2, and class 3 were 1386, 1970 and 1495, respectively. The confusion matrix shows that 1746 measurements were misclassified. The class precision reached for class 1, class 2, and class 3 was 63%, 89.6%, and 68%, respectively. The accuracy of correct classifications overall test optoelectrical signals was 73.53% while the test error was 26.47%. In the case of NB-ACC the accuracy of correct classifications overall test optoelectrical signals was 57.48% while the test error was 42.52%. The confusion matrix shows that 2805 measurements were misclassified.

Table 1. Performance of NB-LPC and NB-ACC.

		Predicted Class NB-LPC					Predicted Class NB-ACC		
		Class1	**Class2**	**Class3**			**Class1**	**Class2**	**Class3**
True	Class1	1386	0	813	True	Class1	1253	894	52
	Class2	0	1970	229		Class2	17	2182	0
	Class3	704	0	1495		Class3	730	1112	357

Table 2 indicates the percentages of correctly and incorrectly classified instances for each true class in this experiment. True positives of class 1, class 2, and class 3 were 2199, 2199, and 2172, respectively. All the measurements belonging to classes 1 and 2 are classified correctly. The confusion matrix shows that 27 measurements were misclassified. The classification accuracy reached for class 1, class 2, and class 3 was 100%, 100%, and 98.8%, respectively. The accuracy of correct classifications overall test optoelectrical signals was 99.59% while the test error was 0.41%. In the case of NB-ACC, the accuracy of correct classifications overall test optoelectrical signals was 74.40% while the test error was 25.60%. The confusion matrix shows that 1689 measurements were misclassified.

Table 2. Performance of SVM-LPC and SVM-ACC.

		Predicted Class SVM-LPC					Predicted Class SVM-ACC		
		Class1	**Class2**	**Class3**			**Class1**	**Class2**	**Class3**
True	Class1	2199	0	0	True	Class1	1212	987	0
	Class2	0	2199	0		Class2	23	2176	0
	Class3	27	0	2172		Class3	228	451	1520

Table 3 shares the percentages of correctly and incorrectly classified instances for each true class and predicted class in this experiment. True positives of class 1, class 2, and class 3 were 2198, 2199, and 2181, respectively. All the measurements belonging to class 2 are classified correctly. The confusion matrix shows that 19 measurements were misclassified. The class precision reached for class 1, class 2, and class 3 was 99.99%, 100%, and 99.2%, respectively. The accuracy of correct classifications overall test optoelectrical signals was 99.71% while the test error was 0.29%. In the case of NB-ACC, the accuracy of correct classifications overall test optoelectrical signals was 99.51% while the test error was 0.49%. The confusion matrix shows that 32 measurements were misclassified.

Table 3. Performance of LDA-LPC and LDA-ACC.

		Predicted Class LDA-LPC					Predicted Class LDA-ACC		
		Class1	**Class2**	**Class3**			**Class1**	**Class2**	**Class3**
True	Class1	2198	1	0	True	Class1	2197	2	0
	Class2	0	2199	0		Class2	9	2190	0
	Class3	18	0	2181		Class3	1	20	2178

Table 4 indicates the percentages of correctly and incorrectly classified instances for each true class and predicted class carried out in this experiment. The configurations KNN-LPC and KNN-ACC achieved the same results. True positives of class 1, class 2, and class 3 were 2199, 2199, and 2199, respectively. The classification accuracy reached for class 1, class 2, and class 3 was 100%, 100%, and 100%, respectively. All the measurements belonging to each class were classified correctly. The accuracy of correct classifications overall test optoelectrical signals was 100% while the test error was 0%.

Table 4. Performance of KNN-LPC and KNN-ACC.

	Predicted Class KNN-LPC					Predicted Class KNN-ACC		
		Class1	Class2	Class3		Class1	Class2	Class3
True	Class1	2199	0	0	True Class1	2199	0	0
	Class2	0	2199	0	Class2	0	2199	0
	Class3	0	0	2199	Class3	0	0	2199

Table 5 shows the percentages of correctly and incorrectly classified instances for each true class and predicted class carried out in this experiment. The configuration NN-LPC achieved the following results. True positives of class 1, class 2, and class 3 were 2199, 0, and 0, respectively. The classification accuracy reached for class 1, class 2, and class 3 was 100%, 0%, and 0%, respectively. All the measurements belonging to classes 2 and 3 were classified incorrectly. The confusion matrix shows that 4398 measurements were misclassified. The accuracy of correct classifications overall test optoelectrical signals was 33.33% while the test error was 66.67%.

Table 5. Performance of NN-LPC and NN-ACC.

	Predicted Class NN-LPC					Predicted Class NN-ACC		
		Class1	Class2	Class3		Class1	Class2	Class3
True	Class1	2199	0	0	True Class1	2199	0	0
	Class2	2199	0	0	Class2	0	2199	0
	Class3	2199	0	0	Class3	0	0	2199

In the case of NN-ACC the accuracy of correct classifications overall test optoelectrical signals was 100% while the test error was 0%. All the measurements belonging to each class were classified correctly.

7.2. Results Entropy Segmentation

Table 6 shows the percentages of correctly and incorrectly classified instances with NB-SH-LPC model. The accuracy of correct classifications overall test optoelectrical signals was 95.33%, while the test error was 4.67%. The confusion matrix shows that 21 measurements were misclassified.

Table 6. Performance of NB-SH-LPC and NB-SH-ACC.

	Predicted Class NB-SH-LPC					Predicted Class NB-SH-ACC		
		Class1	Class2	Class3		Class1	Class2	Class3
True	Class1	143	7	0	True Class1	144	6	0
	Class2	0	150	0	Class2	0	150	0
	Class3	12	2	136	Class3	3	2	145

In the case of SHH-NB-ACC, the accuracy of correct classifications overall test optoelectrical signals was 97.56% while the test error was 2.44%. The confusion matrix shows that 11 measurements were misclassified.

Table 7 shows the percentages of correctly and incorrectly classified instances with the SVM-SH-LPC model. All the measurements belonging to each class were classified correctly. The accuracy of correct classifications overall test optoelectrical signals was 100% while the test error was 0%.

Table 7. Performance of SVM-SH-LPC and SVM-SH-ACC.

		Predicted Class SVM-SH-LPC					Predicted Class SVM-SH-ACC		
		Class1	Class2	Class3			Class1	Class2	Class3
True	Class1	150	0	0	True	Class1	148	0	2
	Class2	0	150	0		Class2	0	150	0
	Class3	0	0	150		Class3	0	0	150

In the case of SVM-SH-ACC, the accuracy of correct classifications overall test optoelectrical signals was 99.56% while the test error was 0.44%. The confusion matrix shows that two measurements were misclassified.

Tables 8 and 9 show the percentages of correctly and incorrectly classified instances with LDA-SH-LPC, LDA-SH-ACC models, KNN-SH-LPC, KNN-SH-ACC models. All the measurements belonging to each class were classified correctly. The accuracy of correct classifications overall test optoelectrical signals was 100% while the test error was 0%.

Table 8. Performance of LDA-SH-LPC and LDA-SH-ACC.

		Predicted Class LDA-SH-LPC					Predicted Class LDA-SH-ACC		
		Class1	Class2	Class3			Class1	Class2	Class3
True	Class1	150	0	0	True	Class1	150	0	0
	Class2	0	150	0		Class2	0	150	0
	Class3	0	0	150		Class3	0	0	150

Table 9. Performance of KNN-SH-LPC and KNN-SH-ACC.

		Predicted Class KNN-SH-LPC					Predicted Class KNN-SH-ACC		
		Class1	Class2	Class3			Class1	Class2	Class3
True	Class1	150	0	0	True	Class1	150	0	0
	Class2	0	150	0		Class2	0	150	0
	Class3	0	0	150		Class3	0	0	150

Table 10 shows the percentages of correctly and incorrectly classified instances with the NN-SH-LPC model. All the measurements belonging to each class were classified correctly. The accuracy of correct classifications overall test optoelectrical signals was 100% while the test error was 0%. In the case of NN-SH-ACC, the accuracy of correct classifications overall test optoelectrical signals was 99.33% while the test error was 0.67%. The confusion matrix shows that three measurements were misclassified.

Table 10. Performance of NN-SH-LPC and NN-SH-ACC.

		Predicted Class NN-SH-LPC					Predicted Class NN-SH-ACC		
		Class1	Class2	Class3			Class1	Class2	Class3
True	Class1	150	0	0	True	Class1	147	1	2
	Class2	0	150	0		Class2	0	150	0
	Class3	0	0	150		Class3	0	0	150

Figure 10, summarizes the performance of the ML models with ACC, LPC, and Shannon's entropy. Without Shannon's entropy segmentation, the overall misclassification reached was 10710. This pipeline needed better results in terms of accuracy for NB-LPC, NB-ACC, SVM-ACC, and NN-LPC. The worst accuracy was 33.33%.

With Shannon's entropy as a segmentation process, the overall misclassification was 37. The accuracy of the ML classifiers was superior to the previous pipeline.

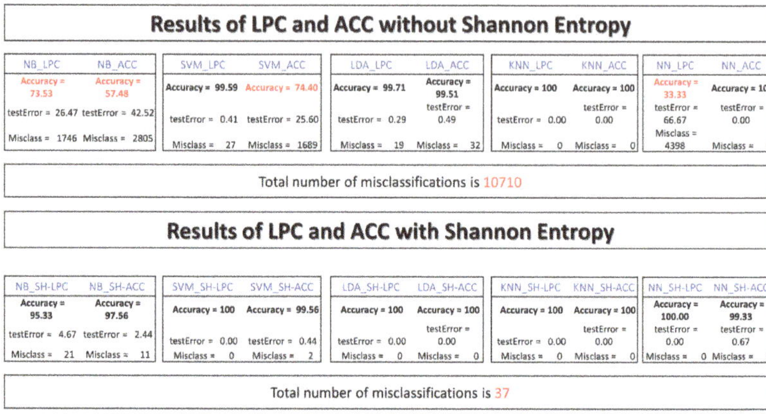

Figure 10. A summary of machine learning model's performance. The first results do not include the Shannon entropy (SH) as a segmentation process. The second results show the performance of ML models with the SH segmentation method.

8. Conclusions

Optical sensors can be viewed as an effective transducer to collect data without physically contacting the object under study. Significant value added for remote sensing such as quantitative or qualitative information, can be gathered. Accessing this information enables us to study the optimal UI parameters, measure many variables, and design early warning systems. Using a phototransistor as a transducer leads to reasonable results in detecting laser beams. The advantage of using this sensor is the internal daylight-blocking filter, which can be bought at low-cost. The main contribution of this study is related to the electronic and physical aspects of sensors in urban sensing systems for SHM tasks and the application of ML methods for its enhancement.

In this paper, we have shown how a TVS can be enhanced using an IT approach. The proposed approach of using an ML framework was implemented to solve the problem related to interference in a real environment. Feature extraction methods were used as a preprocessing stage, and various classifiers were reviewed. The windowing process was integrated into the ML pipeline to split the input signal into temporal segments. The Shannon entropy was used to remove and extract meaningful information from optical patterns.

Results showed significantly better accuracy using Shannon's entropy as a segmentation process. Relevant information was extracted from signals, and ML models were created. Accuracy reached using SVM, LDA, KNN, and NN was over 99%. The accuracy of NB-SH was 95.33% and 97.56% with LPC and ACC, respectively. These results demonstrate Shannon entropy superiority in extracting the optical patterns over frames of complete segments. Without Shannon's entropy segmentation, the worst accuracy was 33.33%.

Practical implementation of this frame can avoid outdoor interferences. In addition, these findings provide additional information about the type of ML techniques that can be used in outdoors environments. This work can be extended to other applications. Three different ECG signals were classified to validate these configurations with Shannon's entropy and LPC-ACC, showing satisfactory results. This indicates that the ML framework with LPC-ACC and the Shannon entropy can solve pattern recognition problems.

Author Contributions: Methodology, J.E.M.-V.; Validation, J.C.R.-Q.; Formal analysis, W.G.-G.; Resources, D.H.-B.; Supervision, W.F.-F.; Project administration, O.S. All authors have read and agreed to the published version of the manuscript.

Funding: This research received no external funding.

Institutional Review Board Statement: Not applicable.

Informed Consent Statement: Not applicable.

Data Availability Statement: Not applicable.

Acknowledgments: This work was made possible thanks to support of the Universidad Autónoma de Baja California and CONACYT, Mexico.

Conflicts of Interest: The authors declare no conflict of interest.

Appendix A. Pseudocode for the Important Parts of the Experimental Design

Figures A1–A4 show a general framework for classifying optical patterns. These procedures were adopted to conduct the experiments.

Appendix A.1. Feature Extraction Pseudocodes

SH_{feat} function

```
Input:  x = raw signal, w_size, Δ_win
Output: A signal segmented with Shannon's entropy SH_feat
Initialization
Create all 1's vector v_ones with length of the w_size
The size of x should be one dimensional, X_size
Obtain the length of x, X_length
Rounds the elements of ((X_size − w_size)/Δ_win) to the nearest integers and add 1, tot_win
create an tot_win-by-X_length matrix of zeros store in Sh_seg
set s_t = 1 and e_n = w_size
        for i = 1 to tot_win
          creates a matrix mat of an 1-by-X_length tiling of copies of   v_ones and multiplies arrays x
          (s_t: e_n, :) and mat.
             apply Ec. (8) to each frame and store in sh_seg(i,:)
                  P(a_n) contains the histogram counts
             s_t = s_t + Δ_win
             e_n = e_n + Δ_win
        end for
Find index greater than mean (sh_seg) and save SH_feat
```

Figure A1. Example of pseudocode to extract features with the Shannon Entropy method. This code is for one-dimensional digital signal processing.

Seg_{feat} function

```
Input:  x = raw signal, frame, ham_win
Output: A signal segmented with hamming window Seg_feat
Initialization
Obtain the length of x, X_length
Rounds the elements of (X_length/ham_win) to the nearest integers minus 2, tot_win
create an tot_win-by- ham_win matrix of zeros,Seg_feat
Create a ham_win -point Hamming window, H

        for i = 0 to tot_win − 1
             s_t = i * frame + 1
             Seg_feat(i + 1,1: ham_win) = (x(s_t: s_t + ham_win − 1).* H)
        end for
```

Figure A2. Example of pseudocode to extract features with a Hamming Window applied to each frame. This is for one-dimensional signals. This code is for one-dimensional digital signal processing.

Appendix A.2. Pseudocodes for Machine Learning Classifiers

Con_{mat} function

```
Input:  Class_n is a cell array with N classes
Output: A confusion matrix Con_mat
Initialization
1: Create a,b,c...n = []
2: length of Class_n, Class_n_length
3: Select w_size and Δ_win for SH_feat function

1:        for i = 1 to Class_n_length
2:                Z_pattern(i) normalized values of Class_n (0 and 1 )
3:                calculate LPC/ACC of SH_feat(Z_pattern(i)), temp
4:                vertical concatenation of matrices 'temp' and 'a', and Do the same for the rest
of the classes
6:        end for
7: Take the features a(:,2:end), b(:,2:end)...n(:,2:end)
8: Split data into training (70%) and testing (30%) a_train,a_test,... n_train, ntest.
9: Fit a multiclass model (SVM,NN,KNN, NB, and LDA)
10:Predict data labels and create confusion matrix, Con_mat
```

Figure A3. This code shows the procedures to build the ML models. ML designers should select LPC or ACC to conduct the experiments. Note that this pseudo code needs the function of SH_{feat}.

$Confusion_{seg}$ function

```
Input:  Class_n is a cell array with N classes
Output: A confusion matrix Confusion_seg
Initialization
1: Create a,b,c...n = []
2: length of Class_n, Class_n_length
3: Select frame and ham_win for Seg_feat function

1:        for i = 1 to Class_n_length
2:                Z_pattern(i) normalized values of Class_n (0 and 1 )
3:                calculate LPC/ACC of Seg_feat(Z_pattern(i)), temp
4:                vertical concatenation of matrices 'temp' and 'a', and Do the same for the rest
of the classes
6:        end for
7: Take the features a(:,2:end), b(:,2:end)...n(:,2:end)
8: Split data into training (70%) and testing (30%) a_train,a_test,... n_train, ntest.
9: Fit a multiclass model (SVM,NN,KNN, NB, and LDA)
10:Predict data labels and create confusion matrix,Confusion_seg
```

Figure A4. This code shows the procedures to build the ML models. ML designers should select LPC or ACC to conduct the experiments. Note that this pseudo code needs the function of Seg_{feat}.

References

1. Dutta, P.; Aoki, P.M.; Kumar, N.; Mainwaring, A.; Myers, C.; Willett, W.; Woodruff, A. Common sense: Participatory urban sensing using a network of handheld air quality monitors. In Proceedings of the 7th ACM Conference on Embedded Networked Sensor Systems, Berkeley, CA, USA, 4–6 November 2009; pp. 349–350. [CrossRef]
2. Calabrese, F.; Ferrari, L.; Blondel, D.D. Urban sensing using mobile phone network data: A survey of research. *ACM Comput. Surv.* **2014**, *47*, 1–20. [CrossRef]
3. Song, G.; Wang, C.; Wang, B. Structural health monitoring (SHM) of civil structures. *Appl. Sci.* **2017**, *7*, 789. [CrossRef]
4. Braunfelds , J.; Senkans, U.; Skels, P.; Janeliukstis, R.; Salgals, T.; Redka, D.; Lyashuk, I.; Porins, J.; Spolitis, S.; Haritonovs, V.; et al. FBG based sensing for structural health monitoring of road infrastructure. *J. Sens.* **2021**, *2021*, 8850368. [CrossRef]
5. Yuan, L.; Fan, W.; Yang, X.; Ge, S.; Xia, C.; Foong, S.Y.; Liew, R.K.; Wang, S.; Van Le, Q.; Lam, S.S. Piezoelectric PAN/$BaTiO_3$ nanofiber membranes sensor for structural health monitoring of real-time damage detection in composite. *Compos. Commun.* **2021**, *25*, 100680. [CrossRef]

6. Siahkouhi, M.; Razaqpur, G.; Hoult, N.; Baghban, M.H.; Jing, G. Utilization of carbon nanotubes (CNTs) in concrete for structural health monitoring (SHM) purposes: A review. *Constr. Build. Mater.* **2021**, *309*, 125137. [CrossRef]
7. Mosleh, A.; Meixedo, A.; Ribeiro, D.; Montenegro, P.; Calçada, R. Early wheel flat detection: An automatic data-driven wavelet-based approach for railways. *Veh. Syst. Dyn.* **2023**, *61*, 1644–1673. [CrossRef]
8. Mosleh, A.; Meixedo, A.; Ribeiro, D.; Montenegro, P.; Calçada, R. Automatic clustering-based approach for train wheels condition monitoring. *Int. J. Rail Transp.* **2022**, 1–26. [CrossRef]
9. Das, S.; Saha, P.; Patro, S. Vibration-based damage detection techniques used for health monitoring of structures: A review. *J. Civ. Struct. Health Monit.* **2016**, *6*, 477–507. [CrossRef]
10. Lakshmi, K.; Rao, A.R.M.; Gopalakrishnan, N. Singular spectrum analysis combined with ARMAX model for structural damage detection. *Struct. Control Health Monit.* **2017**, *24*, e1960. [CrossRef]
11. Bhowmik, B.; Krishnan, M.; Hazra, B.; Pakrashi, V. Real-time unified single-and multi-channel structural damage detection using recursive singular spectrum analysis. *Struct. Health Monit.* **2019**, *18*, 563–589. [CrossRef]
12. Bhowmik, B.; Tripura, T.; Hazra, B.; Pakrashi, V. First-order eigen-perturbation techniques for real-time damage detection of vibrating systems: Theory and applications. *Appl. Mech. Rev.* **2019**, *71*, 060801. [CrossRef]
13. Krishnan, M.; Bhowmik, B.; Hazra, B.; Pakrashi, V. Real time damage detection using recursive principal components and time varying auto-regressive modeling. *Mech. Syst. Signal Process.* **2018**, *101*, 549–574. [CrossRef]
14. Sony, S.; Laventure, S.; Sadhu, A. A literature review of next-generation smart sensing technology in structural health monitoring. *Struct Control Health Monit* **2019**, *26*, e2321. [CrossRef]
15. Kaartinen, E.; Dunphy, K.; Sadhu, A. Lidar-based structural health monitoring: Applications in civil infrastructure systems. *Sensors* **2022**, *22*, 4610. [CrossRef]
16. You, Y.; Wang, Y.; Chao, W.L.; Garg, D.; Pleiss, G.; Hariharan, B.; Campbell, M.; Weinberger, K.Q. Pseudo-lidar++: Accurate depth for 3D object detection in autonomous driving. *arXiv* **2020**, arXiv:1906.06310.
17. Li, Y.; Ibanez-Guzman, J. Lidar for autonomous driving: The principles, challenges, and trends for automotive lidar and perception systems. *IEEE Signal Process. Mag.* **2020**, *37*, 50–61. [CrossRef]
18. Villa, F.; Severini, F.; Madonini, F.; Zappa, F. SPADs and SiPMs arrays for long-range high-speed light detection and ranging (LiDAR). *Sensors* **2021**, *21*, 3839. [CrossRef] [PubMed]
19. Miranda-Vega, J.E.; Rivas-López, M.; Flores-Fuentes, W.; Sergiyenko, O.; Lindner, L.; Rodríguez-Quiñonez, J.C. Reconocimiento de patrones aplicando LDA y LR a senales optoelectrónicas de sistemas de barrido óptico. *Rev. Iberoam. Autom. Inform. Ind.* **2020**, *17*, 401–411. [CrossRef]
20. Flores-Fuentes, W.; Rivas-Lopez, M.; Sergiyenko, O.; Gonzalez-Navarro, F.F.; Rivera-Castillo, J.; Hernandez-Balbuena, D.; Rodríguez-Quiñonez, J.C. Combined application of power spectrum centroid and support vector machines for measurement improvement in optical scanning systems. *Signal Process.* **2014**, *98*, 37–51. [CrossRef]
21. Rivera-Castillo, J.; Flores-Fuentes, W.; Rivas-López, M.; Sergiyenko, O.; Gonzalez-Navarro, F.F.; Rodríguez-Quiñonez, J.C.; Hernández-Balbuena, D.; Lindner, L.; Básaca-Preciado, L.C. Experimental image and range scanner datasets fusion in SHM for displacement detection. *Struct. Control Health Monit.* **2017**, *24*, e1967. [CrossRef]
22. Sergiyenko, O.; Nieto-Hipólito, J.I.; Rodríguez-Quiñones, J.C.; Basaca-Preciado, L.C.; Rivas-López, M.; Starostenko, O.; Tyrsa, V.; Hernández, W. Electromechanical 3D optoelectronic scanners: Resolution constraints and possible ways of improvement. In *Optoelectronic Devices and Properties*; Sergiyenko, O., Ed.; IntechOpen : London, UK, 2011; pp. 549–582. [CrossRef]
23. Tyrsa, V.; Sergiyenko, O.; Burtseva, L.; Bravo-Zanoguera, M.; Devia, L., Rendon, I.; Tyrsa, V. Mobile transport object control by technical vision means. In Proceedings of the Electronics, Robotics and Automotive Mechanics Conference (CERMA'06), Cuernavaca, Mexico, 26–29 September 2006; pp. 74–82. [CrossRef]
24. Sergiyenko, O.; Hernandez, W.; Tyrsa, V.; Cruz, L.F.D.; Starostenko, O.; Peña-Cabrera, M. Remote sensor for spatial measurements by using optical scanning. *Sensors* **2009**, *9*, 5477–5492. [CrossRef] [PubMed]
25. Rivas, M.; Sergiyenko, O.; Aguirre, M.; Devia, L.; Tyrsa, V.; Rendón, I. Spatial data acquisition by laser scanning for robot or SHM task. In Proceedings of the 2008 IEEE International Symposium on Industrial Electronics, Cambridge, UK, 30 June–2 July 2008; pp. 1458–1462. [CrossRef]
26. Lindner, L.; Sergiyenko, O.; Rodríguez-Quiñonez, J.C.; Rivas-Lopez, M.; Hernandez-Balbuena, D.; Flores-Fuentes, W.; Natanael Murrieta-Rico, F.; Tyrsa, V. Mobile robot vision system using continuous laser scanning for industrial application. *Ind. Robot* **2016**, *43*, 360–369. [CrossRef]
27. Sergiyenko, O. Optoelectronic system for mobile robot navigation. *Optoelectron. Instrum. Data Process.* **2010**, *46*, 414–428. [CrossRef]
28. Krishnan, S.; Athavale, Y. Trends in biomedical signal feature extraction. *Biomed. Signal Process. Control* **2018**, *43*, 41–63. [CrossRef]
29. Dave, N. Feature extraction methods LPC, PLP and MFCC in speech recognition. *Int. J. Adv. Res. Eng. Technol.* **2013**, *1*, 1–4.
30. Shannon, C.E. A mathematical theory of communication. *Bell Syst. Tech. J.* **1948**, *27*, 379–423. [CrossRef]
31. Guido, R.C. A tutorial review on entropy-based handcrafted feature extraction for information fusion. *Inf. Fusion* **2018**, *41*, 161–175. [CrossRef]
32. Wickramasinghe, I.; Kalutarage, H. Naive Bayes: Applications, variations and vulnerabilities: A review of literature with code snippets for implementation. *Soft Comput.* **2021**, *25*, 2277–2293. [CrossRef]

33. Leung, K.M. *Naive Bayesian Classifier*; Polytechnic University, Department of Computer Science/Finance and Risk Engineering Hong Kong, 2007; pp. 123–156.
34. Wang, Z.; Xue, X. Multi-class support vector machine. In *Support Vector Machines Applications*; Ma, Y., Guo, G., Eds.; Springer International Publishing: Cham, Switzerland, 2014; pp. 23–48. [CrossRef]
35. Angulo, C.; Parra, X.; Catala, A. K-SVCR. A support vector machine for multi-class classification. *Neurocomputing* **2003**, *55*, 57–77 [CrossRef]
36. Li, T.; Zhu, S.; Ogihara, M. Using discriminant analysis for multi-class classification. In Proceedings of the Third IEEE International Conference on Data Mining, Melbourne, FL, USA, 22 November 2003; pp. 589–592. [CrossRef]
37. Garcia, E.K.; Feldman, S.; Gupta, M.R.; Srivastava, S. Completely lazy learning. *IEEE Trans. Knowl. Data Eng.* **2009**, *22*, 1274–1285 [CrossRef]
38. Triguero, I.; García-Gil, D.; Maillo, J.; Luengo, J.; García, S.; Herrera, F. Transforming big data into smart data: An insight on the use of the k-nearest neighbors algorithm to obtain quality data. *Wiley Interdiscip. Rev. Data Min. Knowl. Discov.* **2019**, *9*, e1289 [CrossRef]
39. Alsouda, Y.; Pllana, S.; Kurti, A. A machine learning driven IoT solution for noise classification in smart cities. *arXiv* **2018**, arXiv:1809.00238.
40. Murugan, P. Feed forward and backward run in deep convolution neural network. *arXiv* **2017**, arXiv:1711.03278.
41. Haldorai, A.; Ramu, A. Canonical correlation analysis based hyper basis feedforward neural network classification for urban sustainability. *Neural Process. Lett.* **2021**, *53*, 2385–2401. [CrossRef]
42. Hemeida, A.M.; Hassan, S.A.; Mohamed, A.A.A.; Alkhalaf, S.; Mahmoud, M.M.; Senjyu, T.; El-Din, A.B. Nature-inspired algorithms for feed-forward neural network classifiers: A survey of one decade of research. *Ain Shams Eng. J.* **2020**, *11*, 659–675. [CrossRef]

entropy

Article

A Low-Illumination Enhancement Method Based on Structural Layer and Detail Layer

Wei Ge *, Le Zhang, Weida Zhan, Jiale Wang, Depeng Zhu and Yang Hong

National Demonstration Center for Experimental Electrical, School of Electronic and Information Engineering, Changchun University of Science and Technology, Changchun 130022, China
* Correspondence: gewei@cust.edu.cn

Abstract: Low-illumination image enhancement technology is a topic of interest in the field of image processing. However, while improving image brightness, it is difficult to effectively maintain the texture and details of the image, and the quality of the image cannot be guaranteed. In order to solve this problem, this paper proposed a low-illumination enhancement method based on structural and detail layers. Firstly, we designed an SRetinex-Net model. The network is mainly divided into two parts: a decomposition module and an enhancement module. Second, the decomposition module mainly adopts the SU-Net structure, which is an unsupervised network that decomposes the input image into a structural layer image and detail layer image. Afterward, the enhancement module mainly adopts the SDE-Net structure, which is divided into two branches: the SDE-S branch and the SDE-D branch. The SDE-S branch mainly enhances and adjusts the brightness of the structural layer image through Ehnet and Adnet to prevent insufficient or overexposed brightness enhancement in the image. The SDE-D branch is mainly denoised and enhanced with textural details through a denoising module. This network structure can greatly reduce computational costs. Moreover, we also improved the total variation optimization model as a mixed loss function and added structural metrics and textural metrics as variables on the basis of the original loss function, which can well separate the structure edge and texture edge. Numerous experiments have shown that our structure has a more significant impact on the brightness and detail preservation of image restoration.

Keywords: low-illumination image enhancement; image decomposition; U-Net; Retinex-Net

Citation: Ge, W.; Zhang, L.; Zhan, W.; Wang, J.; Zhu, D.; Hong, Y. A Low-Illumination Enhancement Method Based on Structural Layer and Detail Layer. *Entropy* **2023**, *25*, 1201. https://doi.org/10.3390/e25081201

Academic Editors: Oleg Sergiyenko, Wendy Flores-Fuentes, Julio Cesar Rodríguez-Quiñonez and Jesús Elías Miranda-Vega

Received: 5 June 2023
Revised: 8 August 2023
Accepted: 10 August 2023
Published: 12 August 2023

1. Introduction

With the development of electronic devices, digital images have played an important role in our lives. They are widely used in fields such as traffic management, medicine [1], satellite remote sensing, and target recognition and tracking. However, the complexity of the shooting environment often leads to low-quality phenomena, such as low recognition, color distortion, and loss of details. Due to the low quality of images, subsequent computer vision tasks become difficult. Because image enhancement can improve the visibility and practicality of low-illumination images, it has important research value.

At present, image enhancement is mainly divided into traditional methods and deep-learning-based methods. Retinex theory, a model for brightness and color perception in human vision and a commonly used low-illumination image enhancement method, was proposed by Land [2] in the 1970s. Afterward, many scholars continued to build on this basis, from the single-scale Retinex (SSR) algorithm to the multi-scale Retinex (MSR) algorithm [3] and then to MSR with color recovery (MSRCR) [4]. However, both SSR and MSR generally exhibit color distortion. Compared with other algorithms, the MSRCR algorithm has a better color restoration ability, but it has high computational complexity and many adjustable parameters that are difficult to adaptively select. In addition, on the basis of Retinex theory, some simple and efficient image enhancement methods based on the Retinex model have been proposed for low-illumination image enhancement [5], such as LIME [6], RobustRetinex [7], and JED [8]. Reference [9] also proposed a convex variational model, which could effectively decompose the gradient field of an image into prominent edges and a relatively smoother illumination field through first- and second-order total change regularization. In summary, although these traditional methods can effectively enhance image brightness and preserve high-frequency information such as edges and corners, they cannot effectively avoid problems such as uneven contrast and color distortion.

In recent years, enhancement methods based on deep learning have gradually developed. The first network based on deep learning to solve low-light image enhancement was LLNet [10]. Paired images, including low-light images and normal images, were input into the network model and trained through automatic encoders to learn the basic signal features in low-light images and adaptively improve the brightness and denoising ability. This method loses details during image reconstruction, resulting in a slightly blurry enhancement of the final image. Afterward, the Retinex-Net network model was proposed by combining Retinex theory with divine-level convolutional networks [11]. Firstly, an image was decomposed into illumination maps and reflection maps through the decomposition network. Secondly, the illumination image was enhanced through an enhancement network, and finally, the enhanced illumination image was multiplied by the decomposed reflection image to obtain the final enhanced image. After Retinex-Net, an author proposed a method to improve the quality of low-light images by analyzing the histogram of the images and utilizing deep learning techniques. For example, the MBLLEN [12] algorithm is a low-light image enhancement algorithm based on a multi-branch network. This algorithm extracts rich image features from different levels, enhances images through multiple sub-networks, and finally generates output images through multi-branch fusion. Another algorithm is the EnlightenGAN [13] algorithm, which improves the quality of low-light images through local discriminators and attention modules. This algorithm has shown good enhancement effects in real scenes, but there are still some shadow areas in some images. Kind [14] used a trainable denoising module for reflectivity recovery. In addition, a learnable mapping function was designed in the lighting adjustment module, where images could be flexibly restored at user-specific lighting levels. Sci [15] adopted a new self-adjustment lighting framework and established a cascaded lighting learning process with weight sharing to achieve fast and flexible image enhancement. These methods all have good enhancement performance. In real-world scenes, unclear details and inappropriate exposure are common. However, the existing methods fail to solve the above problems.

The proposed method draws inspiration from the Retinex theory [16]. The Retinex model can divide an image into two parts: the incident component and the irradiation component. Specifically, the irradiation component reflects the distribution of light in the shooting environment. The reflection component represents the essential properties of an image. In this paper, the image is decomposed into two parts: a structural layer and a detail layer. The structural layer mainly refers to the main contour or global geometric structure information of the image, and the clear boundaries and connected regions are the main reasons for light decay. The detail layer refers to the image containing small scales and details, which are usually periodic and oscillatory. Based on the above ideas, a low-illumination image enhancement method based on a structural layer and a detail layer is proposed. The main contributions of this article include:

(1) This proposed SRetinex-Net model is mainly divided into two parts: a decomposition module and an enhancement module. The decomposition module mainly adopts the SU-Net structure, which decomposes the input image into a structural layer image and a detail layer image. The enhancement module mainly adopts the SDE-Net structure, which is divided into two branches: the SDE-S branch and the SDE-D branch. The SDE-S branch mainly enhances the brightness of the structural layer, while the SDE-D branch enhances the textural detail of the detail layer.

(2) The SU-Net structure is an unsupervised network, which mainly extracts and merges the structural features of input images through a sampling layer and skip connection.

A brightness calibration module was added to the SDE-S branch. After the brightness enhancement of the structural layer image through the Ehnet module, the feature extraction and reconstruction of the enhanced image should be completed through the Adnet module to adjust the image brightness, making the image brightness more balanced and accurate. The SDE-D branch is mainly denoised and enhanced with detailed textures through a denoising module. This network structure greatly improves computational efficiency.

(3) The total variation optimization model was improved as a mixed loss function, and the structure component and texture component were added as variables on the basis of the original loss function, which can make the edge and texture better separated so that the edge of the structural layer image is clear and the details of the detail layer image are more abundant.

(4) Compared with previous methods, the structural layer image structure obtained by decomposing the image is more complete in preservation, and the detail layer image contains more abundant details. In image enhancement, our method does not refer to normal light images. We can adaptively adjust image brightness to better match human visual effects and have conducted extensive experimental comparisons to demonstrate the superiority of our method. Compared with all other methods, we can self-calibrate image brightness, enhance image contrast, and improve image details and visibility.

2. Methods

The low-illumination image enhancement method based on convolutional neural networks makes it difficult to generate complete details during image reconstruction, which can easily lead to slightly blurry enhancement results. To solve this problem, this paper proposes a low-illuminance enhancement method based on decomposing the image into a structural layer and a detail layer. First, the color space of the source image is transformed from RGB to HSV. Then, the V image component is decomposed into a structural layer and a detail layer. Furthermore, the structural layer image's brightness is enhanced through structural branching, while the detail layer image's textural details are enhanced through detail branching. Finally, the enhanced structural layer and detail layer are multiplied to obtain the enhanced V-component image. The enhanced V-component image is combined with the H- and S-component images and transformed into a color space to obtain the final low-illumination-enhanced image.

2.1. Framework of the Proposed Method

Firstly, the color space of the source image is transformed from RGB to HSV. Secondly, the source image is decomposed into the H-, S-, and V-channel components, which can be referred to as I_h, I_s, and I_V, respectively. Finally, the H and S image channels remain unchanged, and the V image channel is extracted as the input to the network. Afterward, the input image I_V can be decomposed into the structural layer I_{Vs} and detail layer I_{Vk} via the decomposition module, which is the input of the enhancement module. Next, the structural layer I_{Vs} is fed into the SDE-S branch to enhance the brightness. The detail layer I_{Vk} is fed into the SDE-D branch to enhance the details. Then, the brightness enhancement image I_{Vs}' and the detail enhancement image I_{Vk}', which are the outputs of the two branches, are multiplied to obtain the enhanced image I_V' of the V-channel component. Finally, the final enhanced image I' is obtained by fusing the components of the I_h, I_s, and I_V' channels and converting it from the HSV space to the RGB space.

As shown in Figure 1, the proposed method can enhance and maintain the detail information of an image while enhancing the brightness and contrast of the image, ensuring the visual quality of the enhanced image.

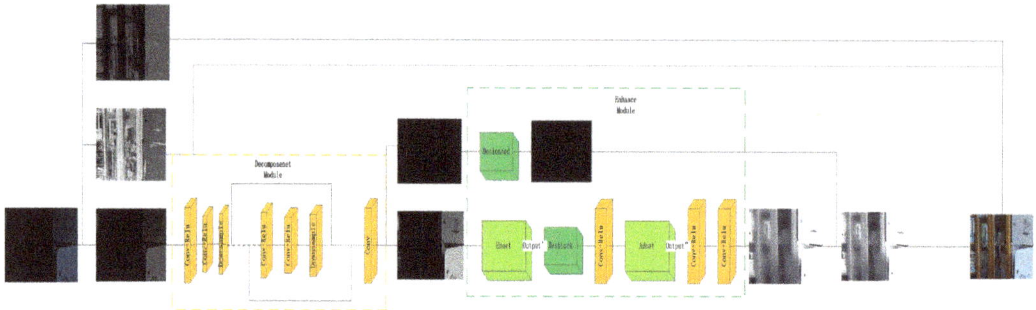

Figure 1. Framework of the proposed method.

2.2. Structure of the Network

2.2.1. Decomposition Module

Compared with the Retinex-based method, the decomposition model decomposes the input image into structural layer images I_{Vs} and detail layer images I_{Vk}, rather than illumination and reflection images. Therefore, we do not need to label images with normal brightness to constrain the network training.

The original U-net structure [17] consumes considerable training time and has the problem of repeated training. Because there are no labels for the structural layers and detail layers of the trained images, we need to retrain each image during training. In order to satisfy this condition, a SU-Net structure is proposed, which mainly uses multiple convolution layers and a nonlinear activation function connection, including an upper sampling layer, lower sampling layer, and skip connection. Because we input a single-channel image, the first layer of convolution has 1 input channel and 64 output channels, the last layer of convolution has 64 input channels and 1 output channel, and the remaining convolution has 64 input and output channels. The entire network completes the feature extraction and reconstruction of images.

2.2.2. Enhancement Module

The enhancement module adopts an SDE-Net structure. The network is divided into two branches: the SDE-S branch and the SDE-D branch. The SDE-S branch mainly enhances the brightness of the structural layer I_{vs} to obtain the enhanced structural layer image I_{Vs}', while the SDE-D branch enhances the textural details of the detail layer I_{vk} to obtain the enhanced detail layer image I_{Vk}'. Ehnet in the SDE-S branch is mainly composed of multiple convolution layers and an activation function, and the size of the convolution kernel is 3×3. The input channel number of the first convolutional layer is 1, and the output channel number of the last convolutional layer is 1. It mainly performs feature extraction and reconstruction on the input image to enhance image brightness. Adnet is a brightness adjustment network that receives a preliminary brightness-enhanced images output via Ehnet, performs feature extraction and reconstruction on the input image to adjust the brightness of the image, prevents image brightness overexposure or insufficient brightness, and makes the brightness of the image more balanced and accurate. Adnet mainly consists of blocks composed of multiple convolutional layers, each with two 3×3, a reduction layer, and an activation function. The number of input channels of the first convolution layer is 1, and the number of output channels of the last convolution layer is 1. The SDE-D branch enhances the texture details of the detail layer I_{vk} through a noise reduction module. This approach can greatly improve computational efficiency, as shown in Figure 2.

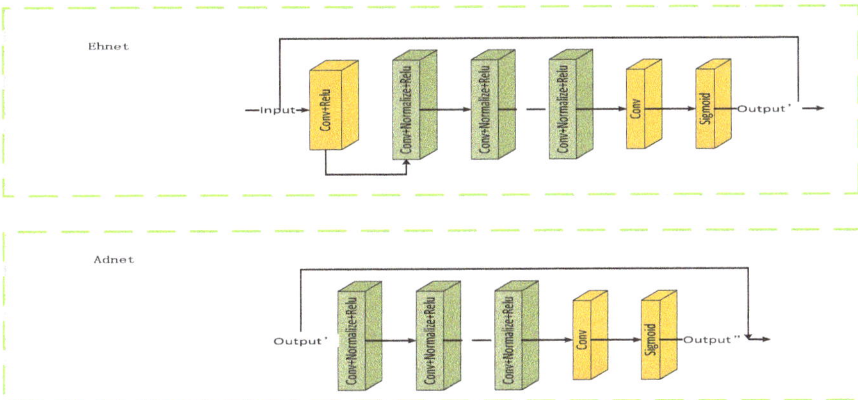

Figure 2. Brightness enhancement network and brightness adjustment network.

2.3. Loss Function

2.3.1. Fully Variational Loss Function

The image can be divided into different image layers based on various methods. For example, the image can be divided into high-frequency and low-frequency signals based on frequency domain methods, and the original image can be decomposed into illumination images and reflection images based on the Retinex algorithm. The image can be decomposed into structural layers and detail layers. The structural layer mainly refers to the main contour or global geometric structural information of the image, with clear boundaries and connected regions. The detail layer refers to a layer that contains small scales and details, which are typically periodic and oscillatory. There are many methods for image decomposition, and image filters can be used for filtering, such as the rolling filter algorithm [18]. A Gaussian filter is used to remove texture, while bilateral filters are used to restore edges, which also causes ringing and artifacts around the edges. Image decomposition can also be achieved using methods such as the TV (total variation) full variation model [19] and the relative total variation (RTV) model [20].

Herein, we use the total variation model as the basis for the optimization framework, and the common total variation objective formula is as follows:

$$S = \underset{S}{\operatorname{argmin}} \sum_i (\| u(i) - I(i) \|_m + \| \beta \nabla u(i) \|_n) \tag{1}$$

We represent the intensity of the input image, its structural layer components, and its detail layer components as I, S, and K, respectively. Our goal is to obtain unknown structural layer images S and detail layer images K from known input images I. Because I = S + K, we only need to estimate one of S and K. In the variational framework, the structural component S is generally obtained by changing the feature metric of the fully variational model, such as in references [21–23], where i represents the pixel intensity at the point and is a balance coefficient, and the subscripts m and n represent the function space of the two terms. In Formula (1), the first term is the fidelity term, which mainly makes the structural layer S infinitely close to the input image I. The second term is the regularization term, which is mainly used to remove the edges in the structural layer diagram.

In order to better decompose the image structural layer and detail layer, considering the anisotropy of image gradients, structural metrics [24] and textural metrics [24] are used to optimize the total variational function.

The structural measurement formula is as follows:

$$G_s(i) = A_J(i)\frac{\| \nabla f(i) \|_1}{R}, \tag{2}$$

where $A_J(i)$ represents the degree of anisotropy in the local area of point i, j represents the positive definite matrix, and a larger value of A indicates that the degree of anisotropy and structural strength at that point is stronger. On the contrary, a smaller value of A indicates a smaller degree of anisotropy and stronger texture details at that point. $\| \nabla f(i) \|_1$ represents the L1 norm of the gradient of the image at point i, and R represents the maximum value.

The texture measurement formula is as follows:

$$G_t(i) = \frac{1}{\{C(i)\}} * \sum_{j \in \{C(i)\}} \cos\theta_{ij} * e^{(-\Phi(h_j,h_i))}, \tag{3}$$

where $\{C(i)\}$ represents the set of domain pixels of point i, j represents the domain position of pixel point i, and $-\Phi(h_j, h_i)$ represents the cross-entropy. $\cos\theta_{ij}$ represents the edge direction positions of pixel points i and j. When i and j are on the same edge, the included angle is 90 degrees. Conversely, when the included angle is 0, it has no effect on the texture measurement. The range of values for G_t is [0, 1].

Therefore, the objective function we utilized is as follows:

$$S = \underset{s}{\operatorname{argmin}} \sum_i \left(\| u(i) - I(i) \|_1^2 + D(i) \| \nabla u(i) \|_1 \right), \tag{4}$$

$$D(i) = [\beta_s(1 - G_s(i)) + \beta_t G_t(i)], \tag{5}$$

where S represents the decomposed structural layer, I represents the original input image, $\| u(i) - I(i) \|_1^2$ refers to the difference between the input image and the output structural layer image, $D(i)$ refers to the i-point structural and textural metrics, $\nabla u(i)$ is the gradient of the structural components at the i-point, and $\| \|_1^2$ is the Lp norm. G_s is a structural metric responsible for filtering out structural edges. When the structural metric value of point i is large, that is, G_s approaches 1, and 1- G_s approaches 0, the gradient regularization term also approaches 0. Therefore, the structural edges of point i can be retained. In this case, the influence of the second term should be reduced; on the contrary, when the texture measurement value of point i is large, that is, G_t tends to 1, and the gradient regularization term is large, the texture edge at point i can be separated from the structure component. At this point, the second main function is to effectively remove textural edges. β_s and β_t are the equilibrium coefficients of G_s and G_t.

We defined the loss function as the objective function (4) and (5) and trained the neural network. The fully variational loss function Formula (6) is as follows:

$$\text{loss} = \sum_i \left(\| u(i) - I(i) \|_1^2 + D(i) \| \nabla u(i) \|_1 \right), \tag{6}$$

Because we do not have the label of the structural layer image, the unknown parameters in the loss function are adjusted with the input image.

The structural layer images and detail layer images obtained using the decomposition module are shown in Figure 3.

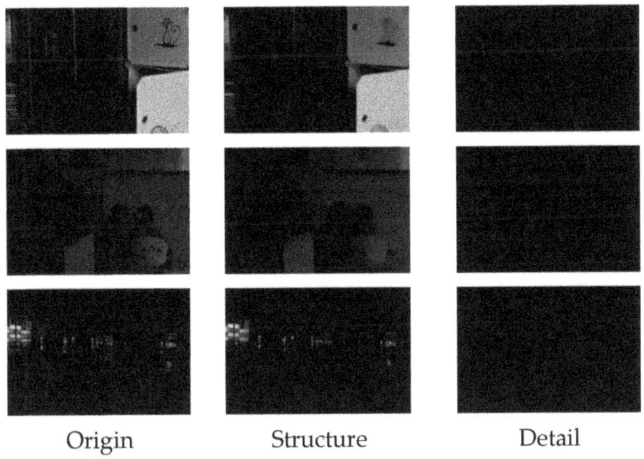

Figure 3. Structural layer images and detail layer images.

2.3.2. Unsupervised Loss Function

This enhancement method can avoid the uncertainty of paired data sets, and we used the unsupervised loss function [15] to achieve this purpose.

$$L_{un} = \gamma_1 L_f + \gamma_2 L_s, \tag{7}$$

where L_f and L_s represent the fidelity loss and smoothing loss. γ_1 and γ_2 are two positive equilibrium parameters.

The fidelity loss function ensures that the estimated illumination is consistent with the pixel level between the inputs of each stage. The specific formula is as follows:

$$L_f = \sum_{t=1}^{T} \| x^t - \left(y + s^{t-1}\right) \|^2, \tag{8}$$

where T represents the total number of stages. In fact, the fidelity loss function uses the redefined input to constrain the output lighting rather than the live scene or normal low-light input photographed artificially. X represents the generated illumination estimation, y represents the low-illumination image to be processed, and s represents the adjustment parameters.

The formula of the illumination smoothing loss function is as follows:

$$L_s = \sum_{i=1}^{N} \sum_{j \in N(i)} w_{i,j} \left| x_i^t - x_j^t \right| \tag{9}$$

where N represents the total number of pixels. I is the i-th pixel. N (i) represents point 5 of i × adjacent pixels in the range of 5. X represents the generated illumination estimation image, and $W_{i,j}$ represents the weight between pixels i and j, which is used to measure the similarity between pixels i and j.

3. Experimental Results and Analysis

To verify the effectiveness of the proposed method, our low-illumination image enhancement method based on structural and detail layers was compared with existing classic algorithms as a comparative experiment, and validation analysis was conducted based on two aspects: subjective visual effects and objective evaluation indicators. In order to verify the generalization of the network, this article used the publicly available LOL dataset and MEF dataset as training datasets. The LOL dataset contains 485 pairs of low-light/normal-light training images and 15 low-light test images. The MEF dataset contains 84 low-light test images.

In the pre-training process of the decomposition module, data preprocessing is performed first. The color space of the source image is transformed from RGB to HSV, and the V-component image is extracted. Then, structural and textural metrics are calculated separately for each V-component image, and the experimental results are saved. Afterward, the network is used for pre-training. The structural metric and detail metric of each training image remain unchanged, and the balance coefficient in the loss function formula is 3.0. A total of 30 iterations are conducted in the pre-training stage. At this time, the network is considered to converge, and the pre-training is complete.

There are two parts to the enhancement module: pre-training and fine-tuning. The structural layers obtained from the decomposition module are pre-trained with 1000 iterations and a learning rate of 0.0003. After approximately 396 iterations, the network converges, and the pre-training ends.

3.1. Subjective Evaluation

In terms of subjective visual effects, as shown in Figure 4, six groups of images are selected, including indoor scenes and natural landscapes. From left to right, there are enhancement images of low-illumination images, the Retinex-Net algorithm, URetinex-Net algorithm, LIME algorithm, Zero DCE++ algorithm, Kind++ algorithm, and the algorithm presented in this article.

(a) (b) (c) (d) (e) (f) (g)

Figure 4. Comparison of low-illumination image enhancement effects of different algorithms. (**a**) Origin images, (**b**) RetinexNet results, (**c**) UretinexNet results, (**d**) LIME results, (**e**) Zero-Dce++ results, (**f**) KinD++ results, and (**g**) our results.

As shown in Figures 4 and 5, it can be seen that the enhanced images of the Retinex-Net algorithm exhibit significant color distortion, with some images exhibiting a noticeable ink sensation. The URetinex-Net algorithm [25] enhances the overall image and has certain defects in image color retention. Many objects tend to have obvious fading phenomena. The LIME algorithm has an excessive enhancement effect on local regions. Zero DCE++ [26] has a poor noise suppression effect and is prone to detail loss. The KinD++ algorithm [27] significantly improves the brightness, but the brightness of the enhanced image cannot maintain the same brightness distribution characteristics as the original image, and there is obvious color distortion. The proposed method in this article has a more reliable enhancement effect, which can work well under different types of lighting conditions, effectively avoiding situations where the overall vision is too high or the enhancement is insufficient. The final enhancement effect is also more natural and realistic.

(a) (b) (c) (d) (e)

Figure 5. Comparison of low-illumination image enhancement effects of different algorithms. (**a**) Origin images, (**b**) Kind results, (**c**) Sci results, (**d**) RUAS results, and (**e**) our results.

The details of Figures 4 and 5 are enlarged in Figures 6 and 7. In the figures, it can be seen that the results obtained with the Retinex-Net algorithm show color distortion and excessive detail enhancement in some areas, such as the bookcase area and cliff area in the image, which are biased toward ink and have artifacts. The URetinex-Net algorithm shows a significant color bias toward white in the enlarged area of the flowerpot and clothing. The LIME algorithm clearly shows the presence of a large amount of noise in the enlarged area of streetlights and swimming pools. The Zero-DCE++ algorithm shows that the contrast enhancement is not sufficient in the enlarged area, resulting in a dim overall color sense in the image and an obvious problem of detail loss. The KinD++ algorithm has the problem of overexposure in the magnified area of the natural landscape. The magnified area of the indoor scene recovers the color distortion, and the brightness recovery is unstable. The proposed method in this article utilizes the advantages of the HSV color space compared with the other methods. While maintaining the structure, it preserves most of the original information of the images and enriches the details of the objects, avoiding color distortion to the greatest extent. At the same time, it can effectively suppress the generation of noise and avoid the presence of artifacts.

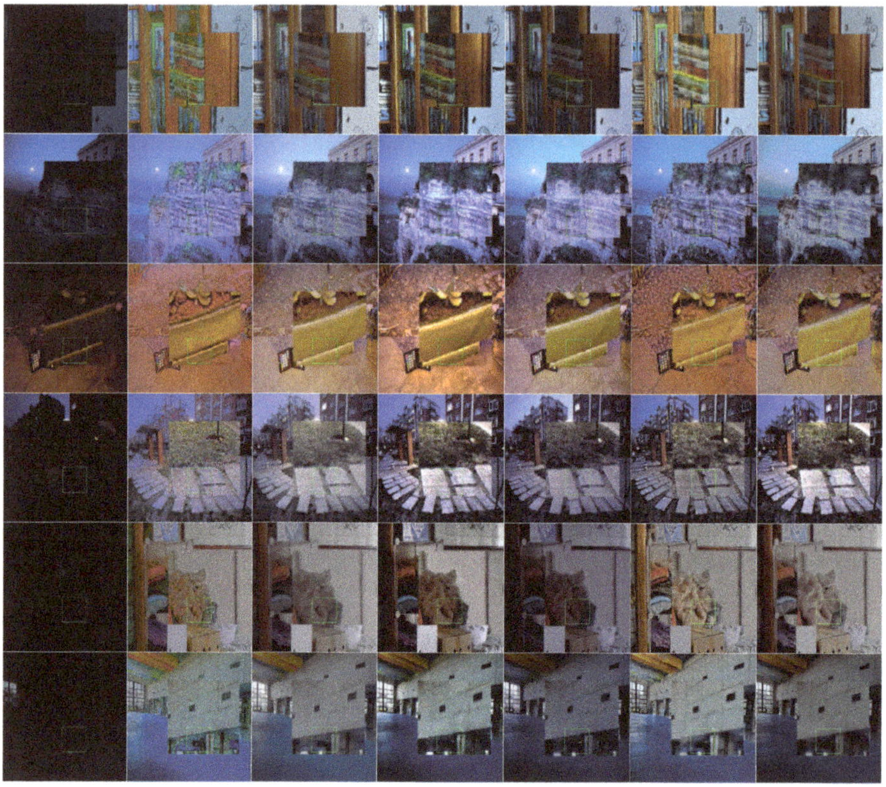

(a) (b) (c) (d) (e) (f) (g)

Figure 6. Comparison of local enlargement details of low-illuminance images using different algorithms. (**a**) Origin images, (**b**) RetinexNet results, (**c**) UretinexNet results, (**d**) LIME results, (**e**) Zero-Dce++ results, (**f**) KinD++ results, and (**g**) our results.

(a) (b) (c) (d) (e)

Figure 7. Comparison of local enlargement details of low-illuminance images using different algorithms. (**a**) Origin images, (**b**) Kind results, (**c**) Sci results, (**d**) RUAS results, and (**e**) our results.

3.2. Objective Evaluation Indicators

In order to better evaluate image quality, this article used the natural image quality evaluator (NIQE) [28], structural similarity index (SSIM), peak signal-to-noise ratio (PSNR), and learned perceptual image patch similarity [29] (LPIPS) to evaluate the resulting images. As an evaluation indicator, the higher the values of the SSIM and PSNR, the better the image quality we will obtain; on the contrary, smaller NIQE and LPIPS values indicate better image quality. BIQI is an image evaluation index without reference images, with values ranging from zero to one. The closer the value is to one, the better the image quality. The EMEE evaluation indicator is used to measure image edge information, especially for images with clear edges. The EMEE value is small, and vice versa. SDME is an image evaluation indicator used to measure the degree of edge change in images. A larger value indicates a more significant edge change in the image. BRISQUE is a five-reference image quality evaluation indicator, with values typically ranging from 50 to 100. The larger the value, the better the image quality. The AME evaluation indicator is suitable for measuring the quality of image edges, and the value is usually positive. Images with clear edges have a higher AME value, while the opposite is true. Visibility is an indicator of image visibility, with larger values indicating clearer targets or details in the image, and vice versa.

The enhanced results of the test datasets are shown in Table 1. ↑ the larger the value, the better the enhancement effect. On the contrary, ↓ the smaller the numerical value, the better the enhancement effect. As shown in the table, our method is better than the other methods in the NIQE, SSIM, PSNR, BIQI, EMEE, BRISQUE, AME, and visibility metrics, except that it performs slightly worse than URetinexNet and LIME in LPIPS and SDME. In summary, we have achieved an effective solution to the existing problems, and the results are excellent.

Table 1. Objective evaluation indicators of different algorithms.

Comparison Algorithm	NIQE ↓	SSIM ↑	PSNR ↑	LPIPS ↓	BIQI ↑	EMEE ↑	SDME ↑	BRISQUE ↓	AME ↑	Visibility ↑
Retinex-Net	7.1888	0.6449	13.7448	2.3146	0.4075	9.1803	89.1120	93.0386	78.9014	1.4980
URetinex-Net	4.7599	0.8238	21.3282	1.3234	0.2692	8.8664	72.2450	94.4427	43.9180	1.3153
SIRE	6.2109	0.4937	10.9447	1.8563	0.3428	8.4146	52.3258	93.3717	37.7913	1.5000
LIME	6.4282	0.7410	16.2744	2.0601	0.3436	7.9899	114.8789	94.8650	83.0246	1.3913
Zero-DCE++	4.3693	0.5479	14.3098	1.8905	0.3604	7.8689	69.8208	94.3531	52.4144	1.4879
KinD++	4.8106	0.7962	15.2666	1.4899	0.3652	8.5482	97.2805	93.3560	73.7956	1.4440
SNR-Aware [30]	5.7982	0.7834	17.3118	1.6384	0.3073	8.6534	68.6665	96.2248	58.0088	1.2607
RUAS [31]	6.2769	0.6075	12.9109	1.9274	0.2815	9.8992	65.1625	95.7833	50.9025	1.4222
OURS	4.3195	0.8321	21.4243	1.3882	0.4394	10.9775	110.7982	92.1687	83.1254	1.5169

3.3. Ablation Experiment

The loss function variable in the decomposition module, the brightness enhancement module, and the adjustment module in the enhancement module of the network model in this paper were ablated. The specific experimental results are shown in the following figure.

As shown in Figure 8, in the ablation experiment for the loss function variables, which was mainly to verify the importance of structural metrics and textural metrics for the generation of structural layers, clarity was used as the key to measure the effects of the variables. Clarity refers to the details and boundaries in an image, and higher values represent more detailed information contained in the image. For the structural layer, the less detailed information we have, the better our final enhancement effect. Therefore, we need to choose variables with smaller clarity values. The red color in the histogram indicates that the loss function variables include both structural metrics and textural metrics; yellow indicates that the loss function variable only contains textural metrics; blue indicates that the loss function variable contains only structural measures. As can be seen in Figure 8, only the loss function containing structural metrics and textural metrics obtains the best structural layer effect.

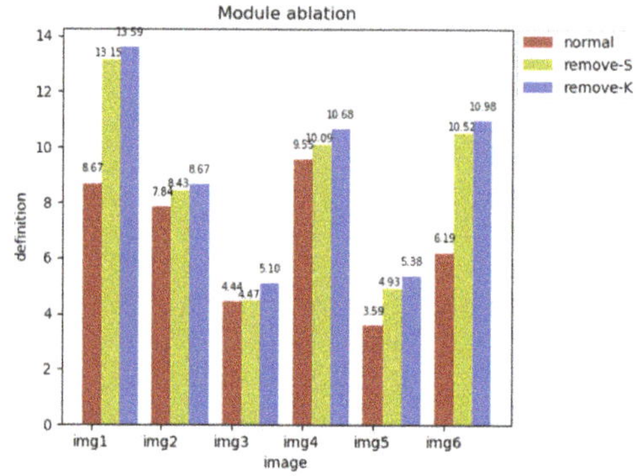

Figure 8. Ablation experiment for loss function variable.

Figure 9 shows the module ablation experiment for the second part of the network structure enhancement module, mainly comparing the basic module, the removed adjustment module, and the removed enhancement module. The peak signal-to-noise ratio is used as the key to measuring the experimental structure. The higher the peak signal-to-noise ratio, the stronger the enhancement effect. Therefore, we need to choose a module with a higher peak signal-to-noise ratio value. As shown in Figure 6, red represents the basic module, yellow represents the removal of the brightness adjustment module, and blue represents the removal of the enhancement module. It can be clearly seen that only the enhancement module and brightness adjustment module coexist, and the network has the best enhancement effect.

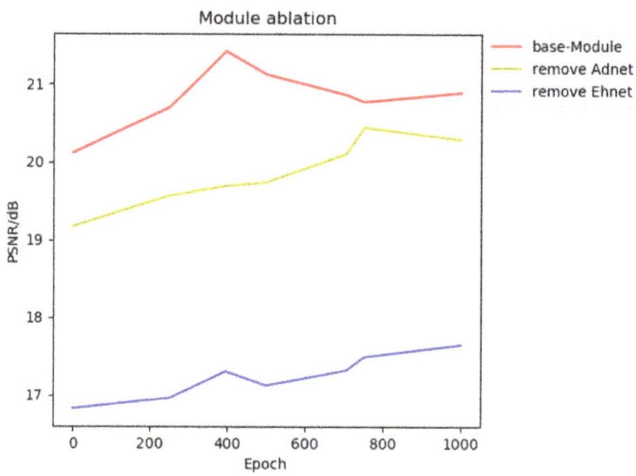

Figure 9. Ablation experiment based on enhancement module.

4. Conclusions

This paper proposed a low-illumination enhancement method based on structural and detail layers. Firstly, we designed an SRetinnex-Net model. The network is mainly divided into two parts: a decomposition module and an enhancement module. Second, the decomposition module mainly adopts the SU-Net structure, and the network decomposes the input image into a structural layer image and detail layer image. Afterward, the enhancement module mainly adopts the SDE-Net structure, which is divided into two branches: the SDE-S branch and the SDE-D branch. The SDE-S branch mainly enhances and adjusts the brightness of the structural layer image through the Ehnet module and the Adnet module, to prevent insufficient or overexposed brightness enhancement in the image. The SDE-D branch is mainly denoised and enhanced with textural details through a denoising module. This network structure can greatly reduce computational costs. Moreover, we also improved the total variation optimization model as a mixed loss function and added structural metrics and textural metrics as variables on the basis of the original loss function, which can well separate the structure edge and texture edge. Numerous experiments have shown that the algorithm proposed in this paper outperforms Retinex-Net, SIRE, LIME, Zero-DCE++, Kind++, RUAS, and other algorithms in evaluation metrics such as the SSIM, PSNR, and NIQE. The algorithm proposed in this article not only improves the brightness of low-illumination images but also has significant advantages in enhancing textural details and color restoration. In the future, the decomposition and enhancement of the entire network play an important role in enhancing low-illumination images, and optimizing the network structure is also a focus of our future research direction. And for low-illumination images without a control group, how to ensure image brightness enhancement without losing image details is a major challenge for us to continue studying low-illumination image enhancement.

Author Contributions: Conceptualization, W.G. and L.Z.; methodology, L.Z.; software, L.Z.; validation, W.G., W.Z. and D.Z.; formal analysis, Y.H.; investigation, L.Z.; resources, W.G.; data curation, J.W.; writing—original draft preparation, L.Z.; writing—review and editing, W.G. and J.W; visualization, D.Z.; supervision, L.Z.; project administration, Y.H. All authors have read and agreed to the published version of the manuscript.

Funding: This research was funded by the Chongnqing Nature Science Foundation (funding number CSTB2002NSCQ-MSX1071).

Institutional Review Board Statement: Not applicable.

Informed Consent Statement: Not applicable.

Data Availability Statement: The data presented in this study are available on request from the corresponding author. The data are not publicly available due to the privacy of the institute.

Conflicts of Interest: The authors declare no conflict of interest.

References

1. Bhutto, J.A.; Tian, L.; Du, Q.; Sun, Z.; Yu, L.; Tahir, M.F. CT and MRI Medical Image Fusion Using Noise-Removal and Contrast Enhancement Scheme with Convolutional Neural Network. *Entropy* **2022**, *24*, 393. [CrossRef] [PubMed]
2. Land, E.H.; McCann, J.J. Lightness and retinex theory. *J. Opt. Soc. Am.* **1971**, *61*, 1–11. [CrossRef]
3. Jobson, D.J.; Rahman, Z.; Woodell, G.A. A multiscale retinex for bridging the gap between color images and the human observation of scenes. *IEEE Trans. Image Process.* **1997**, *6*, 965–976. [CrossRef]
4. Rahman, Z.U.; Jobson, D.J.; Woodell, G.A. Multi-scale retinex for color image enhancement. In Proceedings of the 3rd IEEE International Conference on Image Processing, Lausanne, Switzerland, 19 September 1996; Volume 3, pp. 1003–1006. [CrossRef]
5. Park, S.; Yu, S.; Moon, B.; Ko, S.; Paik, J. Low-light image enhancement using variational optimization-based retinex model. *IEEE Trans. Consum. Electron.* **2017**, *63*, 178–184. [CrossRef]
6. Guo, X.; Yu, L.; Ling, H. LIME: Low-light Image Enhancement via Illumination Map Estimation. *IEEE Trans. Image Process.* **2016**, *26*, 982–993. [CrossRef] [PubMed]
7. Li, M.; Liu, J.; Yang, W.; Sun, X.; Guo, Z. Structure-Revealing Low-Light Image Enhancement Via Robust Retinex Model. *IEEE Trans. Image Process.* **2018**, *27*, 2828–2841. [CrossRef]
8. Ren, X.; Li, M.; Cheng, W.H.; Liu, J. Joint Enhancement and Denoising Method via Sequential Decomposition. In Proceedings of the 2018 IEEE International Symposium on Circuits and Systems (ISCAS), Florence, Italy, 27–30 May 2018.
9. Liang, J.; Zhang, X. Retinex by Higher Order Total Variation L1 Decomposition. *J. Math. Imaging Vis.* **2015**, *52*, 345–355. [CrossRef]
10. Lore, K.G.; Akintayo, A.; Sarkar, S. LLNet: A Deep Autoencoder Approach to Natural Low-light Image Enhancement. *Pattern Recognit.* **2017**, *61*, 650–662. [CrossRef]
11. Wei, C.; Wang, W.; Yang, W.; Liu, J. Deep Retinex Decomposition for Low-Light Enhancement. *arXiv* **2018**, arXiv:1808.04560.
12. Lv, F.; Lu, F.; Wu, J.; Lim, C. MBLLEN: Low-Light Image/Video Enhancement Using CNNs. In Proceedings of the British Machine Vision Conference, Newcastle, UK, 3–6 September 2018.

13. Jiang, Y.; Gong, X.; Liu, D.; Cheng, Y.; Fang, C.; Shen, X.; Yang, J.; Zhou, P.; Wang, Z. EnlightenGAN: Deep Light Enhancement Without Paired Supervision. *IEEE Trans. Image Process.* **2021**, *30*, 2340–2349. [CrossRef] [PubMed]
14. Zhang, Y.; Zhang, J.; Guo, X. Kindling the Darkness: A Practical Low-light Image Enhancer. In Proceedings of the 27th ACM International Conference on Multimedia, Nice, France, 21–25 October 2019.
15. Ma, L.; Ma, T.; Liu, R.; Fan, X.; Luo, Z. Toward Fast, Flexible, and Robust Low-Light Image Enhancement. In Proceedings of the 2022 IEEE/CVF Conference on Computer Vision and Pattern Recognition (CVPR), New Orleans, LA, USA, 18–24 June 2022; pp. 5627–5636. [CrossRef]
16. Jobson, D.J.; Rahman, Z.U.; Woodell, G.A. Properties and performance of a center/surround retinex. *IEEE Trans. Image Process.* **1997**, *6*, 451–462. [CrossRef] [PubMed]
17. Ronneberger, O.; Fischer, P.; Brox, T. *U-Net: Convolutional Networks for Biomedical Image Segmentation*; Springer: Cham, Switzerland, 2015.
18. Zhang, Q.; Shen, X.; Xu, L.; Jia, J. Rolling guidance filter. In Proceedings of the Computer Vision-ECCV 2014: 13th European Conference, Zurich, Switzerland, 6–12 September 2014; Springer International Publishing: Cham, Switzerland, 2014; pp. 815–830.
19. Rudin, L.I.; Osher, S.; Fatemi, E. Nonlinear total variation based noise removal algorithms. *Phys. D Nonlinear Phenom.* **1992**, *60*, 259–268. [CrossRef]
20. Xu, L.; Yan, Q.; Xia, Y.; Jia, J. Structure extraction from texture via relative total variation. *ACM Trans. Graph.* **2012**, *31*, 1–10. [CrossRef]
21. Zhou, F.; Chen, Q.; Liu, B.; Qiu, G. Structure and Texture-Aware Image Decomposition via Training a Neural Network. *IEEE Trans. Image Process.* **2020**, *29*, 3458–3473. [CrossRef] [PubMed]
22. Yin, W.; Goldfarb, D.; Osher, S. A comparison of three total variation based texture extraction models. *J. Vis. Commun. Image Represent.* **2007**, *18*, 240–252. [CrossRef]
23. Aujol, J.F.; Gilboa, G.; Chan, T.; Osher, S. Structure-Texture Image Decomposition—Modeling, Algorithms, and Parameter Selection. *Int. J. Comput. Vis.* **2006**, *67*, 111–136. [CrossRef]
24. Chen, Q.; Liu, B.; Zhou, F. Anisotropy-based image smoothing via deep neural network training. *Electron. Lett.* **2019**, *55*, 1279–1281. [CrossRef]
25. Wu, W.; Weng, J.; Zhang, P.; Wang, X.; Yang, W.; Jiang, J. URetinex-Net: Retinex-based Deep Unfolding Network for Low-light Image Enhancement. In Proceedings of the 2022 IEEE/CVF Conference on Computer Vision and Pattern Recognition (CVPR), New Orleans, LA, USA, 18–24 June 2022; pp. 5891–5900. [CrossRef]
26. Li, C.; Guo, C.; Loy, C.C. Learning to Enhance Low-Light Image via Zero-Reference Deep Curve Estimation. *IEEE Trans. Pattern Anal. Mach. Intell.* **2021**, *44*, 4225–4238. [CrossRef] [PubMed]
27. Zhang, Y.; Guo, X.; Ma, J.; Liu, W.; Zhang, J. Beyond Brightening Low-light Images. *Int. J. Comput. Vis.* **2021**, *129*, 1013–1037. [CrossRef]
28. Mittal, A.; Soundararajan, R.; Bovik, A.C. Making a "Completely Blind" Image Quality Analyzer. *IEEE Signal Process. Lett.* **2013**, *20*, 209–212. [CrossRef]
29. Zhang, R.; Isola, P.; Efros, A.A.; Shechtman, E.; Wang, O. The Unreasonable Effectiveness of Deep Features as a Perceptual Metric. In Proceedings of the 2018 IEEE/CVF Conference on Computer Vision and Pattern Recognition, Salt Lake City, UT, USA, 18–22 June 2018; pp. 586–595. [CrossRef]
30. Xu, X.; Wang, R.; Fu, C.W.; Jia, J. SNR-Aware Low-light Image Enhancement. In Proceedings of the 2022 IEEE/CVF Conference on Computer Vision and Pattern Recognition (CVPR), New Orleans, LA, USA, 18–24 June 2022.
31. Liu, R.; Ma, L.; Zhang, J.; Fan, X.; Luo, Z. Retinex-inspired Unrolling with Cooperative Prior Architecture Search for Low-light Image Enhancement. In Proceedings of the IEEE/CVF Conference on Computer Vision and Pattern Recognition, Nashville, TN, USA, 20–25 June 2021; IEEE: Piscataway, NJ, USA, 2021.

entropy

Article

Deep Learning in Precision Agriculture: Artificially Generated VNIR Images Segmentation for Early Postharvest Decay Prediction in Apples

Nikita Stasenko [1], Islomjon Shukhratov [1], Maxim Savinov [2], Dmitrii Shadrin [1,3] and Andrey Somov [1,*]

1 Skolkovo Institute of Science and Technology, 121205 Moscow, Russia; d.shadrin@skoltech.ru (D.S.)
2 Saint-Petersburg State University of Aerospace Instrumentation (SUAI), 190000 Saint-Petersburg, Russia
3 Department of Information Technology and Data Science, Irkutsk National Research Technical University, 664074 Irkutsk, Russia
* Correspondence: a.somov@skoltech.ru

Abstract: Food quality control is an important task in the agricultural domain at the postharvest stage for avoiding food losses. The latest achievements in image processing with deep learning (DL) and computer vision (CV) approaches provide a number of effective tools based on the image colorization and image-to-image translation for plant quality control at the postharvest stage. In this article, we propose the approach based on Generative Adversarial Network (GAN) and Convolutional Neural Network (CNN) techniques to use synthesized and segmented VNIR imaging data for early postharvest decay and fungal zone predictions as well as the quality assessment of stored apples. The Pix2PixHD model achieved higher results in terms of VNIR images translation from RGB (SSIM = 0.972). Mask R-CNN model was selected as a CNN technique for VNIR images segmentation and achieved 58.861 for postharvest decay zones, 40.968 for fungal zones and 94.800 for both the decayed and fungal zones detection and prediction in stored apples in terms of F1-score metric. In order to verify the effectiveness of this approach, a unique paired dataset containing 1305 RGB and VNIR images of apples of four varieties was obtained. It is further utilized for a GAN model selection. Additionally, we acquired 1029 VNIR images of apples for training and testing a CNN model. We conducted validation on an embedded system equipped with a graphical processing unit. Using Pix2PixHD, 100 VNIR images from RGB images were generated at a rate of 17 frames per second (FPS). Subsequently, these images were segmented using Mask R-CNN at a rate of 0.42 FPS. The achieved results are promising for enhancing the food study and control during the postharvest stage.

Keywords: GAN; CNN; precision agriculture; postharvest decay; fungi; image processing

Citation: Stasenko, N.; Shukhratov, I.; Savinov, M.; Shadrin, D.; Somov, A. Deep Learning in Precision Agriculture: Artificially Generated VNIR Images Segmentation for Early Postharvest Decay Prediction in Apples. *Entropy* **2023**, *25*, 987. https://doi.org/10.3390/e25070987

Academic Editors: Oleg Sergiyenko, Wendy Flores-Fuentes, Julio Cesar Rodriguez-Quinonez and Jesús Elías Miranda-Vega

Received: 5 May 2023
Revised: 19 June 2023
Accepted: 22 June 2023
Published: 28 June 2023

1. Introduction

According to the data provided by United Nations, the human population has grown to 8 billion people [1], and it is expected to increase up to 9.8 billion by 2050 [2]. The growing population will need more sustainable and affordable food sources. It increases the importance of agriculture in the light of sustainable development. In terms of food producing and quality control, agricultural challenges can be divided into preharvesting, harvesting and postharvesting stages [3]. Each stage includes various factors that should be taken into account in order to minimize food losses. During the postharvest stage, farmers primarily concentrate on factors that impact the shelf-life of harvested products during storage and transportation. These factors include temperature [4], humidity [5], as well as the use of gases and chemicals in food containers [6,7]. Each crop has its own number of factors affecting the shelf-life during the postharvest stage, and these factors should be also taken into account [8]. Disparagement of one of these factors or violation during the storage or transportation may result in postharvest losses of food products. Examples of postharvest losses in stored fruits and vegetables include decayed and spoiled areas, often attributed to mishandling, hygiene issues, inadequate humidity control, improper temperature management, and mechanical damages [9]. These factors contribute to the deterioration and loss of quality of stored subjects.

Apple is one of the most popular harvested and cultivated crops. Its global production achieved 93 millions tonnes in 2021 [10]. It is one of the major reasons to monitor apple fruits quality during all the above-mentioned stages to prevent postharvest losses and to avoid potential economic losses. However, there are special factors affecting apple quality during the postharvest stage, e.g., water as loss in apple fruits [11], residual pesticides [12], or concentration of carbon dioxide, ethylene, ethanol or ammonia surrounding apples due to insufficient ventilation in the storage facility [13]. The most common non-destructive methods for preventing postharvest losses include the control of objects using RGB video cameras and sensors [14], near infrared (NIR) data [15], gas sensing spectroscopy [13], fluorescence spectroscopy [16], magnetic resonance imaging (MRI) [17], and even electronic nose [18]. Nevertheless, postharvest losses are still estimated in the range of 40–50% [9]. It should be noted that the control of apple fruits at the postharvest stage is quite comprehensive, making it difficult to monitor each fruit at each step, while any damage may lead to a fungi infection [19] in the stored fruits and also to the formation (and even a rapid growth) of rotten areas which are also known as decayed areas [20]. Moreover, these areas are not well seen visually at early stages, and the decay growth process can be quite dynamic [21].

Artificial intelligence (AI) and its domains, including machine learning (ML) and deep learning (DL), in conjunction with the latest achievements in computer vision (CV), remote sensing, wireless sensing technologies, and Internet of Things (IoT), have provided the added value in a number of application including the space domain [22], medicine [23], power engineering [24], agriculture [25] and food supply [26]. For example, farmers rely on CV for crop quality management, e.g., plant growth monitoring [27], fruit detection [28], disease detection [29] and weed detection [30]. It is necessary for improving the food quality of each plant at preharvest, harvest, and postharvest stages, respectively. Also, there is a set of CV-based approaches for postharvest losses estimation and the evaluation in stored crops [31–33]. However, some postharvest losses, e.g., fungi or postharvest decay zones, should be detected immediately, since the visible decayed or fungi zones (acquired visually or with RGB cameras and sensors) in stored plants may indicate their serious spoilage if we use other types of imaging data, e.g., NIR or thermal imaging, to monitor their quality. This monitoring process requires a special device and equipment, e.g., multispectral or hyperspectral cameras, which are expensive and often not easy to use, given fast detection of defects is still extremely challenging.

In this article, we present an approach based on the application of generative adversarial network (GAN) and convolutional neural network (CNN) for early detection and segmentation of decayed and fungi areas in stored apples at the postharvest stage using visible near-infrared (vis-NIR, or just VNIR) imaging data. We show how artificially generated VNIR imaging data can be used for early postharvest decay detection in stored apples and examine whether GAN- and CNN-based approaches can achieve promising results for image segmentation tasks. The idea of the proposed approach can be divided into two parts:

- Generation of VNIR imaging data containing the stored apples with postharvest decay and fungi zones using the GAN technique.
- Segmentation of generated VNIR images using the CNN technique in order to detect the decayed and fungi zones in the stored apples.

In this research, we study the original and generated VNIR images containing apples of four varieties with several treatments in order to simulate various occasions with apples during the storage. The aim is to present an approach based on the DL techniques combining the GAN and CNN models, for instance, with segmentation of postharvest decay zones and fungi areas. The GAN model will provide the procedure of NIR images synthesis from the input RGB data, while the CNN model is supposed to be used for the instance segmentation of generated images. This is important for the proposed approach, as we aim to train and validate our models to detect the postharvest decay zones and fungi areas separately from each other. For realizing this idea into practice, we propose the following stages.

First, we need to select a GAN based model for the NIR images generation from the input RGB data. There are many available networks, but for the image-to-image translation tasks the following architectures Pix2Pix [34], CycleGAN [35], and Pix2PixHD [36] are mostly applied in agricultural domain [37–43]. We compare Pix2Pix, CycleGAN, and Pix2PixHD models using the dataset containing the paired RGB and NIR images. We are going to work with the images acquired in VNIR range since it includes the full visible spectrum with an abutting portion of the infrared spectrum [44]. The paired images collected in the visible (380–700 nm) and VNIR (400–1100 nm) ranges are required to make sure that the decayed and fungal traits in stored apples are the same for these two ranges. Section 3.1.1, Section 3.1.2, and Section 3.1.3 provide detailed information about the Pix2Pix, CycleGAN, and Pix2PixHD models, respectively.

Second, it is necessary to choose the CNN model for the decayed and fungal areas segmentation in the synthesized VNIR images. In this work, we implement a Mask R-CNN model due to the Feature Pyramid Network (FPN) and ResNet101 backbone, which allow for generating the bounding boxes (object detection) and segmentation masks (instance segmentation). In [45], we have compared the Mask R-CNN to such applied CNN-based models as U-Net [46] and Deeplab [47] for early postharvest decay detection, and Mask R-CNN achieved the highest performance in terms of average precision, namely 67.1% against 59.7% and 56.5%, respectively. Moreover, the Mask R-CNN model generates the bounding boxes and segmentation masks of the postharvest decay and fungal zones separately from each other. This is a so-called 'a tried and tested' method, and that is why we use Mask R-CNN as a CNN-based segmentation model. We discuss the Mask R-CNN model in more detail in Section 3.1.4.

Finally, our plan is to implement the proposed approach and execute it on a Single Board Computer (SBC) with the AI capabilities. This implementation will serve as an evaluation platform for generating segmented VNIR images that highlight any postharvest decay and fungal zones on apples. These zones may be imperceptible to the human eye, but can be detected and selected through our system. We use NVIDIA Jetson Nano as an embedded system with AI capabilities for evaluation. It is a compact and powerful SBC supplied with the accelerated libraries for computer vision and deep learning applications, and is widely used for different real-time problems in agriculture including weed control [48], soil mapping in greenhouse [49], and harvest product detection [50–54]. That is why the presented research is supposed to be an alternative solution for the high-cost NIR hyperspectral devices used for the early postharvest decay detection and prediction for stored food. Figure 1 illustrates the proposed approach.

The contribution of this work is as follows:

- Two experimental testbeds for paired RGB and VNIR imaging data collection under various environmental (temperature and humidity) conditions.
- Application of CNN models, for instance, on the segmentation of decayed and fungi areas in apples at the postharvest stage.
- Separate segmentation of fungi zones and postharvest decay areas in stored apples using the CNN model.
- Application of the trained CNN-based model for the instance segmentation of postharvest decay zones and fungi areas in VNIR images generated by the GAN-based model.
- Implementation of the proposed approach based on the GAN and CNN techniques for postharvest decay detection, segmentation and prediction using generated VNIR imaging data on a low-cost embedded system with the AI capabilities.

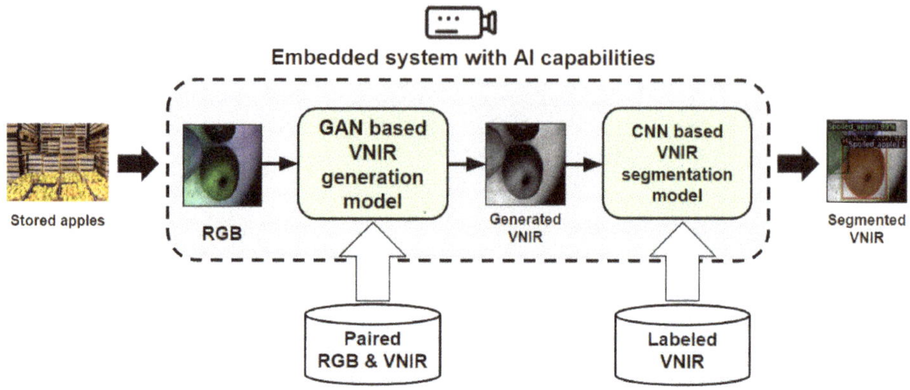

Figure 1. Diagram summarizing the proposed approach for the application of segmented VNIR imagery data via deep learning for early postharvest decay prediction in apples.

This article is organized as follows: Section 2 provides an introduction to relevant research works aimed at early postharvest decay detection and prediction in apples using RGB and VNIR imaging data with the CV and ML methods. Section 3 presents the methods used in this work. Section 3.3 demonstrates the experimental testbeds used for RGB and VNIR imaging data collection and describes the procedure of data annotation. Section 4 shows the results of the comparison of the GAN techniques applied to VNIR images generation from the RGB ones (see Section 4.1). It also presents the application of the CNN technique, for instance, on the segmentation on the generated VNIR images (see Section 4.2), and describes the embedded system running the proposed GAN and CNN (see Section 4.3). Conclusions and discussion of the future work are summarized in Section 5.

2. Related Works

2.1. CV Approaches Based on CNN Models Using RGB Imaging Data

CV techniques with the implementation of ML and DL methods are becoming one of the most useful tools for fruit quality estimation and evaluation at the postharvest stage.

The majority of approaches are based on the collection and analysis of visible morphological traits, such as changes in fruit shape, size, or color during the storage, from stored fruits with CNN models using RGB images as the most acceptable and user-friendly type of data. RGB imagery is closely similar to human vision because red, green and blue are the primary colors in these color models, which makes the process of visible non-destructive quality monitoring and defect detection of stored food production easy and understandable [55]. The majority of cameras and devices for RGB imaging data collection contain a patterned Bayer filter mosaic consisting of squares of four pixels with one red, one blue and two green filters [56]. Usually, the Bayer filter is located on the camera chip.

Generally, a CNN model contains convolutional and pooling layers (added one by one), flatten, fully connected layer and softmax classifier. The convolutional and pooling layers are used in the features extraction part, while the classification part involves the flatten, fully connected layers and softmax classifier. When the image reaches the input layer, a filter in the convolution layer allows it for the selection of feature neurons. An activation function (Sigmoid, Rectified Linear Unit (ReLU), or Softplus) is added to obtain nonlinear results by passing feature neurons through it, and the resulting feature map size is reduced by the pooling layer functions. The flatten layer is the first input layer for the classifier model as it keeps the feature map from the convolution layers. The fully connected layer transforms the obtained feature neurons into a matrix, which performs the classification function with a classification method.

In this way, the CNN structure showed its efficiency in classification, and then in detection and segmentation tasks using RGB imaging data. For example, the automated banana grading system was reported in [57] where a fine-tuned VGG-16 Deep CNN model was applied for banana classification using such traits as skin quality, size, and maturity with the acquired RGB imagery data. A similar approach was proposed in [58] where the VGG-16 model was trained to predict the date of the fruit ripening stage using RGB images with an overall classification of 96.98%.

In [59], the authors developed an automated online carrot grading system, where a lightweight carrot defect detection network (CDDNet) based on ShuffleNet [60] and transfer learning was implemented for carrot quality inspection using RGB and grayscale images. The CDDNet was compared to other CNN models including AlexNet, ResNet50, MobileNet v2, and ShuffleNet, and it demonstrated good performance in terms of detection accuracy and time consuming for binary classification of normal and defective carrots (99.82%), and for classification of normal, bad spots, abnormal, and fibrous root carrots (93.01%). However, the images of carrots contained the carrots of different size and appearance, and the idea of the presented approach was to detect the carrots with visible defects without taking into account the spoilage stage of the defective carrots. Moreover, there was no mention of a possible situation when the carrots are infected, but still there are no visible traits of spoilage.

In [61], the authors report on the implementation of the DeeplabV3+ model [62] with a classical image processing algorithm, e.g., threshold binary segmentation, morphological processing and mask extraction for banana bunches segmentation during sterile bud removal (SBD) on the total of 1500 RGB images. Moreover, YOLOv5-Banana model [63] for the banana fingers segmentation and centroid points extraction, while edge detection and centroid extraction of banana fingers included binarization, morphological opening operation, canny edge detection, and extracting centroid point set. DeeplabV3 was reported to achieve a detection accuracy rate of 86%, mean intersection over union (MIoU) of 0.878 during the debudding period for target segmentation, and the mean pixel precision of 0.936. YOLOv5-Banana achieved 76% detection accuracy rates for the banana bunches during the harvest period. The authors also designed and presented the software to estimate the banana fruit weight during the harvest period.

In [64], several CNN-based models including VGG-16, VGG-19, ResNet50, ResNet101, and ResNet152 were compared to each other for such physiological disorders classification in stored apples as bitter pit, shriveling, and superficial scald. The authors acquired a dataset containing 1080 RGB images (dataset-1) of apples and 4320 augmented images (dataset-2) with the aim to improve data representation during model training and to consider apple position under the monitoring camera and lighting conditions during the storage. The CNN-based models were used and compared for feature extraction, while such classical ML methods as support vector machines (SVM), random forest (RF), k-nearest neighbors algorithm (kNN), and XGBoost were used for the extracted features classification. The highest average accuracy was reported for the VGG-19 model in conjunction with the SVM method in the dataset-1 and dataset-2 with 96.11 and 96.09%.

2.2. Machine Learning and Deep Learning Methods for NIR Data Analysis

NIR spectroscopy covers spectral regions from 780 to 2500 nm that cannot be seen with human eyes, but it allows for obtaining spectral information from ten (generally, referred to as multispectral data [65]) and to more than a hundred wavebands (referred to as hyperspectral data [65]). Measurements performed in the visible (380–700 nm), visible near-infrared (vis-NIR, or just VNIR, 400–1100 nm), and NIR (780–2500 nm) ranges provide the user with more detailed information on the chemical composition of scanned samples. In our case, by samples we mean stored plants, crops and fruits. The state-of-the-art cameras and devices for the hyperspectral data acquisition provide not only spectral information about the scanned samples, but also allow the users to obtain the images of scanned zones in the range of device bands. Spectral information on chemical composition from a wide range of wavebands has simplified the procedure of food quality monitoring and defect detection at the postharvest stage. Moreover, not only the decay zones may occur in stored fruits, but also some fungi like *Sclerotinia sclerotiorum* [66], *Penicillium expansum* [67], *Botrytis cinerea* [68], *Botryosphaeria dothidea* [69] and many others, which should be immediately detected at the early stage. Otherwise, the appearance and growth of decayed and fungi zones may lead to the loss of all stored fruits. It is vital to distinguish various types of postharvest losses, e.g., postharvest decay, and diseases, e.g., various fungi varieties, since each type of loss requires a special type of treatment or removal of spoiled samples from the storage. It should be noted here that the formation of fungal areas may not always lead to the formation of decayed areas. That is why we should detect and identify the fungi and postharvest decay zones separately from each other [70–72].

Both classical ML methods and the DL techniques based on the CNN models are widely used for postharvest losses evaluation in stored plants using VNIR and NIR imaging and spectral data.

In [73], the authors compared several ML methods including linear discriminant analysis (LDA), random forest (RF), support vector machines (SVM), kNN, gradient tree boosting (GTB), and partial least squares-discriminant analysis (PLS-DA) for early *Codling Moth* zones detection in "Gala", "Granny Smith", and "Fuji" stored apples. The research was carried out at the pixel level using NIR hyperspectral reflectance imaging data in the range of 900–1700 nm with an optimal selection of wavelengths. GTB was reported to obtain better results at a pixel level classification with 97.4% of total accuracy for validation dataset.

In [74], the authors implemented the AlexNet model for detecting pesticide residues in postharvest apples using hyperspectral imaging data. There were 12,288 hyperspectral acquired images for the training set and 6144 images for the test set in the 865.11–1711.71 nm range (the camera included 256 bands) and with 3.32 nm spectral resolution. Otsu segmentation algorithm [75] was used for the apples and pesticide residue positioning (they were the regions of interests, or just ROIs), while deep AlexNet [76] provided pesticide category detection. AlexNet was reported to show better results in terms of detection accuracy and time consumption in comparison to the SVM and kNN algorithms (99.09% and 0.0846 s against 74.34% and 11.2301 s, and 43.75% and 0.7645 s, respectively).

As we can see, NIR hyperspectral and multispectral imaging data ensures early disease detection with more details than RGB imaging, but also requires sophisticated equipment, which usually includes a camera with wavebands, imaging spectrograph (or spectrometer), sample stage, illumination lamps and lightning system, as well as supplementary software and devices for processing and capturing NIR data and images [77–79]. However, this is the reason why hyperspectral imaging devices are so expensive and may cost from thousands to ten thousand USD [80]. These high prices reduce the availability and usage of hyperspectral cameras for farmers and food selling companies to perform food quality control at postharvest stages. This issue has raised a demand for developing new approaches for NIR imaging data generation without using high cost hyperspectral systems.

2.3. GAN-Based Models for RGB and NIR Data Analysis

Generative Adversarial Networks (GANs) and, in particular, conditional GAN (cGAN) [81] have demonstrated their effectiveness in a variety of tasks in the agricultural domain including remote sensing [82], image augmentation [83], animal farming [84], and plant phenotyping [85]. The general idea of GAN is based on the usage of two neural network models, where the first network is called generator (generative part, G) and its goal is to create plausible samples, while the second network is called discriminator (adversarial part, D), and it learns to verify whether the created plausible sample is real or fake. GANs are also applied for the so-called image-to-image translation tasks, i.e., where there is a need for high-quality image synthesis from one domain to another. For example, GAN-based models were successfully applied for the multi-channel attention selection in the RGB imagery considering an external semantic guidance in [86,87], MRI data estimation in [88], diffusion models evaluation [89], and NIR imaging generation from the input RGB images in [82,90,91].

Therefore, the approaches based on GAN models allow synthesizing high-quality NIR images from the input RGB images while saving detailed spectral information. At the same time, it is crucial not only to transform the image together with all the relevant information, but also to segment various types of postharvest diseases and defects separately from each other in stored food production in order to choose the specific processing strategy for defected or spoiled food samples. At present, most GAN models provide only the images transformation from one domain to another, but not object detection or instance segmentation operations in the synthesized images. However, as shown in Section 2.1, CNN models demonstrate reasonably good results for the object detection and instance segmentation both for the RGB and the NIR images.

3. Materials and Methods

3.1. DL Techniques

3.1.1. Pix2Pix

The Pix2Pix model [34] is a type of cGAN that has been demonstrated on a range of image-to-image translation tasks, such as converting a satellite image to corresponding maps, or black and white photos to color images. In conditional GANs, the generation of the output image is conditional on the input image. In the case of the Pix2Pix model, the generation process is conditional on the source image. The discriminator covers both the observed source image (*domain A*) and the target image (*domain B*) and must determine whether the target is a plausible transformation of the source image. The generator is trained via the adversarial loss which encourages the generator to make plausible images in the target domain. The generator is also updated via L_1 loss measured between the generated image and the expected output image. This additional loss encourages the generator model to create the plausible translations of the source image. Mathematically, the whole process in Pix2Pix can be defined as:

$$L_{cGAN}(G,D) = \mathbb{E}_{x,y\sim p_{data(x,y)}}[logD(x,y)] + \mathbb{E}_{x,z\sim p_{data(x,z)}}[log(1 - D(x,G(x,z))] \quad (1)$$

where G is the generator, D is the discriminator, x is the observed image, y is the target image, z is the random noise vector, and λ controls the relative importance of the two objectives between *domain A* and *domain B*. The following objective function is used to train the model:

$$G = arg \min_{G} \max_{D} L_{cGAN}(G,D) + \lambda L_{L1}(G) \quad (2)$$

Pix2Pix requires perfectly aligned paired images for the training procedure. In this research, the CNN-based architecture is used both as the generator and the discriminator. Generally, the U-Net model [46] is applied in Pix2Pix as a generator. U-Net trains to generate the images from the images in *domain A* similar to the images in *domain B*. The discriminator is usually a PatchGAN (which is also known as Markovian discriminator [92]), and it trains simultaneously to distinguish the generated images from the real images in *domain B*. The reconstruction loss measures the similarity between the real images and the generated images. Figure 2 shows the block diagram of Pix2Pix.

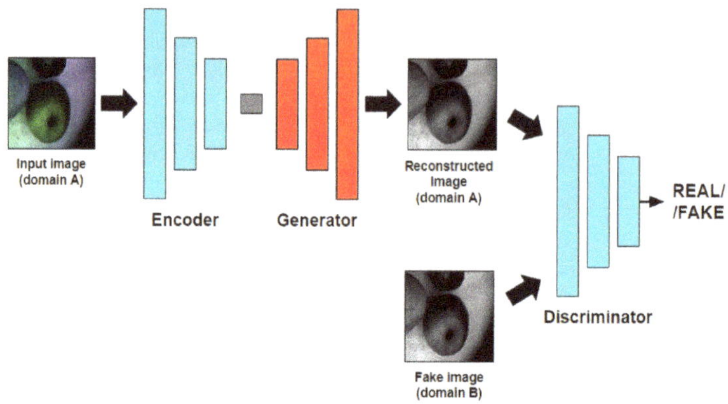

Figure 2. Pix2Pix block diagram.

3.1.2. CycleGAN

The goal of the CycleGAN model [35] is to learn the mapping $G : X \rightarrow Y$ such that the distribution of images from $G(X)$ is indistinguishable from the distribution Y using an unpaired set of image pairs. This mapping is coupled with an inverse mapping $F : Y \rightarrow X$ and a cycle consistency loss introduced to enforce $F(G(X)) \approx X$ and vice versa due to the reason that it is highly underconstrained. For the mapping function $G : X \rightarrow Y$ and its discriminator D_Y

$$L_{GAN}(G, D_Y, X, Y) = \mathbb{E}_y[logD_Y(y)] + \mathbb{E}_x[log(1 - D_Y(G(x)] \tag{3}$$

and the objective is as follows:

$$G, F = arg \min_{G,F} \max_{D_X, D_Y} L(G, F, D_X, D_Y) \tag{4}$$

CycleGAN learns a translation mapping in the absence of aligned paired images. The image generated from *domain A* to *domain B* by the CNN-based generator (G_1) is converted back to *domain A* by another CNN-based generator (G_2), and vice versa, in the attempt to optimize the cycle-consistency loss in addition to the adversarial loss. The block diagram of CycleGAN is shown in Figure 3.

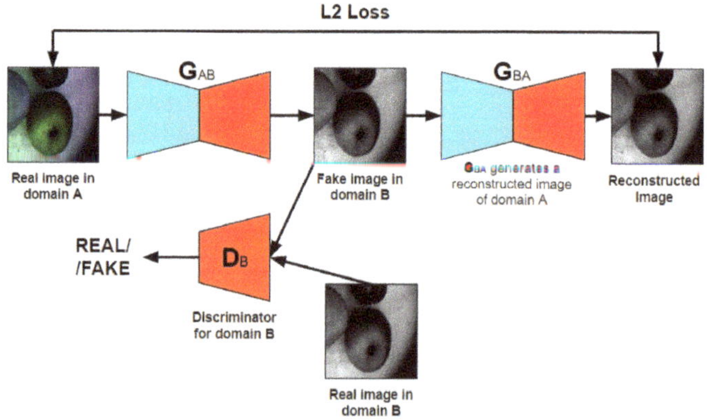

Figure 3. CycleGAN block diagram.

3.1.3. Pix2PixHD

The Pix2PixHD model [36] is a modification of the solution realized in the Pix2Pix model, which includes several improvements including the Coarse-to-Fine generator, multi-scale discriminators, and improved adversarial loss. Pix2PixHD generally consists of global generator G_1 and local enhancer G_2 (see Figure 4, where *** are referred to the residual blocks). Throughout the training process, the global generator is initially trained, followed by the training of the local enhancer in a progressive manner based on their respective resolutions. Subsequently, all the networks are fine-tuned jointly. The purpose of this generator is to efficiently combine global and local information for the task of image synthesis. Three discriminators are used for effective detail capturing on multiple scales.

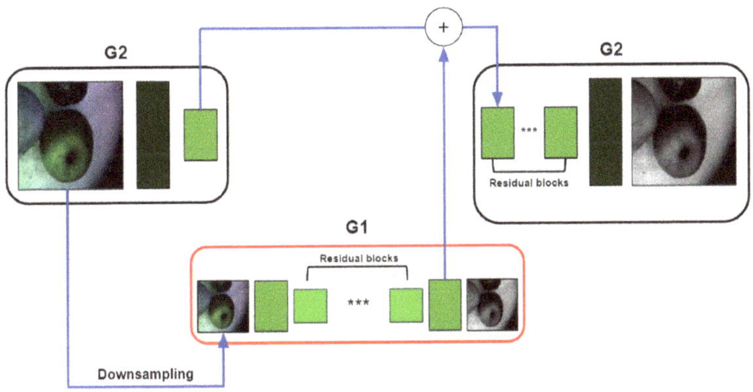

Figure 4. Pix2PixHD generator block diagram.

A significant performance boost was provided by the loss modification and two extra terms, L_{FM}-feature matching loss and perceptual loss, were added L_{VGG} [93] as objective functions. The feature matching loss performs the stabilization of the training. It happens due the point that the generator has to produce natural statistics at multiple scales:

$$L_{FM}(G, D_k) = \lambda_{FM}\mathbb{E}_{y,x}\sum_{i=1}\frac{1}{N_i}[||D_k^{(i)}(y,x) - D_k^{(i)}(y, G(y))||_1] \tag{5}$$

where $D_k^{(i)}$ denotes the output of the i-th layer of the D_k discriminator.

$$L_{VGG} = \lambda_{VGG}\mathbb{E}_{y,x}\sum_{i=1}\frac{1}{M_i}[||F^{(i)}(x) - F^{(i)}(G(y))||_1] \tag{6}$$

where $F^{(i)}$ denotes the i-th layer with M_i elements of the VGG network.

3.1.4. Mask R-CNN

Mask R-CNN [94] is a CNN-based architecture that provides the instance segmentation of various objects in the images. These objects in images are usually called the Regions of Interest (ROIs). This is the latest version of the R-CNN model [95], where R-CNN stands for Regions detected with CNN. Firstly, R-CNN has been improved to Fast R-CNN [96], then to Faster R-CNN [97], and, finally, to Mask R-CNN. As it was mentioned earlier, in R-CNN based models the ROIs are detected with the CNN feature's selective search. In Mask R-CNN, this selective search was improved to Mask R-CNN by adding the Region Proposal Network (RPN) in order to initiate and identify the ROIs and by adding a new branch for the prediction of the mask that covers the found region, i.e., an object in the image. The RPN and ResNet101 backbone allow for making the object detection (bounding boxes generation) and instance segmentation if there are several ROIs in one image and they have different sizes and partially overlap each other. Figure 5 presents a block diagram of Mask R-CNN architecture.

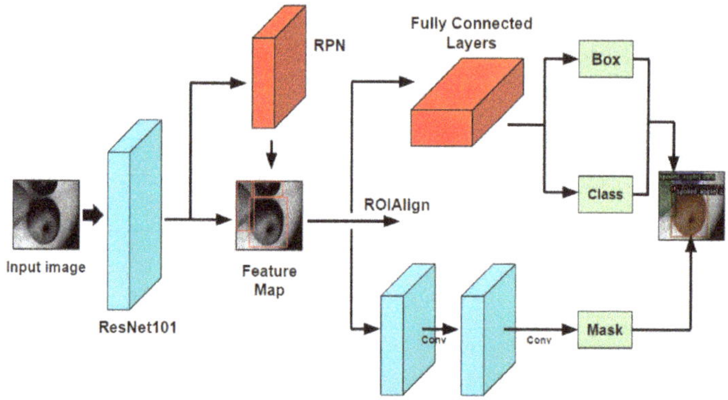

Figure 5. Mask R-CNN block diagram.

3.2. Performance Metrics

In this study, we compare the original VNIR images with the VNIR images generated by the Pix2PixHD model. To perform this, we considered the Mean Average Error (MAE), Mean Average Percentage Error (MAPE), Mean Squared Error (MSE), Root Mean Square Error (RMSE), Peak Signal to Noise Ratio (PSNR), Structural Similarity Index Measure (SSIM), and Feature Similarity Index Measure (FSIM) as follows:

$$\text{MAE} = \frac{1}{n}\sum_{i=1}^{n}|(y_i - x_i)| \tag{7}$$

$$\text{MAPE} = \frac{100\%}{n}\sum_{i=1}^{n}\left|\frac{(y_i - x_i)}{y_i}\right| \tag{8}$$

$$\text{MSE} = \frac{1}{n}\sum_{i=1}^{n}(y_i - x_i)^2 \tag{9}$$

$$\text{RMSE} = \sqrt{\frac{1}{n}\sum_{i=1}^{n}(y_i - x_i)^2} \tag{10}$$

$$\text{PSNR} = 10\log_{10}\left(\frac{R^2}{\frac{1}{n}\sum_{i=1}^{n}(y_i - x_i)^2}\right) \tag{11}$$

$$\text{SSIM} = \left[l(x_i, y_i)^{\alpha} \cdot c(x_i, y_i)^{\beta} \cdot s(x_i, y_i)^{\gamma}\right] \tag{12}$$

$$\text{FSIM} = \left[S_{PC}(x_i, y_i)^{\alpha} \cdot S_{GM}(x_i, y_i)^{\beta}\right] \tag{13}$$

where y_i is the generated or synthesized image, x_i is the original image, n is the number of observations, R is the image maximum possible pixel value, l is the luminance, c is the contrast, s is the structure, α, β, and γ are the weights, S_{PC} is the invariant to light variation in images, and S_{GM} is the computation of image gradient.

We used precision, recall, mean Intersection over Union (IoU), mean Average Precision (mAP), and F1-score to verify the efficiency of the Mask R-CNN model on the synthesized VNIR pictures during the training and validation stages, which are defined as follows:

$$\text{Precision} = \frac{TP}{TP + FP} \tag{14}$$

$$\text{Recall} = \frac{TP}{TP + FN} \tag{15}$$

$$\text{IoU} = \frac{Area\,of\,Overlap}{Area\,of\,Union} \tag{16}$$

$$AP = \sum_{n}(Recall_n - Recall_{n-1})Precision_n \tag{17}$$

$$\text{F1-score} = 2 * \frac{Precision * Recall}{Precision + Recall} \tag{18}$$

Precision and recall are based on True Positives (TP), True Negatives (TN), False Positives (FP), and False Negatives (FN). TP denotes instances in which the model correctly predicts a specific object from a given class in images, TN denotes the instances in which the model correctly predicts an object that does not belong to a given class, and FP denotes the instances in which the model predicts a specific class, but the object does not actually belong to that class. In contrast, FN are the cases in which the model makes no prediction of a particular class, but the object actually belongs to one of the classes. The object classes are described in Section 3.4.

The AP is a region that lies beneath the precision–recall curve. The weighted mean of precisions at each IoU threshold, with the increase in recall from the preceding threshold as the weight, is how AP summarizes a precision–recall curve. It is calculated using (17), where $Precision_n$ and $Recall_n$ are the Precision and Recall at the n-th IoU threshold.

The mAP over all classes or overall IoU thresholds is calculated with the mAP score. AP is averaged over all the classes. There is no distinction between AP and mAP in this case. In our scenario, since AP is averaged across all the classes, there is no difference between AP and mAP. We calculated AP values for IoU = 0.50 (AP_{50}), for IoU = 0.75 (AP_{75}), for the objects with an area less than 32 squared pixels (AP_S), for the objects with an area ranging from 32 to 96 squared pixels (AP_M), and for the objects with an area higher 96 squared pixels (AP_L).

3.3. Experimental Testbeds and Data Acquisition

In this section, we describe the apple fruits used for the experiments and present experimental testbeds for data collection:

(i) The experimental testbed for acquiring the dataset containing paired RGB and VNIR images of stored apples;

(ii) The experimental testbed for stored apple VNIR images collection containing VNIR images acquired by a multispectral camera.

The first testbed is designed for paired RGB and VNIR images collection in order to train and validate the GAN-based DL models for VNIR images translation from RGB images (see Section 3.3.1). The second testbed is used for the stored apples VNIR images collection as well as for the CNN-based model training and validation of postharvest decay zones detection and segmentation in the generated VNIR images (see Section 3.3.2).

3.3.1. Experimental Testbed for Paired RGB and VNIR Imaging Data Collection

We selected 16 apples of four kinds ("Delicious", "Fuji", "Gala", "Reinette Simirenko") and divided them into four rows according to their kind (each row corresponds to each apple kind). Each row contained four apples of different types, where every apple has different treatment from left to right: an apple with no treatment, a thoroughly washed and wiped apple, a mechanically damaged apple, and a shock-frozen apple supercooled under −20°, respectively. The apple without treatment serves as a reference for each kind. A thoroughly washed apple indicates the removal of the natural protective wax layer from an apple. A mechanically damaged apple imitates the wrong storing conditions. A shock-frozen apple simulates the wrong storing conditions. Figure 6 shows these apples.

Figure 6. Apples selected for data collection.

The first testbed is used for data collection under the recommended room storage conditions. The temperature ranges from 25 °C to 32 °C and Relative Humidity (RH) of 34% [98]. The testbed contains aluminum frames and is 1 m in length, 1 m wide, and 1.7 m high. Apples lie on a table with a white tray at the height of 1.3 m above the floor level. We also use SLR camera Canon M50 and the multispectral camera CMS-V1 CMS18100073 (CMS-V) attached at the middle top of the frame and connected to a PC laptop via the USB hub. The distance between the table with the apples on top and the camera is 500 mm. The lamps allowed us to simulate real storage conditions for apples as well as perform the collection of images under full and partial illumination. Detailed information about the acquired dataset and the first experimental testbed is described in [99]. Figure 7 shows the first testbed.

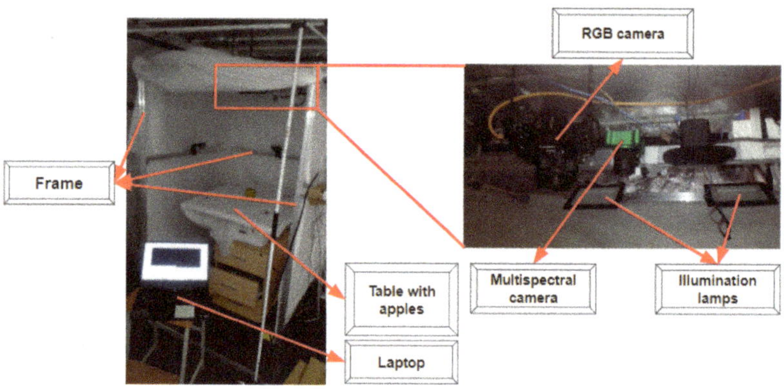

Figure 7. Experimental testbed for paired RGB and VNIR image capturing.

The multispectral camera CMS-V allows acquiring images in the range of 561–838 nm, including the visible and NIR ranges. This camera imager is characterized by the modified Bayer matrix made of a group of 3 × 3 pixels, called macro-pixel, filtering 3 × 3 (9) spectral bands. The raw image delivered by the camera is built of 9 interleaved spectral sub-images (8 colors + 1 Panchromatic) with the 1280 × 1024 pixels resolution. Each RGB image relates to 9 images from the following spectral bands *channel0* = 561 nm, *channel1* = 597 nm, *channel2* = 635 nm, *channel3* = 673 nm, *channel4* = 724 nm, *channel5* = 762 nm, *channel6* = 802 nm, *channel7* = 838 nm, and *channel8* (panchromatic channel) = 0 nm. The resolution of the nine sub-images is 426 × 339 pixels.

We acquired 1305 sequential RGB images and 1305 corresponding VNIR images in 838 nm range to see the decay dynamics in presented apples. The examples of images are shown in Figure 8.

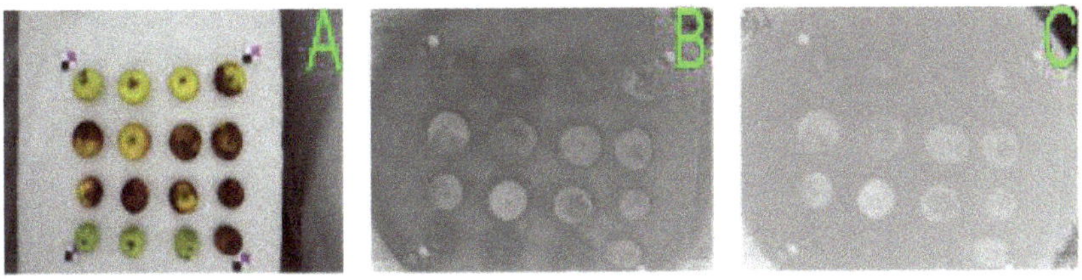

Figure 8. Types of images obtained during the experiments: (**A**)—RGB image of apples acquired under the full illumination; (**B**)—VNIR image of apples acquired under the full illumination (838 nm); (**C**)—VNIR image of apples acquired under the partial illumination (838 nm).

3.3.2. Experimental Testbed for VNIR Imaging Data Collection

In this experiment, we selected 22 apples of the "Alesya", "Fuji", "Golden" and "Reinette Simirenko" seasonal types for data acquisition. The apples were between 8 and 10 cm in diameter, and most of them were multicolor with red and yellow sections. There were also some apples containing fungi zones, i.e., grey-brown moldy areas in apples, as the examples of apples stored under violated storage conditions. These apples were used in order to increase the data representation for early postharvest decay detection tasks in the stored apples using VNIR imaging data. These apples are demonstrated in Figure 9.

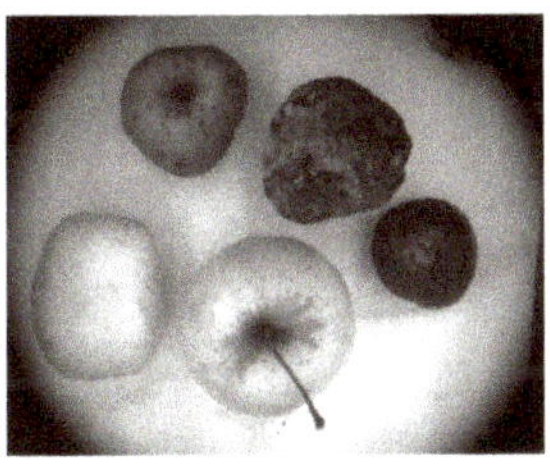

Figure 9. VNIR image of apples selected for data collection.

The second testbed presented in Figure 10 is a greenhouse that includes silicon frames and five shelves, a plastic wrap, a multispectral camera, 10 LED strip lights with red/blue diodes, a power supply (total power is 150 Watt) for controlling the LEDs, a logger, and a pallet with apples. It can be used for the simulation of different processes related to plant breeding in various environmental conditions including extremely dry or wet modes. Temperature and humidity regulation in the testbed is provided with the LED strip lights, the plastic wrap, and several water pallets located on three lower bottom separate shelves.

The silica frames are the basic elements of a presented greenhouse characterized by the following dimensions 170 cm in height, 48 cm in length, and 67 cm in width. Two strip lights were fixed on each shelf while the multispectral camera and the pallet with the apples were fixed on the separate shelves (see Figure 10). Each selected strip has 60 LEDs with the wavelength of 650–660 nm (red light LEDs) and 455–465 nm (blue light LEDs) for highest chlorophyll concentration in plants to provide the most effective photosynthesis processes. This is also fair for crops and plants at the postharvest stages [100]. It is necessary to keep the quality of plant production which is another reason why these LED strip lights are used in the greenhouse. We rely on the power supply (12 V DC, 150 W, IP33) as the energy source for the SMD 5050 LED strip lights, and GL100-N/GL100-WL logger by Graphtech Corporation, supplied with the GS-TH sensor module, for temperature and humidity values registration during the data collection process.

For the VNIR image capturing, the multispectral camera CMS-V described in Section 3.3.1 was also chosen. The camera was connected via USB-A wire to the HP EliteBook 820 G3 Laptop with IntelCore i3-6100 CPU 2.30 GHz, where all the images were acquired and saved as JPG-files with 426 × 339 pixels.

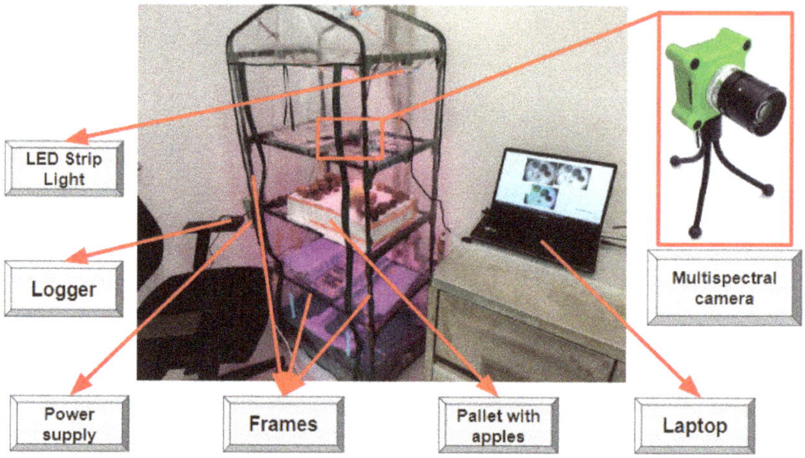

Figure 10. Experimental greenhouse for data acquisition.

We obtained 1029 sequential VNIR images in the 838 nm range collected from CMS-V camera's *channel7*. These images were acquired under the temperature range from 35 °C to 40 °C and RH equal to 70% with the goal to simulate potential violation of the storage process of selected apples. This violation is necessary to speed up the decay processes in apples. We also collected 100 sequential RGB images (see the example in Figure 11) for the CNN-based model training and validation with the aim to demonstrate the up-to-date approach based on the combination of pre-trained GAN-based and CNN-based models. RGB sequential images had the dimensions of 339 pixels × 426 pixels × 3 channels (or simply 339 × 426 × 3).

Figure 11. RGB image of apples selected for data collection.

3.4. Data Annotation

In order to apply a CNN-based deep learning model for the image instance segmentation, we used the Supervisely Ecosystem [101] for annotation and labeling of VNIR imaging data. It is worth reiterating here that we provide this labeling only for the VNIR images acquired with the testbed, described in Section 3.3.2 as these images were specially collected as the sequential VNIR imaging dataset for the DL model training and validation on early postharvest decay detection and segmentation of apples.

Four classes of objects in the images are defined as: *Healthy apple*, *Decay*, *Fungi*, and *Spoiled apple*. By the *Healthy apple* we understand the apples without any visible damages or spoiled zones in the images. The dark gray colored areas with the postharvest decay in apples were indicated as *Decay*. By *Fungi* we indicate white colored moldy zones in apples. Here we distinguish the postharvest decay zones marked as the *Decay* class, and moldy zones marked as the *Fungi* class. If an apple has objects of the *Fungi* class, it means that this apple is supposed to have been stored under the violated storage conditions, e.g., extreme temperature or humidity, which resulted in the apple's full spoilage. The apples with only the postharvest decay zones (*Decay*) can be sent for recycling, while apples with moldy zones (*Fungi*) must be removed from others in order to prevent the spoilage of all samples. We also defined the *Spoiled apples* class: there are stored apples with more than 50 percent of spoiled areas (*Decay* objects) or moldy zones (*Fungi* objects) coverage. Figure 12 illustrates the procedure of image annotation.

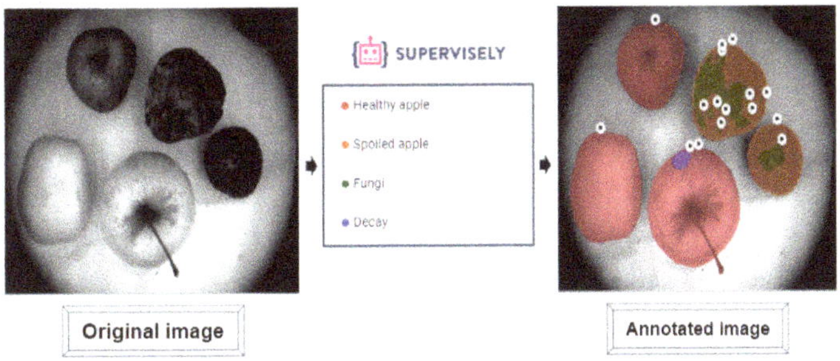

Figure 12. The example of image annotation and objects classes in Supervisely.

4. Results and Discussion

4.1. Image-to-Image Models Comparison for VNIR Images Generation from RGB

In this section, we show the results of deep learning models based on generative adversarial networks comparison for VNIR images translation from RGB images. We provide this comparison on the dataset sequential RGB images and corresponding VNIR images in the 838 nm range presented in Section 3.3.1. To estimate the performance, we split the data into the train set (80%) and the validation set (20%). The augmentation techniques as Random Rotations, Shifts, Zoom, and Flips are implemented to increase the data representativity and to keep the model's efficiency during the training and validation stages. We do not use the transformations such as Contrast/Brightness adjustments because they may lead to the information loss from the acquired VNIR imaging data. Taking into account that the image-to-image translation is also known as the translation from the *domain B* to *domain A* (or just *BtoA*), it was necessary to label *domain B* and *domain A* images from our acquired paired dataset. We identified the RGB images as *domain B* and *domain A* as the VNIR images. All models were evaluated by 200 epochs where the first 100 were implemented with the constant learning rate and the remaining 100 with linearly decreasing to zero. The models training and validation were realized via the Python scripts launched in Google Colab.

For the CycleGAN model, we use ResNet encoder–decoder architecture consisting of two downsampling layers, six ResNet bottleneck blocks and two upsampling layers. We also employ an Adam optimizer with the learning rate of 0.0002 and momentum parameters $\beta_1 = 0.5$ and $\beta_2 = 0.999$.

For the Pix2Pix model training, we fixed the same parameters: batch size = 1, $\beta_1 = 0.5$, $\beta_2 = 0.999$, and learning rate = 0.0002. The U-Net generator had 4 downsampling blocks. Optimization included the generator loss optimization step and the discriminator loss optimization step, respectively. Regularization parameters are as follows: $\lambda_{VGG} = \lambda_{Feat} = 10$, $\lambda_{L_1} = 100$.

For the Pix2PixHD model, we also implement the same parameters: Adam optimizer, batch size = 1, $\beta_1 = 0.5$, $\beta_2 = 0.999$, and learning rate = 0.0002.

Figure 13 shows the discriminator values of CycleGAN (Figure 13a), Pix2Pix (Figure 13b), and Pix2PixHD (Figure 13c) models during the training stage. We show the model's discriminator losses because they show the ability of GAN-based models to identify the quality of synthesized VNIR images by generator in comparison to original VNIR images.

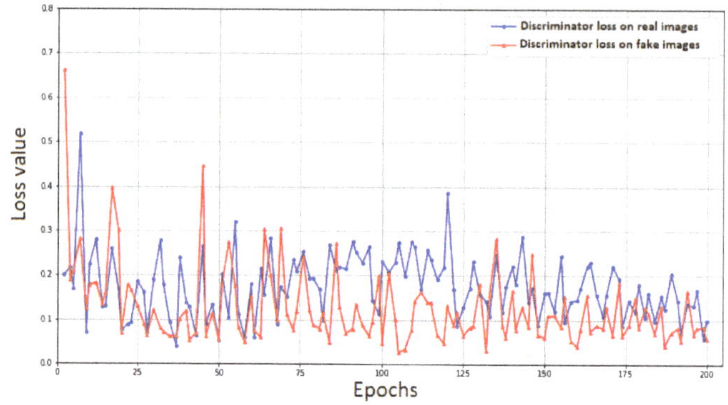

(**a**)

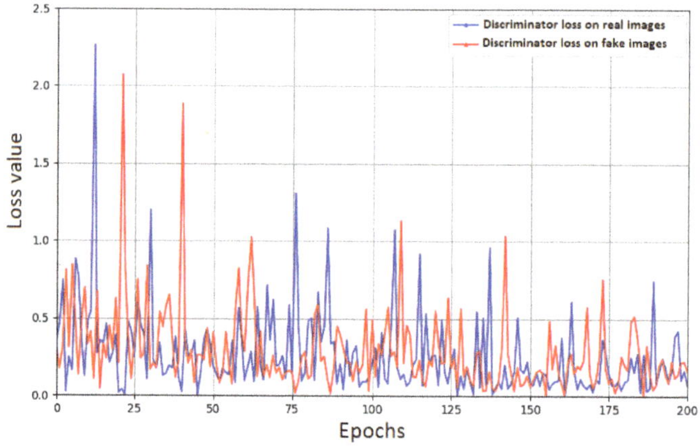

(**b**)

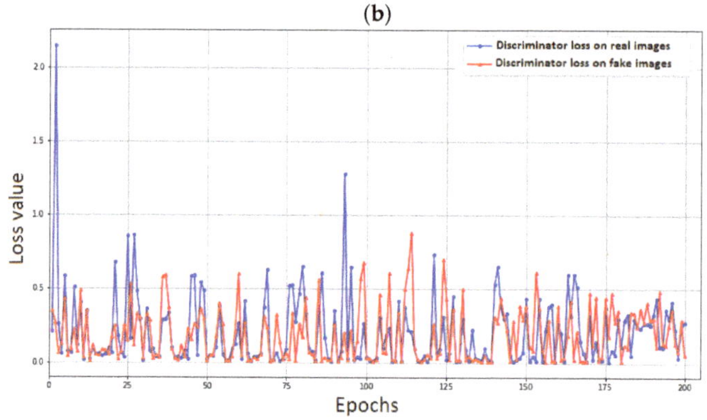

(**c**)

Figure 13. GAN-based models evaluation: (**a**) CycleGAN discriminator loss values during the training; (**b**) Pix2Pix discriminator loss values during the training; and (**c**) Pix2PixHD discriminator loss values during the training.

For selected GAN-based models we see that the training stage is unstable, but the discriminator losses tend to decrease over time. Pix2PixHD shows the lowest loss value in comparison to CycleGAN and Pix2Pix. For the models validation, we reconstructed the VNIR images using model weights acquired during the training. We used MAE, MAPE,

MSE, PSNR and SSIM metrics to estimate the quality of VNIR reconstructed images in comparison with original VNIR images. Figure 14 shows these images (with 'cyclegan', 'pix2pix', 'pix2pixHD' labels, respectively) in comparison to the original VNIR image ('reference' label) via Python visualization tools.

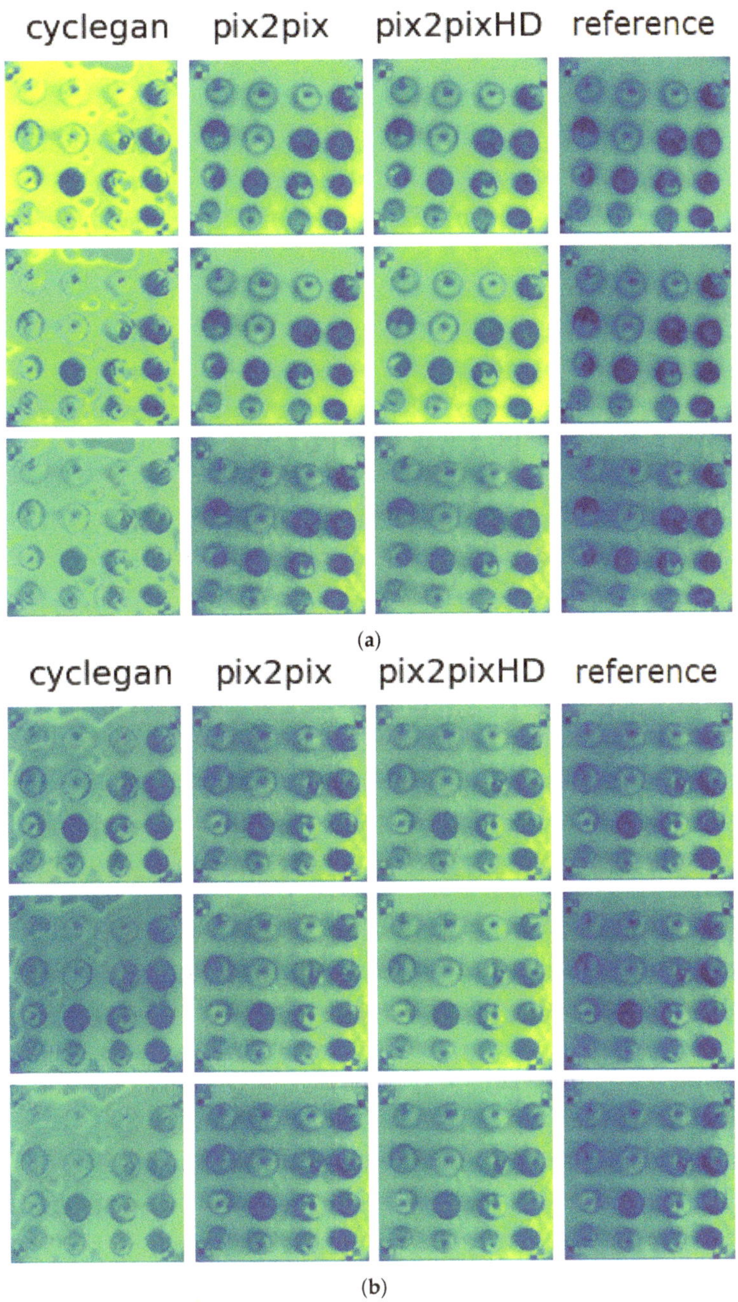

Figure 14. Examples of VNIR generated images in comparison to original VNIR image: (**a**) obtained under full illumination; and (**b**) obtained under partial illumination.

Table 1 summarizes the results of considered models performance, where the results for Pix2PixHD model are highlighted with the black blod. Considering both the pixel-based and the image metrics, one can conclude on the promising results. The generated images look more or less similar to the original ones. The images containing apples, overall light intensity similar to the ground truth and the decay region are mainly preserved However, all the models have particular artifacts. The CycleGAN model has the big stamp-like artifacts and there are a lot of missed decayed zones in the apples. In terms of metrics mentioned in Section 3.2, Pix2Pix and Pix2PixHD models perform the comparable and much better than others, and decay regions preserved relatively well, although the intensity level mismatch can be seen. Pix2PixHD models produce perceptually good images preserving importance for task features and the mean error level is equal to 0.6% In terms of important metrics for the image quality estimation, such as PSNR and SSIM, the Pix2PixHD model showed higher values in comparison to Pix2Pix (46.859 against 46.433, and 0.972 against 0.955, respectively). Taking into account the results of this comparison, we decided to use the Pix2PixHD model for VNIR images generation from RGB during the next stages.

Table 1. Image-to-image models comparison for RGB to VNIR images generation.

Models	MAE	MAPE	MSE	PSNR	SSIM
CycleGAN	0.067	0.105	0.01127	27.375	0.856
Pix2Pix	0.004	0.006	0.00003	46.433	0.955
Pix2PixHD	**0.004**	**0.006**	**0.00003**	**46.859**	**0.972**

4.2. Segmentation of Generated VNIR Images for Early Postharvest Decay Detection in Apples

In this section, we apply the CNN-based models for instance segmentation of generated VNIR images. Based on the results reported in Section 4.1, we use the Pix2PixHD model for the VNIR image generation. The dataset containing 456 images of stored apples (see Section 3.3.2) was used as the input for trained weights of the Pix2PixHD model to generate VNIR images. The examples of synthesized VNIR images from corresponding input RGB images are presented in Figure 15. Comparing the quality of new images with the images that were synthesizing during Pix2PixHD training stage (see Section 4.1), PSNR and SSIM values increased from 46.859 to 52.876 and from 0.972 to 0.994, respectively.

Mask R-CNN is used as the CNN-based model for the images instance segmentation. However, before applying Mask R-CNN to images, synthesized with Pix2PixHD, it was necessary to train Mask R-CNN on real VNIR images to detect and segment the fungi and decayed areas in stored apples. We used the labeled dataset containing 1029 VNIR images (see Section 3.3.2) for Mask R-CNN model training and validation. We report on the object classes used for data labeling in Section 3.4.

In this work, we implemented Mask R-CNN with the L1 as a loss function, ResNet50 as the backbone, Stochastic gradient descent (SGD) as an optimizer, and COCO weights to use Detectron2 library [102]. GaussianNoise, RandomGamma, RandomBrightness, and HorizontalFlip were applied as the data augmentation function to keep the efficiency of the proposed model during the training and validation stages. The model was developed in Python, and all calculations were realized in Google Colab.

In our experiment, we apply the cross-validation for Mask R-CNN model training on the dataset containing VNIR images. Cross-validation is a widespread technique helping avoid the overfitting during the model training on big data. In our case, we deal with the sequential images, i.e., one apple can be located in many images without any changes in position, which may resulted in improving the loss value after decreasing during the training procedure. During cross-validation, the data is usually split into several groups, called folds, where each group is used for the training and validation one by one. For example, if the dataset is separated into three folds, the pipeline is the following: (i) the first fold is a validation set, the second and third folds form the train set; (ii) the first and the third folds are train set, the second fold is a validation set; and (iii) the first and the second folds are training set, the third fold is a validation set. This pipeline is also fair for the cross-validation with four and higher folds distribution. By default, the number of folds, which is also called *k-folds*, is usually set equal to five or ten, but the *k-folds* may be different. In this work, we set the number of folds equal to two, three, six, and nine. We show the mean Average Precision values for each *k-fold* during Mask R-CNN models in Table 2.

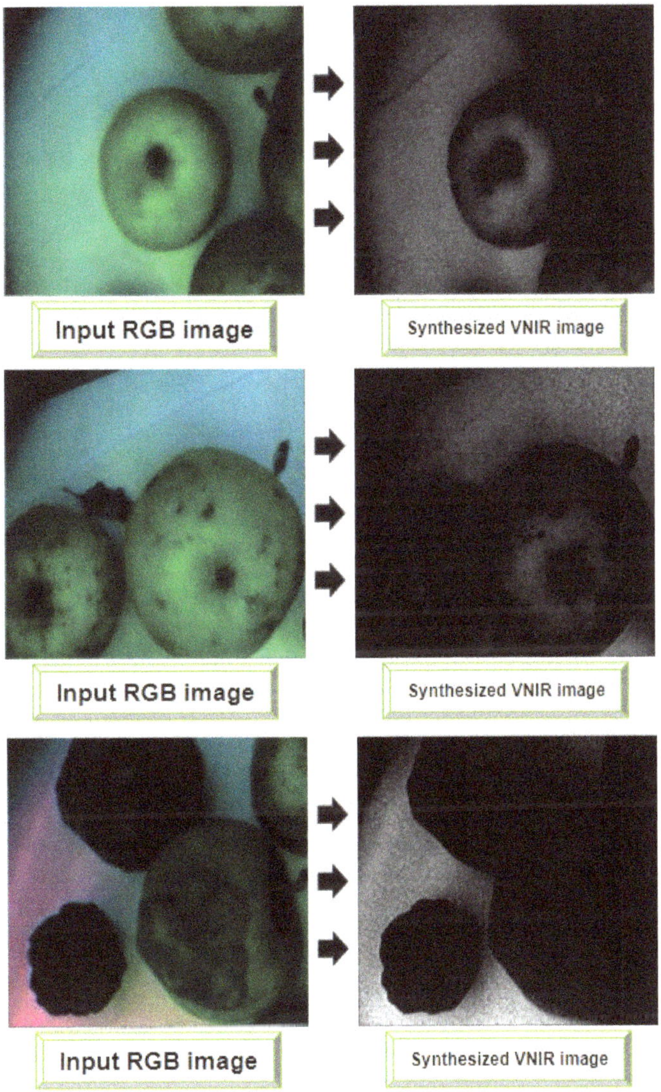

Figure 15. Examples of synthesized VNIR images with Pix2PixHD model weights.

Table 2. Comparison of Average Precision for Mask R-CNN model.

k-Folds	mAP	mAP_{50}	mAP_{75}	mAP_S	mAP_M	mAP_L
2	64.251	90.205	65.606	37.202	75.980	97.412
3	67.652	90.354	65.348	35.400	75.290	96.290
6	67.026	90.950	67.055	38.188	74.609	98.871
9	67.993	91.120	64.871	31.575	75.181	97.257

The results for each object class segmentation (or per-category segmentation) during Mask R-CNN model during all folds are given in Tables 3 and 4). We also used mAP and F1-score metrics to evaluate the segmentation quality during model training for folds distribution. Tables 3 and 4 present the mean mAP and F1-score values for each fold, respectively. As can be seen, the number of folds leads to increasing of the metrics values and segmentation accuracy. This is a demonstration of a cross-validation technique in comparison to ordinary data splitting on the training and validation sets. Figure 16 shows the examples of VNIR images with predicted annotations of object classes (see Section 3.4) acquired during the Mask R-CNN model validation. Here we show the examples of synthesized and annotated images from *k-folds* = 9, as the distribution with the better mAP and F1-score values (see the column for *k-folds* = 9 with black bold in the Tables 3 and 4). Even though the postharvest decay zones (*Decay* object class in Tables 3 and 4) and the fungal areas (*Fungi* object class in Tables 3 and 4) are detected with small values of an F1-score metric (58.861 and 40.968, respectively), a trained Mask R-CNN model allows for the detection and segmentation of spoiled apples (*Spoiled apple* object class), containing either decayed zones or fungal areas, or both, with an F1-score of 94.800, which is promising.

Table 3. Results on per-category segmentation by Mask R-CNN using mAP metric.

Category	mAP			
	k-Folds = 2	*k-Folds* = 3	*k-Folds* = 6	*k-Folds* = 9
Healthy apple	94.785	95.154	93.951	**98.350**
Spoiled apple	87.839	92.567	93.678	**93.997**
Decay	53.509	53.408	54.620	**57.562**
Fungi	31.581	30.609	34.285	**39.967**

Table 4. Results on per-category segmentation by Mask R-CNN using F1-score metric.

Category	F1-Score			
	k-Folds = 2	*k-Folds* = 3	*k-Folds* = 6	*k-Folds* = 9
Healthy apple	95.640	95.589	94.799	**98.375**
Spoiled apple	88.120	93.134	94.689	**94.800**
Decay	53.309	53.213	54.850	**58.861**
Fungi	31.686	37.247	35.126	**40.968**

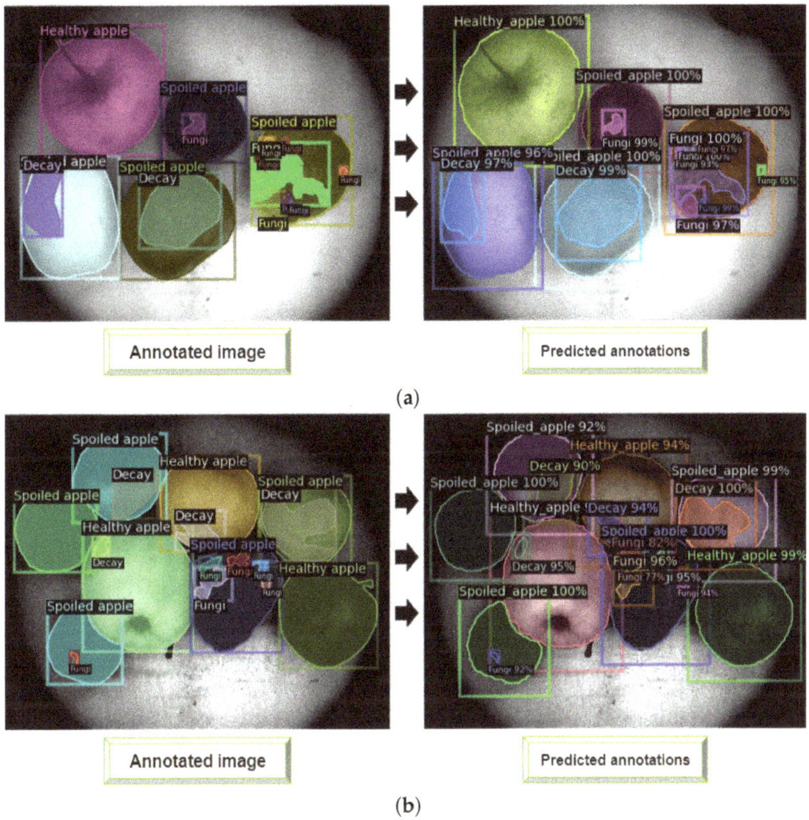

Figure 16. Comparison of object classes annotation in real VNIR images (**a**,**b**, on the left with 'Annotated image' label) to predicted object annotations (**a**,**b**, on the right with 'Predicted annotations' label) during Mask R-CNN model training.

Taking into account the results of Mask R-CNN evaluation on real VNIR imaging data and the results of the Pix2PixHD evaluation in comparison to other GAN-based models (see Section 4.1), we provide the proposed pipeline for segmentation of generated VNIR images. To estimate it we acquired the dataset containing only 456 sequential RGB images without the corresponded VNIR images (see Section 3.4). The images were acquired in the greenhouse (see Section 3.3.2) under the same environmental conditions (temperature range is from 35 °C to 40 °C, and RH is 70%, respectively). In order to simulate possible occasion during the real storage, spoiled apples with the decayed and fungi zones were added to healthy (non-damaged) apples. The concept is as follows: (i) we utilize a set of RGB images as input data; (ii) these RGB images are passed through a GAN-based model (specifically, Pix2PixHD with pre-trained weights in our case); (iii) VNIR images are generated from the input RGB images using Pix2PixHD; and (iv) the generated VNIR images are fed into a CNN-based model (specifically, Mask R-CNN with pre-trained weights) to obtain these images with predicted annotation masks. Figure 17 shows the examples of images which were synthesized and segmented with the proposed pipeline. As it can be seen in Figure 17b,c, the proposed approach helps detect and segment the decayed zones separately from the fungi zones in the stored apples. All computations were also provided in Google Colab.

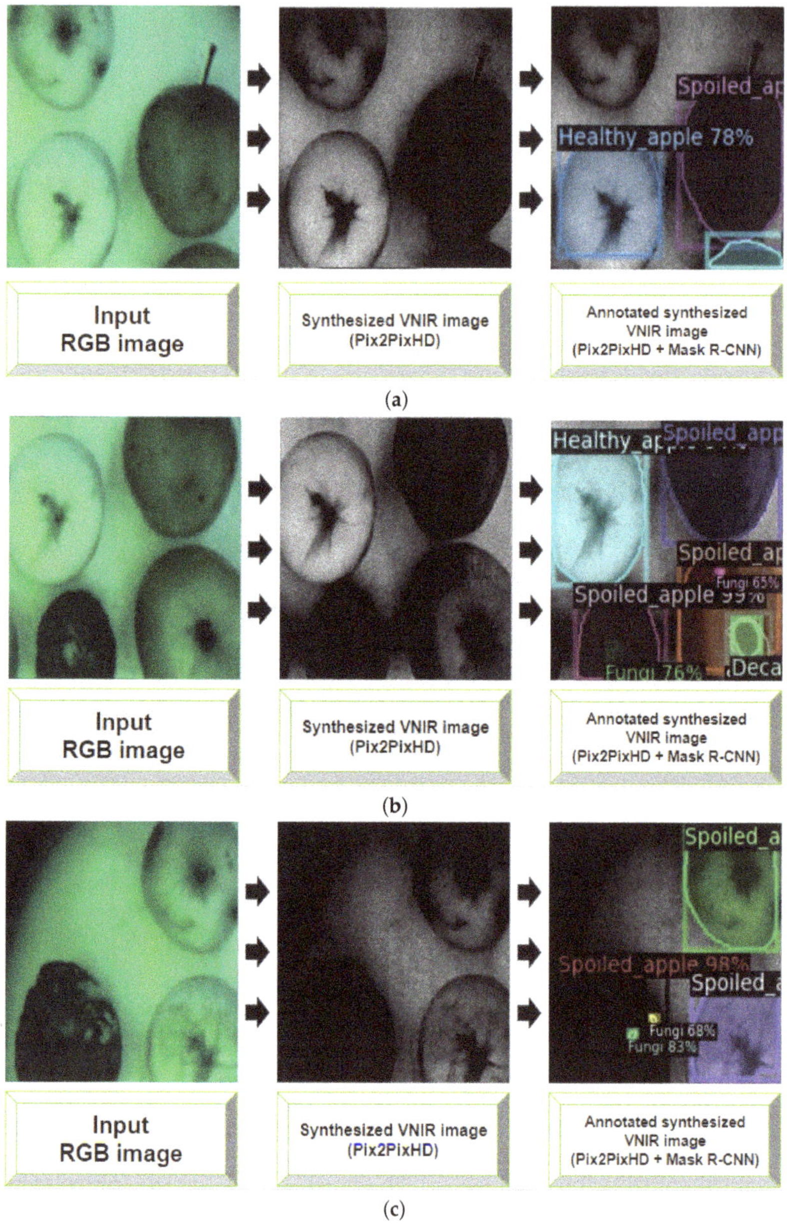

Figure 17. Synthesized VNIR images (**a**–**c**) segmentation with Mask R-CNN model.

4.3. Early Postharvest Decay Detection in Stored Apples Using Generated VNIR Imaging Data on an Embedded System

To evaluate the applicability of a GAN- and CNN-based models in real-life scenarios we conduct an experiment using the NVIDIA Jetson Nano embedded system [103]. The goal of the experiment is to validate the model's ability to handle video streams with varying frames per second (FPS).

We used 100 RGB images. Input RGB images are characterized by the size of 256 pixels. A GAN model was used to generate VNIR images from input images and processed over 100 images at an average rate of 17 FPS. The generated images were then tested with Mask R-CNN, resulting in an average rate of 0.420 FPS. Low FPS in Mask R-CNN can be attributed to its complexity compared to Pix2PixHD. As the two-stage detection model that performs instance segmentation by detecting objects and generating pixel-level masks for each object, it requires more computational resources. Figure 18 shows the examples of VNIR images generated and segmented using the NVIDIA Jetson Nano based on the input RGB data.

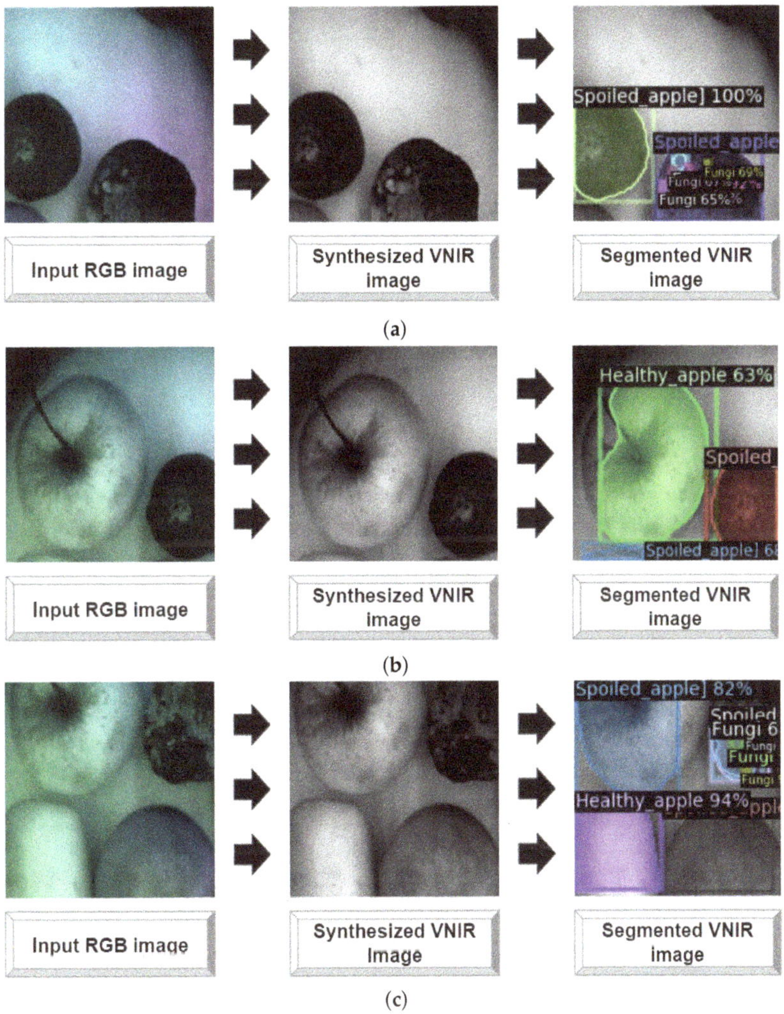

Figure 18. Generated and segmented VNIR images (**a**–**c**) using Jetson Nano.

4.4. Discussion

In this section, we compare our results with other relevant research works in the field of application of NIR imaging data and deep learning techniques for early postharvest decay and fungal zones prediction in stored apples. The proposed approach is based on the joint application of GAN and CNN techniques for artificial generation and subsequent segmentation of VNIR images. However, in order to segment the decayed and fungal zones in artificially generated VNIR images, we had to train and validate a CNN technique on the real VNIR images containing these zones in stored apples. To perform this, we acquired the dataset of VNIR images (see Section 3.3) and then trained and validated the Mask R-CNN model (see Section 4.2).

Taking into the account the ability of Mask R-CNN to provide the multi-class instance and semantic segmentation (see Section 3.1.4), we trained the model not only to detect and identify the quality of apple (*Healthy apple* or *Spoiled apple*, see Section 3.4), but also to detect and predict the decayed and fungal zones separately from each other. Novelty is that the model is trained and validated to identify the quality of stored apples by taking into account the presence of decayed and fungal areas in the apples themselves. In this context, an apple is classified as *Spoiled apple* if it contains the decayed or fungal zones, whether they are separate or combined. Conversely, if an apple does not exhibit any decayed or fungal zones prior to storage stage, i.e., during the VNIR image collection, it is classified as a *Healthy apple*. However, if the decayed and/or fungal zones emerge in the apple during the storage stage, its classification transitions from a *Healthy apple* to *Spoiled apple*.

Relevant works in this area can be classified into three main groups according to main tasks: (i) defective apples detection based on the internal quality parameters [104,105]; (ii) early defect detection in apples [104,106]; and (iii) early fungi detection in apples [73,107,108]. Table 5 presents a comparative study of these works.

Table 5. Comparative table of relevant research works.

References	Task	NIR Images Range, nm	Technique	Metric	Value
[104]	Real-time apple defect inspection	850	YOLO v4	F1	92.000
[105]	Apples surface defect segmentation	460–842	U-Net	F1-score	87.000
[105]	Apples surface defect segmentation	460–842	the improved U-Net	F1-score	91.000
[106]	Early bruise detection in apples	900–2350	Faster R-CNN	mAP	96.900
[106]	Early bruise detection in apples	900–2350	YOLO v3-Tiny	mAP	99.100
[106]	Early bruise detection in apples	900–2350	YOLO 5s	mAP	99.600
[107]	Moldy core detection in apples	400–850	CARS-PLS-DA model	Accuracy	87.880
[73]	Codling Moth detection in apples	900–1700	Gradient tree boosting	F1-score	97.000
[108]	Moldy core detection in apples	200–1100	BP-ANN	Accuracy	95.000

The authors applied various tools and methods based on machine learning for detecting the defected and diseased zones in wide NIR ranges (400–2350 nm, globally) with detailed spectral information on the diseased zones. The most relevant and similar approach to the current research is reported in [104], where a YOLO v4 model in sorting machine for real-time detection of defects in "Red Fuji", "Golden Delicious", and "Granny Smith" apples is implemented. The authors used the RGB and corresponded NIR images in the range of 850 nm of the apples in the machine's sorting line. Moreover, the ability of trained YOLO v4 models to detect with bounding box 'calyx' and 'stem' zones separately from 'defect' zones was demonstrated. In this work, we applied the Mask R-CNN not only to detect (with bounding box) and segment (with mask) the decayed and the fungal areas in stored apples, but also to identify the quality of apples as diseased (*Spoiled apple*) if such zones are detected by the model. F1-score and mAP values for *Decay* and *Fungi* zones are not that high. These problems can be fixed in our future work by obtaining more VNIR images containing the fungal and the decayed areas in order to increase the data representation during the model validation. On the other hand, the results for *Spoiled apple* (apple contains *Fungi* and/or *Decay* zones) segmentation are 98.350 and 98.375, respectively, which is promising. Finally, the proposed approach is for an apple quality control during the storage stage, i.e., before sending the stored apples to the fruit sorting machine. The system, which could generate VNIR images without a multispectral or hyperspectral camera based only on the input RGB images with segmented fungal and decayed zones, if they occur in stored apples, can be applied as an additional stage for the fruit and vegetable control before sending them to a sorting machine.

In [106], the authors compared several Faster R-CNN, YOLO v3-Tiny, and YOLO 5s models for early decay (or bruise) detection in apples. The approach proposed in this work showed promising results in terms of the mAP metric (98.350 for Mask R-CNN validation, in our case, against 96.900 for Faster R-CNN, 99.100 for YOLO v3-Tiny, and 96.600 for YOLO 5s), and the selected model was trained to segment the decayed and fungal zones in apples, while authors in [106] trained the models to identify and predict the apples without ('No bruise'), with a small ('Mild bruise') and significant ('Severe bruise') decayed areas in apples. The authors also acquired the NIR images in spectral range of 900–2350 nm, while in this work the images from 838 nm range were used in order to make sure that the diseased zones in VNIR images are visible in the RGB images as well.

In [105], the authors trained and validated U-Net and the improved U-Net model for the defect segmentation in VNIR images of apples. In this work, we have demonstrated the semantic segmentation of decayed and fungal areas with an advanced experimental methodology. We simulated ordinary and extreme storage conditions during the paired RGB and VNIR images collection procedures. Taking this into account, we achieved a relevant value for the diseased apples segmentation in terms of the F1-score metric.

We have demonstrated the potential for the postharvest decay and fungi prediction for stored apples. However, it can be scaled to other crops that are widely used in food production, e.g., carrots, tomatoes, cucumber, fruits or bananas. For example, the system that allows the generation and segmentation of VNIR images can be applied for segmentation and prediction of such fungi as *Sclerotinia sclerotiorum* or *Botrytis cinerea*. 'Sclerotinia' and 'Botrytis' fungal zones have similar morphology and, if they occur in plants, it is a nontrivial task to identify one fungi variety from another one using only RGB imagery or visual estimation of the internal fungal traits with human eyes [109]. The system supplied with the trained and validated DL technique based on the GAN and CNN models can assist the user with the additional spectral information about each fungi acquired from the generated VNIR images. It is useful for more precise antifungal activities during the food quality control.

Another potential scenario is the application of the proposed research for the preharvest diseases and the defect detection for the plants both growing in natural environments and in artificially controlled systems. For example, it can be a robot moving platform or unmanned aerial vehicle without a hyperspectral camera, but with an embedded system that may generate and segment the NIR imaging data from the input RGB one. However, DL technique should be trained, tested and validated precisely, as the proposed system has to detect and segment not only the diseased plants from the healthy ones, but also to detect the kind of defect (damage, decay, fungi variety) with the following suggestion of spoiled fruit processing.

5. Conclusions

NIR imagery provides detailed information about the diseased areas in stored fruits, which is why the hyperspectral cameras containing thousands of bands are used for food quality monitoring at postharvest stages. However, hyperspectral devices are expensive and are not friendly for the farmers and sellers' usage. In this article, we have presented the approach based on the GAN and CNN DL techniques for early postharvest decay zones and fungi areas detection and prediction in stored apples using synthesized and segmented VNIR images.

The conclusions of this work are as follows:

- The analysis of Pix2Pix, CycleGAN, and Pix2PixHD models, which are widely used GAN techniques, and their application to a dataset containing paired 1305 sequential RGB images and 1305 sequential VNIR images of stored apples of different varieties and various pre-treatments. The images were acquired under the full and partial illumination with the goal to simulate real storage conditions.
- Comparison of the real VNIR images with the VNIR images synthesized by selected GAN based models. The VNIR images generated via Pix2PixHD a 0.972 score for the SSIM metric.
- The training and test of Mask R-CNN on another dataset containing only 1029 sequential VNIR images of apples under violated storage conditions. Within this test, an F1-score of 58.861 is achieved for the postharvest decay zones and F1-score 40.968 for the fungal zones detection. The spoiled apples with the decayed and fungal zones are detected and segmented with F1-score 94.800.
- Testing of the proposed solution on an embedded system with AI capabilities. We used 100 RGB images of stored apples as an input data for NVIDIA Jetson Nano, and the time processing of VNIR images generation by Pix2PixHD showed 17 FPS. The detection and segmentation by Mask R-CNN achieved 0.42 FPS.

The proposed approach is a promising solution able to substitute expensive hyperspectral imaging devices for early postharvest decay prediction tasks in postharvest food quality control.

Author Contributions: Conceptualization, N.S., D.S. and A.S.; methodology, N.S.; software, N.S. ans I.S.; validation, N.S., I.S. and M.S.; formal analysis, N.S. and D.S.; investigation, N.S.; resources, N.S.; data curation, N.S.; writing—original draft preparation, N.S., I.S.; writing—review and editing, N.S., I.S., M.S., D.S. and A.S.; visualization, N.S. and I.S.; supervision, A.S. All authors have read and agreed to the published version of the manuscript.

Funding: This research received no external funding

Institutional Review Board Statement: Not applicable.

Data Availability Statement: Not applicable.

Conflicts of Interest: The authors declare no conflict of interest.

Abbreviations

The following abbreviations are used in this manuscript:

NIR	Near Infrared Image
VNIR	Visible Near Infrared Image
AI	Artificial Intelligence
CV	Computer Vision
ML	Machine Learning
SVM	Support Vector Machines
RF	Random Forest
kNN	K-Nearest Neighbors Algorithm
GTB	Gradient Tree Boosting
DL	Deep Learning
CNN	Convolutional Neural Network
GAN	Generative Adversarial Network
ROI	Regions of Interests
SBC	Single Board Computer
RH	Relative Humidity

References

1. United Nations Data about Current World Population. Available online: https://www.worldometers.info/world-population/ (accessed on 26 June 2023).
2. United Nations Data on Current and Prospected World Population. Available online: https://population.un.org/wpp/Graphs/Probabilistic/POP/TOT/900 (accessed on 26 June 2023).
3. Ullah, S.; Hashmi, M.; Lee, J.; Youk, J.H.; Kim, I.S. Recent Advances in Pre-harvest, Post-harvest, Intelligent, Smart, Active, and Multifunctional Food Packaging. *Fibers Polym.* **2022**, *23*, 2063–2074. [CrossRef]
4. Coradi, P.C.; Maldaner, V.; Lutz, É.; da Silva Daí, P.V.; Teodoro, P.E. Influences of drying temperature and storage conditions for preserving the quality of maize postharvest on laboratory and field scales. *Sci. Rep.* **2020**, *10*, 22006. [CrossRef] [PubMed]
5. Mohammed, M.; Alqahtani, N.; El-Shafie, H. Development and evaluation of an ultrasonic humidifier to control humidity in a cold storage room for postharvest quality management of dates. *Foods* **2021**, *10*, 949. [CrossRef] [PubMed]
6. Sun, X.; Baldwin, E.; Bai, J. Applications of gaseous chlorine dioxide on postharvest handling and storage of fruits and vegetables—A review. *Food Control* **2019**, *95*, 18–26. [CrossRef]
7. Yahia, E.M.; Fonseca, J.M.; Kitinoja, L. Postharvest losses and waste. In *Postharvest Technology of Perishable Horticultural Commodities*; Elsevier: Amsterdam, The Netherlands, 2019; pp. 43–69.
8. Palumbo, M.; Attolico, G.; Capozzi, V.; Cozzolino, R.; Corvino, A.; de Chiara, M.L.V.; Pace, B.; Pelosi, S.; Ricci, I.; Romaniello, R.; et al. Emerging Postharvest Technologies to Enhance the Shelf-Life of Fruit and Vegetables: An Overview. *Foods* **2022**, *11*, 3925. [CrossRef]
9. Elik, A.; Yanik, D.K.; Istanbullu, Y.; Guzelsoy, N.A.; Yavuz, A.; Gogus, F. Strategies to reduce post-harvest losses for fruits and vegetables. *Strategies* **2019**, *5*, 29–39.
10. FAO Data on Global Apple Production. Available online: https://www.fao.org/faostat/en/#data/QCL/visualize (accessed on 26 June 2023).
11. Harker, F.; Feng, J.; Johnston, J.; Gamble, J.; Alavi, M.; Hall, M.; Chheang, S. Influence of postharvest water loss on apple quality: The use of a sensory panel to verify destructive and non-destructive instrumental measurements of texture. *Postharvest Biol. Technol.* **2019**, *148*, 32–37. [CrossRef]
12. de Andrade, J.C.; Galvan, D.; Effting, L.; Tessaro, L.; Aquino, A.; Conte-Junior, C.A. Multiclass Pesticide Residues in Fruits and Vegetables from Brazil: A Systematic Review of Sample Preparation Until Post-Harvest. *Crit. Rev. Anal. Chem.* **2021**, 1–23. Available online: https://www.tandfonline.com/doi/abs/10.1080/10408347.2021.2013157 (accessed on 26 June 2023).
13. Bratu, A.M.; Petrus, M.; Popa, C. Monitoring of post-harvest maturation processes inside stored fruit using photoacoustic gas sensing spectroscopy. *Materials* **2020**, *13*, 2694. [CrossRef]
14. Sottocornola, G.; Baric, S.; Nocker, M.; Stella, F.; Zanker, M. Picture-based and conversational decision support to diagnose post-harvest apple diseases. *Expert Syst. Appl.* **2022**, *189*, 116052. [CrossRef]
15. Malvandi, A.; Feng, H.; Kamruzzaman, M. Application of NIR spectroscopy and multivariate analysis for Non-destructive evaluation of apple moisture content during ultrasonic drying. *Spectrochim. Acta Part A Mol. Biomol. Spectrosc.* **2022**, *269*, 120733. [CrossRef]
16. Schlie, T.P.; Dierend, W.; Koepcke, D.; Rath, T. Detecting low-oxygen stress of stored apples using chlorophyll fluorescence imaging and histogram division. *Postharvest Biol. Technol.* **2022**, *189*, 111901. [CrossRef]
17. Wang, L.; Huang, J.; Li, Z.; Liu, D.; Fan, J. A review of the polyphenols extraction from apple pomace: Novel technologies and techniques of cell disintegration. *Crit. Rev. Food Sci. Nutr.* **2022**, 1–14. [CrossRef] [PubMed]
18. Wu, X.; Fauconnier, M.L.; Bi, J. Characterization and Discrimination of Apples by Flash GC E-Nose: Geographical Regions and Botanical Origins Studies in China. *Foods* **2022**, *11*, 1631. [CrossRef] [PubMed]
19. Biasi, A.; Zhimo, V.Y.; Kumar, A.; Abdelfattah, A.; Salim, S.; Feygenberg, O.; Wisniewski, M.; Droby, S. Changes in the fungal community assembly of apple fruit following postharvest application of the yeast biocontrol agent Metschnikowia fructicola. *Horticulturae* **2021**, *7*, 360. [CrossRef]
20. Bartholomew, H.P.; Lichtner, F.J.; Bradshaw, M.; Gaskins, V.L.; Fonseca, J.M.; Bennett, J.W.; Jurick, W.M. Comparative Penicillium spp. Transcriptomics: Conserved Pathways and Processes Revealed in Ungerminated Conidia and during Postharvest Apple Fruit Decay. *Microorganisms* **2022**, *10*, 2414. [CrossRef]
21. Morales-Cedeno, L.R.; del Carmen Orozco-Mosqueda, M.; Loeza-Lara, P.D.; Parra-Cota, F.I.; de Los Santos-Villalobos, S.; Santoyo, G. Plant growth-promoting bacterial endophytes as biocontrol agents of pre-and post-harvest diseases: Fundamentals, methods of application and future perspectives. *Microbiol. Res.* **2021**, *242*, 126612. [CrossRef]
22. Nikparvar, B.; Thill, J.C. Machine learning of spatial data. *ISPRS Int. J. Geo-Inf.* **2021**, *10*, 600. [CrossRef]
23. Zhang, Y.; Liu, M.; Yu, F.; Zeng, T.; Wang, Y. An o-shape neural network with attention modules to detect junctions in biomedical images without segmentation. *IEEE J. Biomed. Health Inform.* **2021**, *26*, 774–785. [CrossRef]
24. Zhao, S.; Blaabjerg, F.; Wang, H. An overview of artificial intelligence applications for power electronics. *IEEE Trans. Power Electron.* **2020**, *36*, 4633–4658. [CrossRef]
25. Meshram, V.; Patil, K.; Meshram, V.; Hanchate, D.; Ramkteke, S. Machine learning in agriculture domain: A state-of-art survey. *Artif. Intell. Life Sci.* **2021**, *1*, 100010. [CrossRef]
26. Kakani, V.; Nguyen, V.H.; Kumar, B.P.; Kim, H.; Pasupuleti, V.R. A critical review on computer vision and artificial intelligence in food industry. *J. Agric. Food Res.* **2020**, *2*, 100033. [CrossRef]
27. Rasti, S.; Bleakley, C.J.; Holden, N.; Whetton, R.; Langton, D.; O'Hare, G. A survey of high resolution image processing techniques for cereal crop growth monitoring. *Inf. Process. Agric.* **2022**, *9*, 300–315. [CrossRef]

28. Tang, Y.; Qiu, J.; Zhang, Y.; Wu, D.; Cao, Y.; Zhao, K.; Zhu, L. Optimization strategies of fruit detection to overcome the challenge of unstructured background in field orchard environment: A review. *Precis. Agric.* **2023**, *24*, 1183–1219. [CrossRef]
29. Ouhami, M.; Hafiane, A.; Es-Saady, Y.; El Hajji, M.; Canals, R. Computer vision, IoT and data fusion for crop disease detection using machine learning: A survey and ongoing research. *Remote Sens.* **2021**, *13*, 2486. [CrossRef]
30. Wu, Z.; Chen, Y.; Zhao, B.; Kang, X.; Ding, Y. Review of weed detection methods based on computer vision. *Sensors* **2021**, *21*, 3647. [CrossRef] [PubMed]
31. Mendigoria, C.H.; Aquino, H.; Concepcion, R.; Alajas, O.J.; Dadios, E.; Sybingco, E. Vision-based postharvest analysis of musa acuminata using feature-based machine learning and deep transfer networks. In Proceedings of the 2021 IEEE 9th Region 10 Humanitarian Technology Conference (R10-HTC), Bangalore, India, 30 September–2 October 2021; pp. 1–6.
32. Bucio, F.; Isaza, C.; Gonzalez, E.; De Paz, J.Z.; Sierra, J.R.; Rivera, E.A. Non-Destructive Post-Harvest Tomato Mass Estimation Model Based on Its Area via Computer Vision and Error Minimization Approaches. *IEEE Access* **2022**, *10*, 100247–100256. [CrossRef]
33. Ropelewska, E. Postharvest Authentication of Potato Cultivars Using Machine Learning to Provide High-Quality Products. *Chem. Proc.* **2022**, *10*, 30.
34. Isola, P.; Zhu, J.Y.; Zhou, T.; Efros, A.A. Image-to-Image Translation with Conditional Adversarial Networks. *arXiv* **2018**, arXiv:1611.07004.
35. Zhu, J.Y.; Park, T.; Isola, P.; Efros, A.A. Unpaired Image-to-Image Translation using Cycle-Consistent Adversarial Networks. *arXiv* **2020**, arXiv:1703.10593.
36. Wang, T.C.; Liu, M.Y.; Zhu, J.Y.; Tao, A.; Kautz, J.; Catanzaro, B. High-Resolution Image Synthesis and Semantic Manipulation with Conditional GANs. *arXiv* **2018**, arXiv:1711.11585.
37. Christovam, L.E.; Shimabukuro, M.H.; Galo, M.d.L.B.; Honkavaara, E. Pix2pix conditional generative adversarial network with MLP loss function for cloud removal in a cropland time series. *Remote Sens.* **2022**, *14*, 144. [CrossRef]
38. de Lima, D.C.; Saqui, D.; Mpinda, S.A.T.; Saito, J.H. Pix2pix network to estimate agricultural near infrared images from rgb data. *Can. J. Remote Sens.* **2022**, *48*, 299–315. [CrossRef]
39. Farooque, A.A.; Afzaal, H.; Benlamri, R.; Al-Naemi, S.; MacDonald, E.; Abbas, F.; MacLeod, K.; Ali, H. Red-green-blue to normalized difference vegetation index translation: A robust and inexpensive approach for vegetation monitoring using machine vision and generative adversarial networks. *Precis. Agric.* **2023**, *24*, 1097–1115. [CrossRef]
40. Bertoglio, R.; Mazzucchelli, A.; Catalano, N.; Matteucci, M. A comparative study of Fourier transform and CycleGAN as domain adaptation techniques for weed segmentation. *Smart Agric. Technol.* **2023**, *4*, 100188. [CrossRef]
41. Jung, D.H.; Kim, C.Y.; Lee, T.S.; Park, S.H. Depth image conversion model based on CycleGAN for growing tomato truss identification. *Plant Methods* **2022**, *18*, 83. [CrossRef] [PubMed]
42. van Marrewijk, B.M.; Polder, G.; Kootstra, G. Investigation of the added value of CycleGAN on the plant pathology dataset. *IFAC-PapersOnLine* **2022**, *55*, 89–94. [CrossRef]
43. Yang, J.; Zhang, T.; Fang, C.; Zheng, H. A defencing algorithm based on deep learning improves the detection accuracy of caged chickens. *Comput. Electron. Agric.* **2023**, *204*, 107501. [CrossRef]
44. Tsuchikawa, S.; Ma, T.; Inagaki, T. Application of near-infrared spectroscopy to agriculture and forestry. *Anal. Sci.* **2022**, *38*, 635–642. [CrossRef]
45. Stasenko, N.; Savinov, M.; Burlutskiy, V.; Pukalchik, M.; Somov, A. Deep Learning for Postharvest Decay Prediction in Apples. In Proceedings of the IECON 2021—47th Annual Conference of the IEEE Industrial Electronics Society, Toronto, ON, Canada, 13–16 October 2021; pp. 1–6.
46. Ronneberger, O.; Fischer, P.; Brox, T. U-net: Convolutional networks for biomedical image segmentation. In Proceedings of the International Conference on Medical Image Computing and Computer-Assisted Intervention, Munich, Germany, 5–9 October 2015; Springer: Berlin/Heidelberg, Germany, 2015; pp. 234–241.
47. Yurtkulu, S.C.; Şahin, Y.H.; Unal, G. Semantic Segmentation with Extended DeepLabv3 Architecture. In Proceedings of the 2019 27th Signal Processing and Communications Applications Conference (SIU), Sivas, Turkey, 24–26 April 2019; pp. 1–4.
48. Assunção, E.; Gaspar, P.D.; Mesquita, R.; Simões, M.P.; Alibabaei, K.; Veiros, A.; Proença, H. Real-Time Weed Control Application Using a Jetson Nano Edge Device and a Spray Mechanism. *Remote Sens.* **2022**, *14*, 4217. [CrossRef]
49. Saddik, A.; Latif, R.; Taher, F.; El Ouardi, A.; Elhoseny, M. Mapping Agricultural Soil in Greenhouse Using an Autonomous Low-Cost Robot and Precise Monitoring. *Sustainability* **2022**, *14*, 15539. [CrossRef]
50. de Aguiar, A.S.P.; dos Santos, F.B.N.; dos Santos, L.C.F.; de Jesus Filipe, V.M.; de Sousa, A.J.M. Vineyard trunk detection using deep learning–An experimental device benchmark. *Comput. Electron. Agric.* **2020**, *175*, 105535. [CrossRef]
51. Mazzia, V.; Khaliq, A.; Salvetti, F.; Chiaberge, M. Real-time apple detection system using embedded systems with hardware accelerators: An edge AI application. *IEEE Access* **2020**, *8*, 9102–9114. [CrossRef]
52. Beegam, K.S.; Shenoy, M.V.; Chaturvedi, N. Hybrid consensus and recovery block-based detection of ripe coffee cherry bunches using RGB-D sensor. *IEEE Sens. J.* **2021**, *22*, 732–740. [CrossRef]
53. Zhang, W.; Liu, Y.; Chen, K.; Li, H.; Duan, Y.; Wu, W.; Shi, Y.; Guo, W. Lightweight fruit-detection algorithm for edge computing applications. *Front. Plant Sci.* **2021**, *12*, 740936. [CrossRef]
54. Vilcamiza, G.; Trelles, N.; Vinces, L.; Oliden, J. A coffee bean classifier system by roast quality using convolutional neural networks and computer vision implemented in an NVIDIA Jetson Nano. In Proceedings of the 2022 Congreso Internacional de Innovación y Tendencias en Ingeniería (CONIITI), Bogota, Colombia, 5–7 October 2022; pp. 1–6.

55. Fan, K.J.; Su, W.H. Applications of Fluorescence Spectroscopy, RGB-and MultiSpectral Imaging for Quality Determinations of White Meat: A Review. *Biosensors* **2022**, *12*, 76. [CrossRef]
56. Zou, X.; Zhang, Y.; Lin, R.; Gong, G.; Wang, S.; Zhu, S.; Wang, Z. Pixel-level Bayer-type colour router based on metasurfaces. *Nat. Commun.* **2022**, *13*, 3288. [CrossRef]
57. Rivero Mesa, A.; Chiang, J. Non-invasive grading system for banana tiers using RGB imaging and deep learning. In Proceedings of the 2021 7th International Conference on Computing and Artificial Intelligence, Tianjin, China, 23–26 April 2021; pp. 113–118.
58. Nasiri, A.; Taheri-Garavand, A.; Zhang, Y.D. Image-based deep learning automated sorting of date fruit. *Postharvest Biol. Technol.* **2019**, *153*, 133–141. [CrossRef]
59. Deng, L.; Li, J.; Han, Z. Online defect detection and automatic grading of carrots using computer vision combined with deep learning methods. *LWT* **2021**, *149*, 111832. [CrossRef]
60. Zhang, X.; Zhou, X.; Lin, M.; Sun, J. ShuffleNet: An extremely efficient convolutional neural network for mobile devices. *arXiv* **2017**, arXiv:1707.01083.
61. Wu, F.; Yang, Z.; Mo, X.; Wu, Z.; Tang, W.; Duan, J.; Zou, X. Detection and counting of banana bunches by integrating deep learning and classic image-processing algorithms. *Comput. Electron. Agric.* **2023**, *209*, 107827. [CrossRef]
62. Baheti, B.; Innani, S.; Gajre, S.; Talbar, S. Semantic scene segmentation in unstructured environment with modified DeepLabV3+. *Pattern Recognit. Lett.* **2020**, *138*, 223–229. [CrossRef]
63. Wu, F.; Duan, J.; Ai, P.; Chen, Z.; Yang, Z.; Zou, X. Rachis detection and three-dimensional localization of cut off point for vision-based banana robot. *Comput. Electron. Agric.* **2022**, *198*, 107079. [CrossRef]
64. Buyukarikan, B.; Ulker, E. Classification of physiological disorders in apples fruit using a hybrid model based on convolutional neural network and machine learning methods. *Neural Comput. Appl.* **2022**, *34*, 16973–16988. [CrossRef]
65. Li, J.; Zheng, K.; Yao, J.; Gao, L.; Hong, D. Deep unsupervised blind hyperspectral and multispectral data fusion. *IEEE Geosci. Remote Sens. Lett.* **2022**, *19*, 1–5. [CrossRef]
66. Liang, J.; Li, X.; Zhu, P.; Xu, N.; He, Y. Hyperspectral reflectance imaging combined with multivariate analysis for diagnosis of Sclerotinia stem rot on Arabidopsis thaliana leaves. *Appl. Sci.* **2019**, *9*, 2092. [CrossRef]
67. Vashpanov, Y.; Heo, G.; Kim, Y.; Venkel, T.; Son, J.Y. Detecting green mold pathogens on lemons using hyperspectral images. *Appl. Sci.* **2020**, *10*, 1209. [CrossRef]
68. Fahrentrapp, J.; Ria, F.; Geilhausen, M.; Panassiti, B. Detection of gray mold leaf infections prior to visual symptom appearance using a five-band multispectral sensor. *Front. Plant Sci.* **2019**, *10*, 628. [CrossRef]
69. Wan, L.; Li, H.; Li, C.; Wang, A.; Yang, Y.; Wang, P. Hyperspectral Sensing of Plant Diseases: Principle and Methods. *Agronomy* **2022**, *12*, 1451. [CrossRef]
70. Błaszczyk, U.; Wyrzykowska, S.; Gąstoł, M. Application of Bioactive Coatings with Killer Yeasts to Control Post-Harvest Apple Decay Caused by Botrytis cinerea and Penicillium italicum. *Foods* **2022**, *11*, 1868. [CrossRef]
71. Amaral Carneiro, G.; Walcher, M.; Baric, S. Cadophora luteo-olivacea isolated from apple (Malus domestica) fruit with post-harvest side rot symptoms in northern Italy. *Eur. J. Plant Pathol.* **2022**, *162*, 247–255. [CrossRef]
72. Ghooshkhaneh, N.G.; Golzarian, M.R.; Mollazade, K. VIS-NIR spectroscopy for detection of citrus core rot caused by Alternaria alternata. *Food Control* **2023**, *144*, 109320. [CrossRef]
73. Ekramirad, N.; Khaled, A.Y.; Doyle, L.E.; Loeb, J.R.; Donohue, K.D.; Villanueva, R.T.; Adedeji, A.A. Nondestructive detection of codling moth infestation in apples using pixel-based nir hyperspectral imaging with machine learning and feature selection. *Foods* **2022**, *11*, 8. [CrossRef]
74. Jiang, B.; He, J.; Yang, S.; Fu, H.; Li, T.; Song, H.; He, D. Fusion of machine vision technology and AlexNet-CNNs deep learning network for the detection of postharvest apple pesticide residues. *Artif. Intell. Agric.* **2019**, *1*, 1–8. [CrossRef]
75. Huang, C.; Li, X.; Wen, Y. AN OTSU image segmentation based on fruitfly optimization algorithm. *Alex. Eng. J.* **2021**, *60*, 183–188. [CrossRef]
76. Krizhevsky, A.; Sutskever, I.; Hinton, G.E. Imagenet classification with deep convolutional neural networks. *Commun. ACM* **2017**, *60*, 84–90. [CrossRef]
77. Zhang, D.; Zhou, X.; Zhang, J.; Lan, Y.; Xu, C.; Liang, D. Detection of rice sheath blight using an unmanned aerial system with high-resolution color and multispectral imaging. *PLoS ONE* **2018**, *13*, e0187470. [CrossRef]
78. Sun, Y.; Xiao, H.; Tu, S.; Sun, K.; Pan, L.; Tu, K. Detecting decayed peach using a rotating hyperspectral imaging testbed. *LWT* **2018**, *87*, 326–332. [CrossRef]
79. Li, J.; Luo, W.; Wang, Z.; Fan, S. Early detection of decay on apples using hyperspectral reflectance imaging combining both principal component analysis and improved watershed segmentation method. *Postharvest Biol. Technol.* **2019**, *149*, 235–246. [CrossRef]
80. Hyperspectral Imaging Systems Market Size Report. Available online: https://www.grandviewresearch.com/industry-analysis/hyperspectral-imaging-systems-market (accessed on 26 June 2023).
81. Mirza, M.; Osindero, S. Conditional generative adversarial nets. *arXiv* **2014**, arXiv:1411.1784.
82. Illarionova, S.; Shadrin, D.; Trekin, A.; Ignatiev, V.; Oseledets, I. Generation of the nir spectral band for satellite images with convolutional neural networks. *Sensors* **2021**, *21*, 5646. [CrossRef]
83. Lu, Y.; Chen, D.; Olaniyi, E.; Huang, Y. Generative adversarial networks (GANs) for image augmentation in agriculture: A systematic review. *Comput. Electron. Agric.* **2022**, *200*, 107208. [CrossRef]

84. Khatri, K.; Asha, C.; D'Souza, J.M. Detection of Animals in Thermal Imagery for Surveillance using GAN and Object Detection Framework. In Proceedings of the 2022 International Conference for Advancement in Technology (ICONAT), Goa, India, 21–22 January 2022; pp. 1–6.
85. Valerio Giuffrida, M.; Scharr, H.; Tsaftaris, S.A. Arigan: Synthetic arabidopsis plants using generative adversarial network. In Proceedings of the IEEE International Conference on Computer Vision Workshops, Venice, Italy, 22–29 October 2017; pp. 2064–2071
86. Tang, H.; Xu, D.; Yan, Y.; Corso, J.J.; Torr, P.H.; Sebe, N. Multi-channel attention selection gans for guided image-to-image translation. *arXiv* **2020**, arXiv:2002.01048.
87. Guo, Z.; Shao, M.; Li, S. Image-to-image translation using an offset-based multi-scale codes GAN encoder. *Vis. Comput.* **2023**, 1–17. [CrossRef]
88. Fard, A.S.; Reutens, D.C.; Vegh, V. From CNNs to GANs for cross-modality medical image estimation. *Comput. Biol. Med.* **2022**, *146*, 105556. [CrossRef] [PubMed]
89. Saharia, C.; Chan, W.; Chang, H.; Lee, C.; Ho, J.; Salimans, T.; Fleet, D.; Norouzi, M. Palette: Image-to-image diffusion models. In Proceedings of the ACM SIGGRAPH 2022 Conference Proceedings, Vancouver, BC, Canada, 7–11 August 2022; pp. 1–10.
90. Kshatriya, B.S.; Dubey, S.R.; Sarma, H.; Chaudhary, K.; Gurjar, M.R.; Rai, R.; Manchanda, S. Semantic Map Injected GAN Training for Image-to-Image Translation. In Proceedings of the Satellite Workshops of ICVGIP 2021, Gandhinagar, India, 8–10 December 2022; Springer: Berlin/Heidelberg, Germany, 2022; pp. 235–249.
91. Sa, I.; Lim, J.Y.; Ahn, H.S.; MacDonald, B. deepNIR: Datasets for generating synthetic NIR images and improved fruit detection system using deep learning techniques. *Sensors* **2022**, *22*, 4721. [CrossRef] [PubMed]
92. Li, C.; Wand, M. Precomputed real-time texture synthesis with markovian generative adversarial networks. In Proceedings of the European Conference on Computer Vision, Amsterdam, The Netherlands, 11–14 October 2016; Springer: Berlin/Heidelberg, Germany, 2016, pp. 702–716.
93. Simonyan, K.; Zisserman, A. Very deep convolutional networks for large-scale image recognition. *arXiv* **2014**, arXiv:1409.1556.
94. He, K.; Gkioxari, G.; Dollár, P.; Girshick, R. Mask r-cnn. In Proceedings of the IEEE International Conference on Computer Vision, Venice, Italy, 22–29 October 2017; pp. 2961–2969.
95. Girshick, R.; Donahue, J.; Darrell, T.; Malik, J. Rich feature hierarchies for accurate object detection and semantic segmentation. In Proceedings of the IEEE Conference on Computer Vision and Pattern Recognition, Washington, DC, USA, 23–28 June 2014; pp. 580–587.
96. Girshick, R. Fast r-cnn. In Proceedings of the IEEE International Conference on Computer Vision, Santiago, Chile, 7–13 December 2015; pp. 1440–1448.
97. Ren, S.; He, K.; Girshick, R.; Sun, J. Faster r-cnn: Towards real-time object detection with region proposal networks. *Adv. Neural Inf. Process. Syst.* **2015**, *28*, 91–99. [CrossRef] [PubMed]
98. Saletnik, B.; Zaguła, G.; Saletnik, A.; Bajcar, M.; Słysz, E.; Puchalski, C. Method for Prolonging the Shelf Life of Apples after Storage. *Appl. Sci.* **2022**, *12*, 3975. [CrossRef]
99. Nesteruk, S.; Illarionova, S.; Akhtyamov, T.; Shadrin, D.; Somov, A.; Pukalchik, M.; Oseledets, I. XtremeAugment: Getting More From Your Data Through Combination of Image Collection and Image Augmentation. *IEEE Access* **2022**, *10*, 24010–24028. [CrossRef]
100. Martínez-Zamora, L.; Castillejo, N.; Artés-Hernández, F. Postharvest UV-B and photoperiod with blue+ red LEDs as strategies to stimulate carotenogenesis in bell peppers. *Appl. Sci.* **2021**, *11*, 3736. [CrossRef]
101. Supervisely Data Annotator. Available online: https://app.supervise.ly (accessed on 26 June 2023).
102. Wu, Y.; Kirillov, A.; Massa, F.; Lo, W.Y.; Girshick, R. Detectron2. 2019. Available online: https://github.com/facebookresearch/detectron2 (accessed on 26 June 2023).
103. NVIDIA. Jetson Modules Technical Specificatons. 2023. Available online: https://developer.nvidia.com/embedded/jetson-modules (accessed on 26 June 2023).
104. Fan, S.; Liang, X.; Huang, W.; Zhang, V.J.; Pang, Q.; He, X.; Li, L.; Zhang, C. Real-time defects detection for apple sorting using NIR cameras with pruning-based YOLOV4 network. *Comput. Electron. Agric.* **2022**, *193*, 106715. [CrossRef]
105. Tang, Y.; Bai, H.; Sun, L.; Wang, Y.; Hou, J.; Huo, Y.; Min, R. Multi-Band-Image Based Detection of Apple Surface Defect Using Machine Vision and Deep Learning. *Horticulturae* **2022**, *8*, 666. [CrossRef]
106. Yuan, Y.; Yang, Z.; Liu, H.; Wang, H.; Li, J.; Zhao, L. Detection of early bruise in apple using near-infrared camera imaging technology combined with deep learning. *Infrared Phys. Technol.* **2022**, *127*, 104442. [CrossRef]
107. Zhang, Z.; Pu, Y.; Wei, Z.; Liu, H.; Zhang, D.; Zhang, B.; Zhang, Z.; Zhao, J.; Hu, J. Combination of interactance and transmittance modes of Vis/NIR spectroscopy improved the performance of PLS-DA model for moldy apple core. *Infrared Phys. Technol.* **2022**, *126*, 104366. [CrossRef]
108. Hu, Q.X.; Tian, J.; Fang, Y. Detection of moldy cores in apples with near-infrared transmission spectroscopy based on wavelet and BP network. *Int. J. Pattern Recognit. Artif. Intell.* **2019**, *33*, 1950020. [CrossRef]
109. Sadek, M.E.; Shabana, Y.M.; Sayed-Ahmed, K.; Abou Tabl, A.H. Antifungal activities of sulfur and copper nanoparticles against cucumber postharvest diseases caused by Botrytis cinerea and Sclerotinia sclerotiorum. *J. Fungi* **2022**, *8*, 412. [CrossRef]

entropy

MDPI

Article

SCFusion: Infrared and Visible Fusion Based on Salient Compensation

Haipeng Liu [1], Meiyan Ma [1], Meng Wang [1,2,*], Zhaoyu Chen [1] and Yibo Zhao [1]

1 Faculty of Information Engineering and Automation, Kunming University of Science and Technology, Kunming 650500, China; 20212204010@stu.kust.edu.cn (Z.C.)
2 Yunnan Province Key Laboratory of Computer, Kunming University of Science and Technology, Kunming 650500, China
* Correspondence: 20212204021@stu.kust.edu.cn

Abstract: The aim of infrared and visible image fusion is to integrate the complementary information of the two modalities for high-quality fused images. However, many deep learning fusion algorithms have not considered the characteristics of infrared images in low-light scenes, leading to the problems of weak texture details, low contrast of infrared targets and poor visual perception in the existing methods. Therefore, in this paper, we propose a salient compensation-based fusion method that makes sufficient use of the characteristics of infrared and visible images to generate high-quality fused images under low-light conditions. First, we design a multi-scale edge gradient module (MEGB) in the texture mainstream to adequately extract the texture information of the dual input of infrared and visible images; on the other hand, the salient tributary is pre-trained by salient loss to obtain the saliency map based on the salient dense residual module (SRDB) to extract salient features, which is supplemented in the process of overall network training. We propose the spatial bias module (SBM) to fuse global information with local information. Finally, extensive comparison experiments with existing methods show that our method has significant advantages in describing target features and global scenes, the effectiveness of the proposed module is demonstrated by ablation experiments. In addition, we also verify the facilitation of this paper's method for high-level vision on a semantic segmentation task.

Keywords: image fusion; salient compensation; infrared and visible images; deep learning

Citation: Liu, H.; Ma, M.; Wang, M.; Chen, Z.; Zhao, Y. SCFusion: Infrared and Visible Fusion Based on Salient Compensation. *Entropy* **2023**, *25*, 985. https://doi.org/10.3390/e25070985

Academic Editors: Oleg Sergiyenko, Wendy Flores-Fuentes, Julio Cesar Rodriguez-Quinonez and Jesús Elías Miranda-Vega

Received: 1 June 2023
Revised: 20 June 2023
Accepted: 23 June 2023
Published: 27 June 2023

1. Introduction

It is difficult to obtain high quality images during image acquisition due to weather, environment, etc. [1,2]. To improve image quality, researchers have proposed various image processing technology methods [3,4], and image fusion, as an image enhancement technique, can synthesize the complementary information between images to maximize the details of the imaged scene [5]. Among them, infrared and visible image fusion has become a hot research topic in the field of image processing due to its applications in the military and other fields [6]. Visible images usually contain a large number of texture details, but they are susceptible to environmental effects; in contrast, infrared images have the feature of highlighting thermal targets, but infrared images have problems such as a lack of background information, noise, and low resolution [7]. Therefore, the complementary characteristics of infrared and visible images enable their fusion to comprehensively describe the imaging scene, thus providing more feature information for subsequent advanced vision tasks such as pedestrian detection [8], image segmentation [9], etc.

Most of the existing methods for infrared and visible image fusion include some traditional methods and deep learning methods. The traditional methods mainly include multi-scale-decomposition-based methods [10,11], sparse representation-based methods [12,13], subspace-based methods [14], saliency-based methods [15], and hybrid methods [16].

However, most traditional methods achieve image fusion by complex mathematical transformations and hand-designed fusion rules; therefore, they cannot adapt to increasingly complex fusion scenarios. Additionally, due to the powerful deep feature extraction ability of deep learning, it has received more and more attention from scholars in the field of image fusion. Deep learning-based fusion methods are divided into three main categories: auto-encoder(AE)-based methods [17–20], convolutional neural network (CNN)-based methods [21–24], and generative adversarial network (GAN)-based methods [25–28].

Although deep-learning-based image fusion methods have been able to generate satisfactory fused images in recent years, there are still some pressing challenges in the field of image fusion. On the one hand, existing fusion algorithms [22,23,27] have a prerequisite: the assumption that infrared images provide salient target information and visible images provide background texture information, which holds under certain conditions (when visible images contain more information), but when conditions such as poor lighting of the visible image imaging scene are poor, this assumption leads to loss of background information in the fused image and the problem of target contrast degradation. On the other hand, there are some self-encoder-based methods [17–20] that use hand-designed fusion strategies to fuse depth features; however, the depth features tend to be uninterpretable, and the hand-designed fusion strategies are not able to assign appropriate weights to the depth features so that they are not better able to fuse the features. In contrast, some end-to-end methods [22,23,29] use feature fusion by simply cascading the feature information at the end of the feature extraction network, which is susceptible to the loss of shallow detailed texture feature information. In addition, existing methods do not pay attention to the infrared region at the target level when constructing the loss function, which cannot target the saliency of the infrared target in the fused image, resulting in the inevitable weakening of the infrared target in the fused image.

To solve the above problems, we propose a salient-compensation-based fusion framework for infrared, and visible images, called SCFusion. We will describe our approach in detail in Section 3. Overall, our main contributions are four-fold:

- It is presented a saliency-compensated infrared and visible image fusion framework consisting of a multi-scale edge gradient block (MEGB), a salient dense residual module (SRDB), and a spatial bias module (SBM). The fused images have significantly enhanced target information and rich scene descriptions.
- A scene texture mainstream consisting of multi-scale edge gradient blocks (MEGB) is designed to effectively extract the scene texture features of the source image, and the visible and infrared images can complement each other as scene texture information in different scenes, effectively solving the limitation of visible images by low-light scenes.
- A salient tributary trained individually by salient loss is designed, which uses the salient dense residual module (SRDB) to extract saliency targets, improving the target capture capability of the fusion network and eliminating the problem of low contrast in target regions of existing methods.
- A spatial bias module (SBM) is designed to compensate infrared features into texture features at different stages, where information extraction and fusion compensation are performed simultaneously, without the need to design additional fusion strategies.

The remainder of this paper is organized as follows. Section 2 briefly describes the related works of image fusion. In Section 3, we introduce our proposed SCFusion in detail, including network architecture and loss function. Section 4 illustrates the impressive performance of our method in comparison with other alternatives, followed by some concluding remarks in Section 5.

2. Related Work

2.1. Infrared and Visible Fusion

Deep learning has been sufficiently applied in computer vision tasks including image fusion due to its powerful capability of adaptation, numerous methods based on deep learning have been proposed, which are broadly classified into the following three main categories:

AE-based image fusion: Most of the self-encoder-based methods pre-train on large datasets to obtain encoders and decoders to implement the process of feature extraction and reconstruction, followed by feature fusion using manually designed fusion rules. DenseFuse [17] consists of a convolutional layer, a fusion layer, and a dense block, while the fusion layer is implemented by simple addition and parametrization. To further improve the feature extraction, NestFuse [18] and RFN-Nest [30] introduced nested connections and residual dense blocks in the network. Later, in order to make the network pay attention to specific regions of the source image, Jian et al. [31] employed an attention mechanism to focus on salient targets and texture details of the source image. Xu [20] et al., applied dissociative representation learning to a self-encoder approach considering the interpretability of feature extraction.

GAN-based image fusion: Generative adversarial networks (GANs) are able to effectively model data distribution even without supervised information, making the network remarkably compatible with infrared and visible image fusion tasks. FusionGAN [25] is the first approach to implement GANs into infrared and visible image fusion tasks, which defines the fusion task as an adversarial game between generators and discriminators. However, with a single discriminator, it is susceptible to a break in the balance of the data distribution between infrared and visible images; therefore, Ma et al., proposed DDcGAN [26], which proposes a dual-discriminator adversarial generative network. AttentionGAN [32] incorporates an attention mechanism based on DDcGAN [26], which intends to have the network retain the target information of infrared images and background information. Additionally, later, Zhou et al. [27] proposed an approach to generate adversarial networks with gradient and intensity discriminators as multi-task fusion, which imported gradient and intensity into the GAN to make the network pay more attention to the gradient and intensity of infrared and visible images.

CNN-based image fusion: Infrared and visible image fusion methods based on convolutional neural networks (CNN) achieve end-to-end feature extraction, fusion, and reconstruction by designing network structures and loss functions. RXDNFuse [33] combines the advantages of DenseNet [17] and ResNet [34] to propose residual dense networks for a more comprehensive extraction of features at different scales. SeAFusion [29] proposed an approach to drive the fusion task with semantic loss to better integrate the fusion task with subsequent advanced vision tasks. Li et al. [35] proposed a dual-attention-based feature fusion module based on the theory of meta-learning, in which the network accepts source image inputs of different resolutions. STDFusionNet [22] proposed the use of target masks to assist in extracting the target of the visible image and the background of the visible image as a way to improve the fusion effect, but the labeling of the mask is manually labeled, which results in a large preliminary workload. PIAFusion [7] considers the lighting conditions, although it embeds the lighting probability into the loss function, which is prone to the problem of overexposure to the background of the daytime scene.

2.2. The High-Level Vision Tasks

As one of the important methods in the field of computer vision, semantic segmentation aims to predict the semantic category of each pixel in an image; it has crucial importance in the field of autonomous driving [36]. However, many semantic segmentation methods are designed based on the conditions of good illumination, while the performance of these methods decreases when the image has poor illumination conditions or is occluded. Therefore, it has become a new problem in the field of semantic segmentation to improve the accuracy of segmentation networks when the visible images are contaminated. Some researchers have started to experiment with semantic segmentation methods that combine infrared images with visible images, and most of these methods also involve the process of infrared and visible image fusion. RTFNet [37] employs ResNet to extract the features of two source images separately as an encoder; multimodal fusion is implemented by accumulating the feature blocks of RGB and Thermal encoder paths over the elements, with an upception block designed to recover the feature map resolution. AFNet [38] computes the infrared image and visible image by designing the attention fusion module to the spatial correlation between feature maps while guiding the fusion of features from different modalities in the process. AMFuse [39] was designed specifically for multimodal fusion with an add–multiply fusion block fusing common and complementary features of infrared and visible images, with an attention module and a spatial pyramid pool module added to the module to enhance the information in multi-scale contexts.

However, infrared and visible image fusion methods ignore the variation in complementary information of infrared and visible images in normal light and low-light environments. Therefore, we propose a new fusion method that is able to sufficiently exploit the features of infrared and visible images under different lighting conditions, so as to retain more meaningful information.

3. Methods

3.1. Network Architecture

In order to balance the background texture details of the infrared and visible images without limiting the light conditions of the input image and to enhance the contrast between the infrared target and the scene, we designed the saliency-compensated fusion network, whose overall network is shown in Figure 1. The framework mainly consists of the multiscale edge gradient block (MEGB), the salient dense residual module (SRDB) and the spatial bias module (SBM). The visible and infrared images are integrated into the texture mainstream together to obtain enhanced texture features, while the infrared images are integrated into the salient mainstream to obtain enhanced salient features, both of which are effectively fused with global and local information by the spatial bias module (SBM). The relevant modules will be described in detail below.

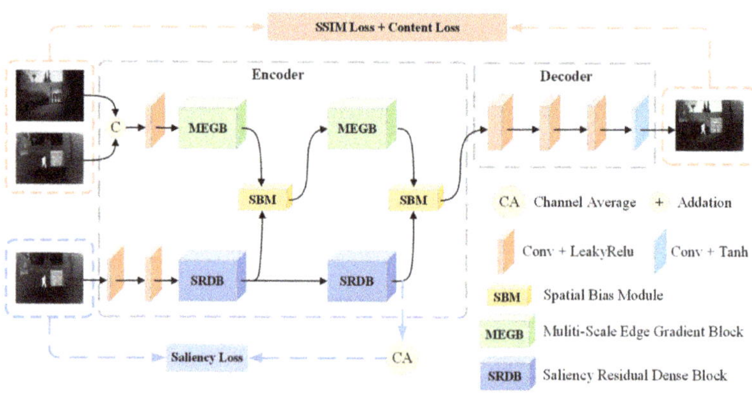

Figure 1. Overall framework for SCFusion. It consists of multiscale edge gradient block (MEGB),salient dense residual block (SRDB), and spatial bias block (SBM). The saliency map generated by the saliency tributary is pre-trained by saliency loss, which is then sent to the main network to generate the fused image with the texture features obtained by MEGB under the joint training of structural similarity loss and content loss.

3.1.1. Multiscale Edge Gradient Block (MEGB)

The specific structure is shown in Figure 2, which consists of multiscale mainstream and residual gradient streams. Most networks use convolutional layers of the same size convolutional kernel to extract features, which is difficult to perceive the information comprehensively. So, the multiscale mainstream is added with branches of convolutional layers of different sizes of convolutional kernels to increase the perceptual field. To reduce the information loss in the multi-scale features, different convolutional computations are not added with pooling layers, while the residual gradient flow is combined with the Sobel operator to maintain the strong texture rationality of the features. The multiscale output is then combined with the output of the residual gradient flow to complete the texture detail enhancement.

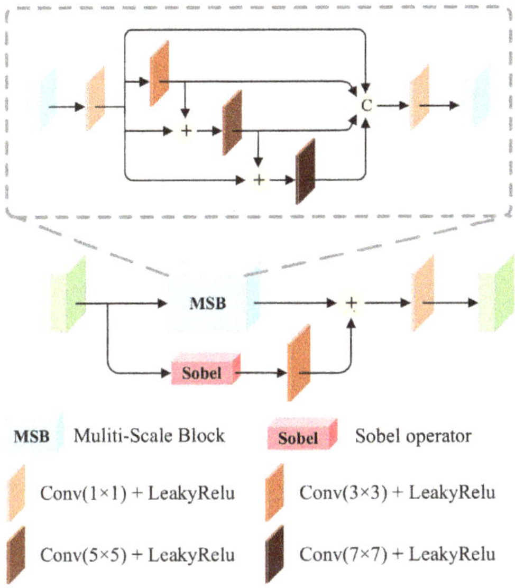

Figure 2. Multiscale edge gradient block (MEGB). It accomplishes texture detail enhancement by combining the output of the multi-scale with the output of the residual gradient flow.

Specifically, in the feature mainstream, we are given a pair of strictly aligned infrared images I_{ir} and visible images I_{vi}, which are approximated by a shallow convolutional layer for modal differences and then joined in the channel dimension to obtain Φ_H In the tributary stream, the infrared images I_{ir} are passed through a shallow convolutional layer to obtain Φ_C Φ_H is directly input to MEGB, and MSB uses different convolutional kernels to extend the perceptual field of the network, and multi-scale features Φ_D cascade to enhance the feature description. The module MSB output feature Φ_M can be expressed as:

$$\Phi_M = Conv(C(\Phi_D)), n \in \{1,3,5,7\} \tag{1}$$

The texture extraction of the hybrid features is also performed using the Sobel operator to enhance the features' fine-grained representation, and the above process can be expressed as follows:

$$\Phi_{T_1} = Conv(Conv(\nabla_{Sobel}\Phi_H) \oplus \Phi_M) \tag{2}$$

where $Conv(\cdot)$ denotes the convolution operation, $C(\cdot)$ denotes the cascade on the channel dimension, ∇_{Sobel} denotes the Sobel operator, and $\oplus$ denotes element-wise summation.

In summary, MEGB breaks the limitation of texture extraction from lighting conditions by combining multi-scale features and Sobel texture features in parallel to maximize texture details in infrared and visible images.

3.1.2. Salient Dense Residual Block (SRDB)

The specific structure is shown in Figure 3, which integrates dense connectivity [17], residual streams [35], and channel attention (CAB). To obtain comprehensive feature information, we introduce dense connectivity in the mainstream, but to address the high memory cost and energy consumption due to feature reuse, it is replaced by aggregating the features of all previous layers in the last layer of dense connectivity. Densely connected features are input to attention in order to make the network more focused on the attention region. It is remarkable that we generate salient target images in the training phase, while the infrared salient target features are input directly into the subsequent network in the inference phase.

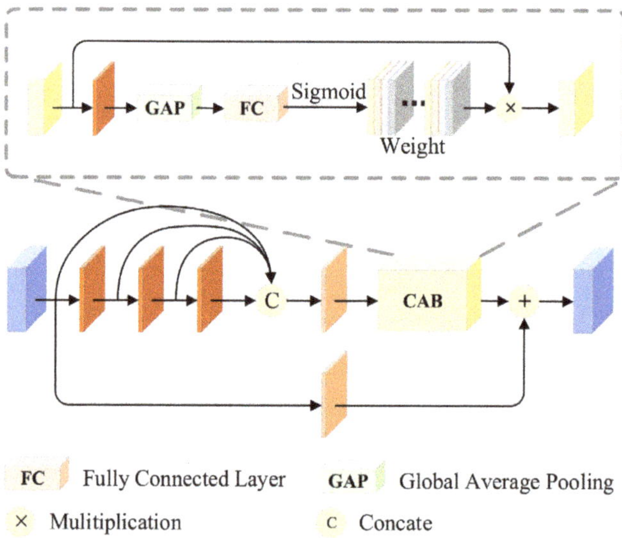

Figure 3. Salient dense residual block (SRDB). It achieves contrast enhancement by combining attentional features with residual flow features.

Specifically, we send Φ_C into the SRDB, and after feature reuse, feature Φ_E can be represented as:

$$\Phi_E = C\Big(\Phi_C, Conv(\Phi_C), Conv^2(\Phi_C), Conv(\Phi_C)\Big) \quad (3)$$

The attention first passes through a 3 × 3 convolutional layer, followed by a global average pooling to obtain the global feature vector, a fully connected layer to learn the importance of each channel, and then a sigmoid activation function to obtain the weights and assign higher weights to the features with higher contrast, and multiply the weights with the original input features to obtain the attention feature V_C.

Finally, the contrast enhancement is achieved by adding the attention features with the residual stream features to highlight the salient targets, and the above process can be defined as:

$$V_C = Sigmoid(FC(GAP(Conv(\Phi_E)))) \cdot \Phi_E \quad (4)$$

$$\Phi_{S_1} = Conv(\Phi_C) \oplus V_C \quad (5)$$

where $GAP(\cdot)$ denotes the global average pooling, $FC(\cdot)$ denotes the fully connected layer, $Sigmoid(\cdot)$ denotes the activation function, and Φ_{S_1} is the final output feature of SRDB.

In a nutshell, SRDB calculates the contrast of features on the basis of channel attention to achieve contrast enhancement, which further preserves the high contrast of infrared targets.

3.1.3. Spatial Bias Block (SBM)

The specific structure of the module is shown in Figure 4. The module has two inputs, a texture feature from the mainstream and a salient feature from the tributary. In the salient tributary we focus on the infrared target; meanwhile, we also need to learn the relationship between different distant targets, i.e., the global information to enhance the semantic information of the image, but the simple convolutional layer has the problem of not being able to learn the long-range dependencies due to the limited perceptual field, so we learn the global information by adding a spatial bias channel to the texture tributary. This module is lightweight, unlike the self-attention operation which is too burdensome. The spatial bias term B can be expressed as:

$$B = Relu(BN(\Phi_{S_1}, SB)) \tag{6}$$

where $B(\cdot)$ denotes the output of the significant features after adding the spatial bias term, SB denotes the spatial bias, and BN and $Relu$ denote the batch normalization and nonlinear activation layers, respectively.

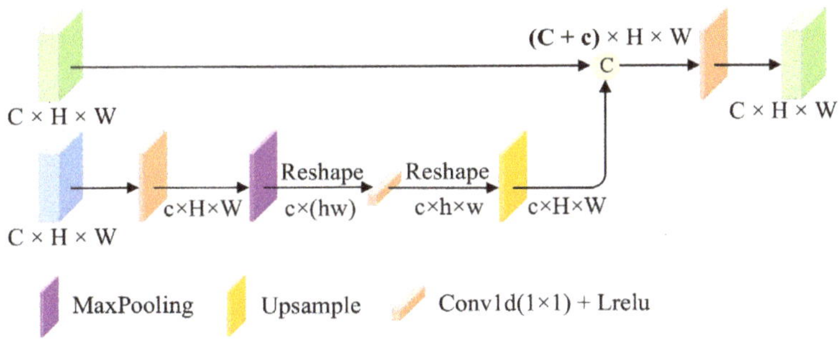

Figure 4. Spatial bias block (SBM). It allows the network to learn both local and global information by connecting spatially biased features with texture features in channel cascades.

Instead, textures are represented by the grayscale distribution of pixels and their surrounding spatial domains, i.e., local information. By cascading spatial bias features with texture features in the channel direction, the network can learn both local and global information. In order to aggregate global knowledge in the feature map, we use 1×1 convolution in the passband dimension. Finally, texture feature Φ_{T_1} is spliced with saliency feature Φ_{S_1} to complete the process of asymptotic fusion, which can be expressed as

$$\Phi'_{T_1} = Conv(C(\Phi_{T_1}, B(\Phi_{S_1}))) \tag{7}$$

In conclusion, a simple and efficient fusion rule is the key to image fusion, and SBM utilizes lightweight spatial bias terms to fuse local and global information without increasing the complexity of the network.

3.2. Loss Function

We know that under different lighting conditions, image texture information may exist in either visible or infrared images; the salient targets are more prominent in infrared images. Therefore, our method aims to fully extract texture details in both infrared and visible images from the texture mainstream while enhancing the salient targets weakened by the mainstream from the saliency tributaries. Therefore, our method is a two-stage model trained by the mainstream loss function and the tributary loss function, and its training process is shown in Algorithm 1.

Algorithm 1: Training procedure

Input: Infrared images I_{ir} and visible images I_{vi}
Output: Fused images I_f

For M_1 epoch **do**
    For p_1 step **do**
        Select n infrared images $\{I_{ir}^1, I_{ir}^2, \dots, I_{ir}^{n_1}\}$;
        Use the salient branch to extract salient feature maps Φ_s;
        Calculate the salient loss $L_{salient}$ according to Eq.(16);
        Update the parameters by Adam Optimizer;
    end
    Save weights of the salient branch for the M_1 epoch;
end
For M_2 epoch **do**
    For p_2 step **do**
        Select n infrared images $\{I_{ir}^1, I_{ir}^2, \dots, I_{ir}^{n_2}\}$;
        Select n Visible images $\{I_{vi}^1, I_{vi}^2, \dots, I_{vi}^{n_2}\}$;
        Generate n fused images $\{I_f^1, I_f^2, \dots, I_f^{n_2}\}$ by our fusion network;
        Calculate the total loss L_{total} according to Eq.(10);
        Update the parameters by Adam Optimizer;
    end
    Save weights of the network model for the M_2 epoch;
end

3.2.1. Mainstream Loss

The mainstream branch aims to make the fused image retain rich texture details and improve the visual quality and evaluation index, so we design the structural similarity loss and content loss to guide the network to generate the fused image; the formula of fusion loss is as follows:

$$L_F = \lambda_1 L_{SSIM} + \lambda_2 L_{Content} \tag{8}$$

where λ_1, λ_2 are the weighting factors to balance the two losses. The two loss functions are described in detail below.

Structural Similarity Loss

For the fusion task, we want to close the similarity between the fused image and the source image to improve its fusion performance so that the visual effect of the image is more in line with the visual effect perceived by human eyes. Structural similarity (SSIM) can effectively evaluate the similarity between the source and fused images, which consists of three components: luminance similarity, contrast similarity, and structural similarity. The loss of structural similarity is formulated as follows:

$$L_{SSIM} = 1 - \frac{SSIM\left(I_f, I_{VI}\right) + SSIM\left(I_f . I_{ir}\right)}{2} \tag{9}$$

$$L_{SSIM}(x, y) = \frac{(2\mu_x\mu_y + C_1)(2\sigma_{xy} + C_2)}{\left(\mu_x^2 + \mu_y^2 + C_1\right)\left(\sigma_x^2 + \sigma_y^2 + C_2\right)} \tag{10}$$

where I_f denotes the fused image, I_{ir}, and I_{vi} denote the infrared image and visible image, respectively; $SSIM(x,y)$ indicates the calculation of the structural similarity between two images; μ_x and μ_y are the averages of all pixels in the two source images; σ_x and σ_y are the variances of the pixel values of the two source images; and C_1 and C_2 are constants to ensure the stability of the function.

Content Loss

In addition, our texture mainstream expects the fused image to retain abundant texture details while maintaining the best intensity distribution, so the content loss is introduced, which consists of two parts: intensity loss and texture loss. The content loss is defined as follows:

$$L_{Content} = L_{Int} + \alpha L_{Grad} \tag{11}$$

where L_{Int} denotes the intensity loss, L_{Grad} denotes the gradient loss, and ∂ is used to obtain a balance between the strength loss and texture loss.

The intensity loss measures the intensity distribution between the fused image and the source image at the pixel level, so the intensity loss is defined as follows:

$$L_{Int} = \frac{1}{HW}||I_f - Max\,(I_{ir}, I_{vi})\,||_1 \tag{12}$$

where H and W are the height and width of the input image, respectively, and $||\cdot||_1$ represents $l_1 - norm$.

In addition, to encourage clearer texture details, we expect the gradient of the fused image to be close to the gradient maximum of the visible and infrared images, so the texture loss is defined as follows:

$$L_{Grad} = \frac{1}{HW}||\left|\nabla_{Sobel} I_f\right| - Max\,(|\nabla_{Sobel} I_{ir}|, |\nabla_{Sobel} I_{vi}|)\,||_1 \tag{13}$$

where ∇_{Sobel} denotes the Sobel gradient operator, which measures the gradient texture of the image; $|\cdot|$ denotes the absolute operation.

3.2.2. Salient Loss

The purpose of the fusion task is to serve the subsequent advanced vision task, and the salient target is crucial for the subsequent task, so in order to preserve the salient target of the fused image, we use the target mask to construct the intermediate salient loss, which is defined as follows:

$$L_{Salient} = \frac{1}{HW}||I_m \cdot I_{ir} - CA\,(\Phi_{ir})\,||_1 \tag{14}$$

where I_m denotes the target mask, and CA denotes the channel average.

In summary, our network of significant target compensation is able to obtain ideal texture details with significant targets guided by structural similarity loss, content loss, and salient loss, and can round-the-clock fuse the meaningful information of source images.

4. Experimental Validation

4.1. Experimental Configurations

In this paper, we conducted extensive qualitative and quantitative experiments on three datasets, including TNO [40], MSRS [7], and M3FD [28], to comprehensively evaluate our approach and validate the generalization of our method. In addition, we selected seven methods such as DenseFuse [17], RFN-Nest [30], Fusiongan [25], SDNet [41], U2Fusion [23], FLFuse [24], and PIAFusion [7] for comparison with our method.

The experimental results of visualization are subjective, in this paper, we introduce the standard deviation (SD), visual information fidelity (VIF), and the average gradient (AG). The difference correlation sum of SD is based on statistical concepts to evaluate the distribution and contrast of fused images, and VIF is based on the human visual system designed to measure the fidelity of information from the perspective of human visual perception. SCD measures the correlation between the information of the fused image and the corresponding source image, EN evaluates the amount of information contained in the fused image from an information-theoretic perspective, and SF evaluates the texture details contained in the fused image by calculating the row frequency and column frequency. All the above evaluation metrics are of higher values, indicating better image quality.

This paper presents a two-stage model, so we train the textured main stream and the salient tributary in turn. In the first stage, we train the salient tributaries: epoch = 10. After that, the output features of SRDB are supplemented as mainstream saliency features. Then train the fusion network: epoch = 8. In the training phase of the experiments, a data augmentation method was used to address the problem of small existing visible and infrared image fusion datasets, and a common dataset of aligned visible and infrared images, MSRS was used as the training set. For the hyper-parameter setting: $\lambda_1 = 1$, $\lambda_2 = 15$, $\alpha = 3$. Additionally, we leverage the Adam optimizer with a batch size of 64. The learning rate is 1×10^{-4}. The test set was selected from the public datasets TNO, RoadScene, MSRS and M3FD for infrared and visible image fusion, and 42, 20, 361 and 300 pairs of images each were selected for algorithm comparison experiments. The experiments in this paper were conducted on a GeForce RTX 2080Ti 11GB with PyTorch as the deep learning framework. All comparison algorithms in the experiments were experimented with in the original thesis setup.

4.2. Comparison Experiments

4.2.1. Qualitative Results

The visualization results for eight image pairs in the three datasets are given in Figures 5–7.

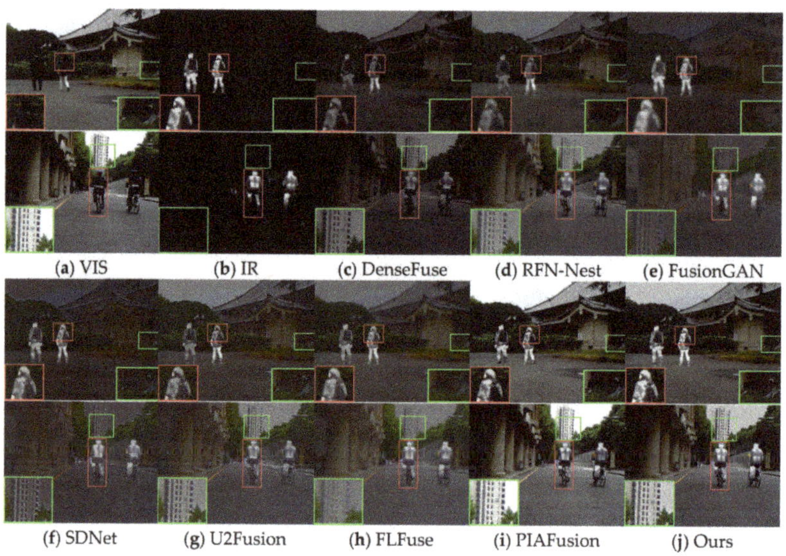

Figure 5. Vision quality comparison on the MSRS dataset. Areas with large differences are highlighted by RED and GREEN boxes, and enlarged images of RED boxes are in the lower right or left corner.

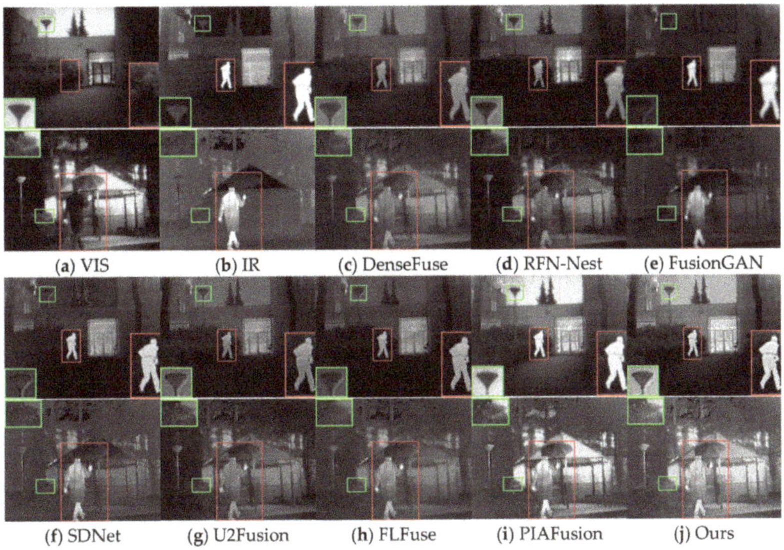

Figure 6. Vision quality comparison on the TNO dataset. Areas with large differences are highlighted by RED and GREEN boxes, and enlarged images of RED boxes are in the lower right or left corner.

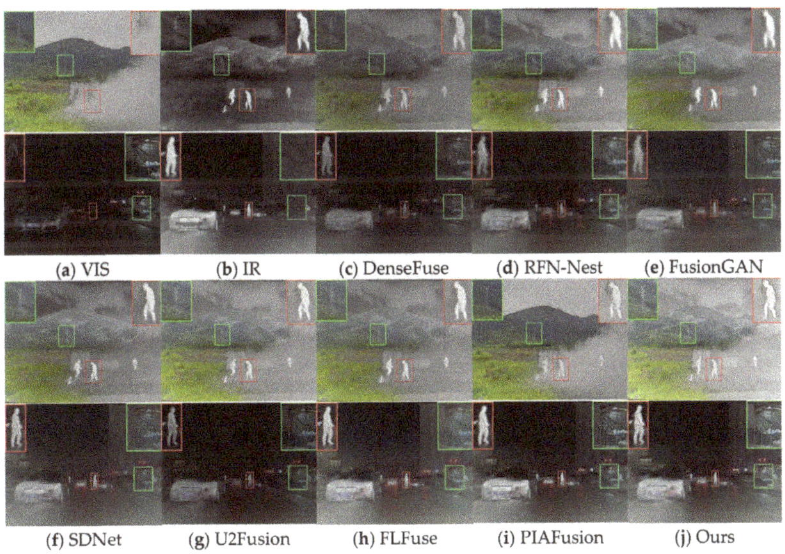

Figure 7. Vision quality comparison on the M3FD dataset. Areas with large differences are highlighted by RED and GREEN boxes, and enlarged images of RED boxes are in the lower right or left corner.

In the daytime scene, as shown in Figure 5, DenseFuse and RFN-Nest weaken the infrared target, and FusionGAN causes the problem of blurred edge texture, while SDNet and FLFuse weaken the background texture detail of the image, as seen in the green box; only PIAFusion and the method in this paper can integrate the effective information.

In the night scene as shown in Figure 6, the visible image contains only a small amount of texture information, while the infrared image has background texture detail information in addition to the prominent target. Many methods focus excessively on the information of one of the modal images, and it is difficult to achieve good results in different scenes. Among them, the infrared targets in DenseFuse, RFN-Nest, U2Fusion and FLFuse are weakened, and the fused images of FusionGAN and SDNet are more towards the infrared images, resulting in blurred background information. Since PIAFusion adds light perception coefficients to the loss function, the method in this paper fully extracts the details contained in both images in the texture mainstream and uses saliency tributaries to supplement the weakened salient targets, so it can effectively fuse the complementary information in low-light scenes.

In the scenes where the visible image targets are obscured as shown in Figure 7, the method in this paper can mine the salient targets hidden in smoke because the method in this paper uses intermediate salient loss to guide the tributaries to enhance their strong contrast. Among the seven comparison algorithms, DenseFuse can retain texture information but ignores the salient contrast of the target, while background information is smoothed to different levels in RFN-Nest, FusionGAN, SDNet, U2Fusion, and FLFuse. In contrast, although PIAFusion can better preserve the high contrast of salient targets, it is easy to lose the IR modal information of obscured objects such as sky and smoke due to the smoothness of light perception loss.

In summary, our method has both comprehensive scene information and retains rich contrast information and texture details of the target region.

4.2.2. Quantitative Results

We performed a quantitative evaluation on three datasets, TNO, MARS, and M3FD.The comparison of the metrics of different methods is shown in Table 1 below. The best values of AG and SF indicate that our fusion method has richer contrast information and also contains richer texture details; the best value of EN indicates that our method retains sufficient edge information; and the best value of SCD indicates that our fusion results contain more realistic information. SD and VIF perform optimal or suboptimal on the three datasets, indicating that our method has richer contrast information and generates fused images that are more consistent with the human visual system. In addition, six metrics are optimal or suboptimal on three datasets indicating that our method has superior generalization performance and can be applied to different types of datasets. In conclusion, our method is able to mine effective information in low-light and occluded scenes and integrate the information into the fused images with the help of spatially paranoid blocks. Therefore, our method has a greater advantage over other methods to obtain high-quality fused images.

Table 1. Quantitative results of six metrics on TNO, MSRS and M3FD datasets. Bold: best. Underline and italic: second best.

Dataset	Algorithm	Evaluation Methods					
		SD	VIF	AG	SCD	EN	SF
TNO	DenseFuse	8.5765	0.6704	2.4895	1.5916	6.3422	0.0248
	RFN-Nest	*9.3153*	0.8103	2.6109	*1.7711*	*6.9285*	0.0226
	FusionGAN	8.6058	0.6457	2.3636	1.3688	6.5199	0.0240
	SDNet	9.0398	0.7523	*4.5252*	1.5488	6.6670	*0.0448*
	U2Fusion	8.8553	0.6787	3.4891	1.5862	6.4230	0.0327
	FLFuse	9.2156	0.7986	3.2772	1.7172	6.6307	0.0329
	PIAFusion	9.1093	**0.8835**	4.4265	1.6540	6.8937	0.0447
	Ours	**9.7039**	*0.8121*	**5.5097**	**1.8117**	**7.0620**	**0.0502**
MSRS	DenseFuse	7.0692	0.6752	2.0412	1.3296	5.8397	0.0235
	RFN-Nest	6.9939	0.5364	1.5376	1.2881	5.7514	0.0181
	FusionGAN	5.4307	0.4234	1.2258	0.7948	5.2179	0.0146
	SDNet	5.3143	0.3745	2.1439	0.8298	4.8852	0.0270
	U2Fusion	5.6231	0.3967	2.0100	1.0034	4.7525	0.0256
	FLFuse	6.4837	0.4837	1.7743	1.1090	5.5079	0.0193
	PIAFusion	**7.9268**	**0.9072**	**3.6801**	*1.7395*	**6.4304**	**0.0444**
	Ours	*7.7783*	*0.7354*	*3.3791*	**1.8057**	*6.4044*	*0.0421*

Table 1. *Cont.*

Dataset	Algorithm	Evaluation Methods					
		SD	VIF	AG	SCD	EN	SF
M3FD	DenseFuse	8.6130	0.6694	2.6528	1.5051	6.4264	0.0298
	RFN-Nest	9.0712	0.7338	2.5848	*1.6352*	6.7151	0.0274
	FusionGAN	8.8489	0.5154	2.3610	1.1257	6.4690	0.0274
	SDNet	8.8867	0.6321	4.0228	1.3912	6.6134	0.0454
	U2Fusion	9.0141	0.7061	3.8500	1.5488	6.6285	0.0408
	FLFuse	8.7556	0.6969	2.1329	1.4934	6.5734	0.0233
	PIAFusion	**10.1639**	**0.9300**	*4.9702*	1.3363	*6.8036*	*0.0575*
	Ours	*9.4840*	*0.7894*	**5.4374**	**1.7589**	**6.9482**	**0.0606**

4.3. Application of Semantic Segmentation

In this section we validate the facilitation of this paper's approach for advanced vision on a semantic segmentation task [29]. Specifically, we train the semantic segmentation algorithm [42] on the source and fused images, respectively. We selected 1000 images as the training set and tested the segmentation performance of different models on 360 images, and the qualitative and quantitative results are shown in Figure 8 and Table 2.

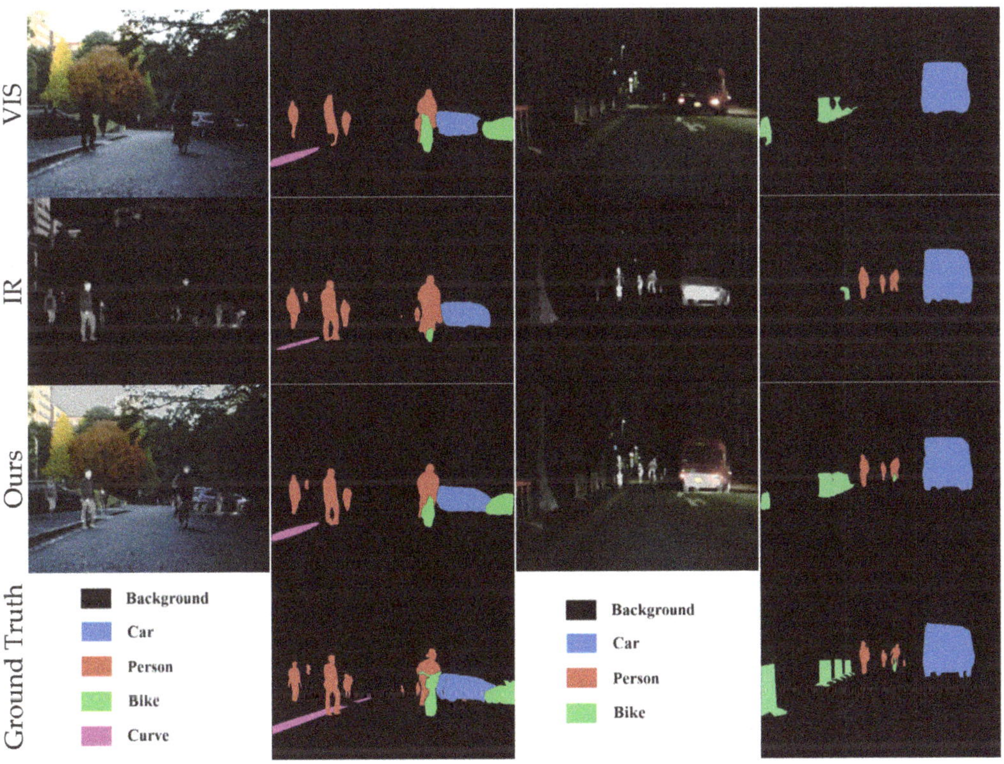

Figure 8. Vision quality comparison of the segmentation results.

Table 2. Segmentation performance (mIoU) of visible, infrared, and fused images at different times in the same scene. (Bold: best.).

Label Class		Background	Car	Person	Bike	Curve	Car Stop	Guardrail	Color Cone	Bump	Mean
Day	VIS	0.9800	0.8906	0.5556	0.7260	**0.5798**	0.4824	**0.8090**	**0.6508**	**0.5669**	0.6934
	IR	0.9482	0.5470	0.6564	0.0847	0.1032	0.1268	0.0368	0.0087	0.1304	0.2936
	Ours	**0.9834**	**0.9074**	**0.7332**	**0.7347**	0.5469	**0.5395**	0.7588	0.6335	0.5534	**0.7101**
Night	VIS	0.9652	0.6960	0.1305	0.5889	0.2750	0.1762	**0.3666**	0.3792	0.1943	0.4191
	IR	0.9593	0.4680	0.7103	0.0873	0.2599	0.0292	0.0000	0.0223	0.1945	0.3034
	Ours	**0.9763**	**0.7902**	**0.7205**	**0.6057**	**0.4419**	**0.2881**	0.3390	**0.4354**	**0.2233**	**0.5356**
All	VIS	0.9726	0.7933	0.3431	0.6575	0.4274	0.3293	**0.5878**	0.5150	0.3806	0.5563
	IR	0.9538	0.5075	0.6834	0.0860	0.1816	0.0780	0.0184	0.0155	0.1625	0.2985
	Ours	**0.9799**	**0.8488**	**0.7269**	**0.6702**	**0.4944**	**0.4138**	0.5489	**0.5345**	**0.3884**	**0.6229**

In the daytime scene as shown in columns one and two of Figure 8, the visible images contain a large amount of information, so the segmentation accuracy for visible images is high as shown in the second row of Table 2. However, some detection of people is lost due to the lack of guidance of infrared targets in the visible image. Additionally, the infrared image lacks the complement of the visible image background, and the segmentation accuracy of the bicycle is low as shown in the sixth column of the third row of Table 2.

In the night scene, as shown in Figure 8, columns three and four, the visible image cannot capture enough information due to the lack of light, so the segmentation network has a low segmentation accuracy for people in the scene, as shown in Table 2, fifth row, fifth column. While the infrared image captures the thermal target so the segmentation accuracy for people is higher as shown in the fifth column of the sixth row of Table 2; however, the infrared image reduces the segmentation accuracy of the bicycle.

Our method is shown in row three of Figure 8. Since the inclusion of the spatial bias term enables the network to perceive long-distance information and enhances the semantic information of the images, our method fully integrates the useful information of both source images, so our method outperforms the segmentation accuracy of pedestrians and bicycles than unimodal images in both daytime and nighttime scenes.

4.4. Ablation Experiment

In this section, we qualitatively and quantitatively analyze the effectiveness of the loss functions and modules in the method of this paper through ablation studies. The results are shown in Table 3 and Figure 9.

Table 3. Quantitative evaluation results of ablation study. (Bold: best).

Experiment	Evaluation Methods					
	SD	VIF	AG	SCD	EN	SF
Ls + SRDB + SBM + MEGB	7.7783	0.7354	**3.3791**	**1.8057**	**6.4044**	**0.0421**
W/O $L_{Salient}$	7.7368	**0.7765**	3.2551	1.6536	6.1237	0.0420
W/O $L_{Content}$	5.9613	0.6719	1.9179	0.9957	5.4447	0.0239
W/O L_{SSIM}	6.9871	0.6764	3.1844	1.4672	5.9284	0.0400
W/O SRDB	7.6180	0.7489	3.3681	1.5740	6.1327	0.0414
W/O SBM	7.4222	0.5957	3.3583	1.2744	5.9838	0.0403
W/O MEGB	**7.8551**	0.4782	3.1330	0.7154	6.0242	0.0385

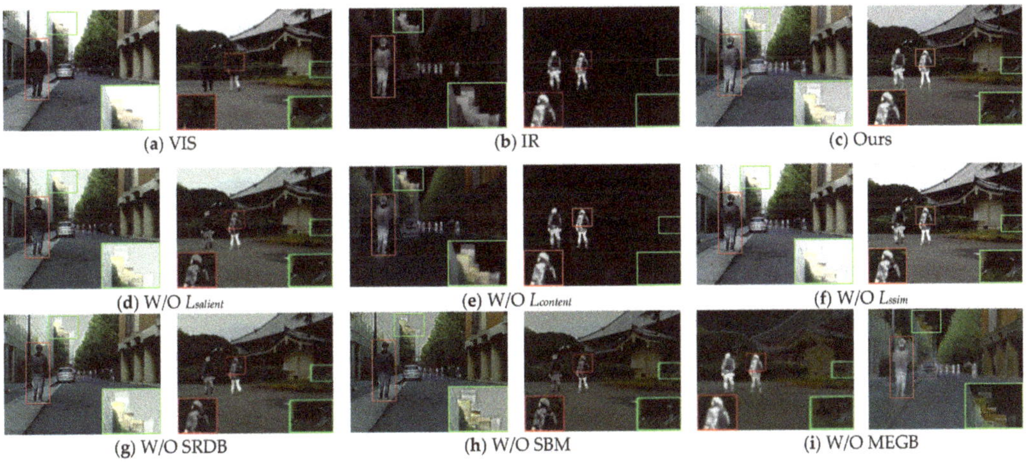

Figure 9. Vision quality comparison of the ablation study on important loss functions and modules.

4.4.1. Loss of Salience

The salient loss guides the tributary network to retain the high contrast of the infrared targets, aiming to compensate for the salient target features towards the feature mainstream. As shown in Figure 9d, the contrast of the targets marked in the red boxes significantly decreases after removing the salient loss, and the SD values (evaluated contrast) in Table 3 decrease, indicating that there is no salient loss, and the network's infrared targets are weakened.

4.4.2. Loss of Content

Content loss uses intensity loss and gradient loss jointly to constrain the network to maintain the optimal intensity distribution while retaining abundant texture detail. As shown in Figure 9e, after removing the content loss, it is obvious that a significant decrease in background texture detail and a significant decrease in various metrics can be seen in the fused image biased toward the infrared image, which shows that the content loss has an important role in the overall network to synthesize the characteristics of the infrared and visible images.

4.4.3. Structural Similarity Loss

The structural similarity loss aims to measure the similarity of the fused image to the source image. As shown in Figure 9f, when the structural similarity loss is removed, over-exposure is perpetuated in the visible image overexposure region for the fused image. On the other hand, the values of SD and EN vary greatly, indicating that the fused image contains less information with lower image contrast.

4.4.4. Salient Dense Residual Block

SRDB utilizes attention to enable network features to extract a strong pixel distribution in the attention channel. As shown in Figure 9g, after removing the saliency-dense residual blocks, we can notice a significant decrease in the saliency of the fused image targets. The value of SD in Table 3 significantly decreases, indicating that the attention block is critical to the strong pixel distribution.

4.4.5. Spatial Bias Block

The SBM effectively completes the progressive fusion process by adding information from the salient tributaries to the main stream. In Figure 9h and Table 3, it can be seen that the overall brightness of the fused image becomes darker and the target contrast decreases after removing the spatial bias block (SBM). On the other hand, the values of VIF, SCD and SD decrease significantly, which shows that adding spatial bias terms to the tributary can both effectively enhance the IR target and fused image more in line with the human visual system.

4.4.6. Multiscale Edge Gradient Block

MEGB can fully extract the texture information of the image by using multiscale feature extraction with gradient operator embedding. As shown in Figure 9i, when we exclude the multiscale edge gradient block, the overall scene is relatively smoother with less gradient variation. Additionally, the values of AG and SF in Table 3 drop significantly, indicating that the module does enhance the representation of network texture details.

In summary, our designed module not only facilitates the fusion image visually, but also improves significantly in terms of metrics, so our designed module facilitates the maintenance of both texture and salient targets.

5. Summary

This paper proposed a saliency-compensated infrared and visible image fusion method, SCFusion. On the one hand, MEGB helps the extraction and retention of texture gradients of the overall network, which enhances the ability of the fused image to describe the global scene information. On the other hand, SRDB is designed to extract salient targets of infrared images and generate salient maps guided by salient loss. Finally, the information fusion is completed by compensating the saliency features of the tributaries into the main stream using SBM blocks. The experiments comparing the qualitative and quantitative aspects of this paper's method with existing methods show the effectiveness of this paper's method, and the fusion experiments with different lighting scenes also show that this paper's method can effectively help to fully fuse the information of infrared and visible images in low-light scenes. Moreover, experiments on our semantic segmentation task validate the facilitation of our approach for subsequent high-level vision tasks. However, there are limitations to our method. Although our method can mitigate the loss of fused image scene information when the visible image is obscured by smoke to some extent, our method cannot remove the overexposure effect caused by strong light interference. We will further investigate the combination of low-light enhancement and image fusion tasks to solve the problem of strong light interference in the future.

Author Contributions: Conceptualization, M.M., Z.C and Y.Z.; methodology, M.M. and Z.C.; software, M.M., Y.Z. and Z.C; validation, M.M., H.L. and Z.C.; formal analysis, H.L.; investigation, M.W.; resources, M.W.; data curation, M.W.; writing—original draft preparation, M.M.; writing—review and editing, M.M., M.W. and Z.C.; visualization, M.M. and Z.C.; supervision, H.L. and M.W.; project administration, M.W.; funding acquisition, M.W. All authors have read and agreed to the published version of the manuscript.

Funding: This research was funded by the National Natural Science Foundation of China (62062048, 62263017), and the Yunnan Department of Science and Technology Project (202201AT070113).

Institutional Review Board Statement: Not applicable.

Data Availability Statement: Not applicable.

Acknowledgments: We thank all the editors and reviewers in advance for their valuable comments that will improve the presentation of this paper.

Conflicts of Interest: The authors declare no conflict of interest.

References

1. Zhang, H.; Xu, H.; Tian, X.; Jiang, J.; Ma, J. Image fusion meets deep learning: A survey and perspective. *Inf. Fusion* **2021**, *76*, 323–336. [CrossRef]
2. Chen, J.; Li, X.; Luo, L.; Ma, J. Multi-focus image fusion based on multi-scale gradients and image matting. *IEEE Trans. Multimed.* **2021**, *24*, 655–667. [CrossRef]

3. Parihar, A.S.; Singh, K.; Rohilla, H.; Asnani, G. Fusion-based simultaneous estimation of reflectance and illumination for low-light image enhancement. *IET Image Process* **2021**, *15*, 1410–1423. [CrossRef]
4. Shi, Z.; Guo, B.; Zhao, M.; Zhang, C. Nighttime low illumination image enhancement with single image using bright/dark channel prior. *EURASIP J. Image Video Process* **2018**, *2018*, 13. [CrossRef]
5. Zhang, X. Benchmarking and comparing multi-exposure image fusion algorithms. *Inf. Fusion* **2021**, *74*, 111–131. [CrossRef]
6. Karim, S.; Tong, G.; Li, J.; Qadir, A.; Farooq, U.; Yu, Y. Current advances and future perspectives of image fusion: A comprehensive review. *Inf. Fusion* **2022**, *90*, 185–217. [CrossRef]
7. Tang, L.; Yuan, J.; Zhang, H.; Jiang, X.; Ma, J. PIAFusion: A progressive infrared and visible image fusion network based on illumination aware. *Inf. Fusion* **2022**, *83*, 79–92. [CrossRef]
8. Zhao, Y.; Cheng, J.; Zhou, W.; Zhang, C. Infrared pedestrian detection with converted temperature map. In Proceedings of the 2019 Asia-Pacific Signal and Information Processing Association Annual Summit and Conference (APSIPA ASC), Lanzhou, China, 18–21 November 2019; pp. 2025–2031.
9. Zhou, S.; Yang, P.; Xie, W. Infrared image segmentation based on Otsu and genetic algorithm. In Proceedings of the 2011 International Conference on Multimedia Technology, Hangzhou, China, 26–28 July 2011; pp. 5421–5424.
10. Li, G.; Lin, Y.; Qu, X. An infrared and visible image fusion method based on multi-scale transformation and norm optimization. *Inf. Fusion* **2021**, *71*, 109–129. [CrossRef]
11. Chen, J.; Li, X.; Luo, L.; Mei, X.; Ma, J. Infrared and visible image fusion based on target-enhanced multiscale transform decomposition. *Inf. Sci.* **2020**, *508*, 64–78. [CrossRef]
12. Liu, Y.; Wang, Z. Simultaneous image fusion and denoising with adaptive sparse representation. *IET Image Proc.* **2015**, *9*, 347–357. [CrossRef]
13. Yin, H. Sparse representation with learned multiscale dictionary for image fusion. *Neurocomputing* **2015**, *148*, 600–610. [CrossRef]
14. Liu, Y.; Liu, S.; Wang, Z. A general framework for image fusion based on multi-scale transform and sparse representation. *Inf. Fusion* **2015**, *24*, 147–164. [CrossRef]
15. Li, H.; Wu, X.-J.; Kittler, J. MDLatLRR: A novel decomposition method for infraredand visible image fusion. *IEEE Trans. Image Process* **2020**, *29*, 4733–4746. [CrossRef]
16. Ma, J.; Zhou, Z.; Wang, B.; Zong, H. Infrared and visible image fusion based on visual saliency map and weighted least square optimization. *Infrared Phys. Technol.* **2017**, *82*, 8–17. [CrossRef]
17. Li, H.; Wu, X.J. DenseFuse: A fusion approach to infrared and visible images. *IEEE Trans. Image Process.* **2018**, *28*, 2614–2623. [CrossRef]
18. Li, H.; Wu, X.J.; Durrani, T. NestFuse: An infrared and visible image fusion architecture based on nest connection and spatial/channel attention models. *IEEE Trans. Instrum. Meas.* **2020**, *69*, 9645–9656. [CrossRef]
19. Xu, H.; Gong, M.; Tian, X.; Huang, J.; Ma, J. CUFD: An encoder–decoder network for visible and infrared image fusion based on common and unique feature decomposition. *Comput. Vis. Image Underst.* **2022**, *218*, 103407. [CrossRef]
20. Xu, H.; Wang, X.; Ma, J. DRF: Disentangled representation for visible and infrared image fusion. *IEEE Trans. Instrum. Meas.* **2021**, *70*, 1–13. [CrossRef]
21. Tang, L.; Deng, Y.; Ma, Y.; Huang, J.; Ma, J. SuperFusion: A versatile image registration and fusion network with semantic awareness. *IEEE/CAA J. Autom. Sin.* **2022**, *9*, 2121–2137. [CrossRef]
22. Ma, J.; Tang, L.; Xu, M.; Zhang, H.; Xiao, G. STDFusionNet: An infrared and visible image fusion network based on salient target detection. *IEEE Trans. Instrum. Meas.* **2021**, *70*, 1–13. [CrossRef]
23. Xu, H.; Ma, J.; Jiang, J.; Guo, X.; Ling, H. U2Fusion: A unified unsupervised image fusion network. *IEEE Trans. Pattern Anal. Mach. Intell.* **2020**, *44*, 502–518. [CrossRef]
24. Xue, W.; Wang, A.; Zhao, L. FLFuse-Net: A fast and lightweight infrared and visible image fusion network via feature flow and edge compensation for salient information. *Infrared Phys. Technol.* **2022**, *127*, 104383. [CrossRef]
25. Ma, J.; Yu, W.; Liang, P.; Li, C.; Jiang, J. FusionGAN: A generative adversarial network for infrared and visible image fusion. *Inf. Fusion* **2019**, *48*, 11–26. [CrossRef]
26. Xu, H.; Liang, P.; Yu, W.; Jiang, J.; Ma, J. Learning a Generative Model for Fusing Infrared and Visible Images via Conditional Generative Adversarial Network with Dual Discriminators. In Proceedings of the Twenty-Eighth International Joint Conference on Artificial Intelligence (IJCAI), Macao, China, 10–16 August 2019; pp. 3954–3960.
27. Zhou, H.; Hou, J.; Zhang, Y.; Ma, J.; Ling, H. Unified gradient-and intensity-discriminator generative adversarial network for image fusion. *Inf. Fusion* **2022**, *88*, 184–201. [CrossRef]
28. Liu, J.; Fan, X.; Huang, Z.; Wu, G.; Liu, R.; Zhong, W.; Luo, Z. Target-aware dual adversarial learning and a multi-scenario multi-modality benchmark to fuse infrared and visible for object detection. In Proceedings of the IEEE/CVF Conference on Computer Vision and Pattern Recognition, New Orleans, LA, USA, 18–4 June 2022; pp. 5802–5811.
29. Tang, L.; Yuan, J.; Ma, J. Image fusion in the loop of high-level vision tasks: A semantic-aware real-time infrared and visible image fusion network. *Inf. Fusion* **2022**, *82*, 28–42. [CrossRef]
30. Li, H.; Wu, X.J.; Kittler, J. RFN-Nest: An end-to-end residual fusion network for infrared and visible images. *Inf. Fusion* **2021**, *73*, 72–86. [CrossRef]
31. Jian, L.; Yang, X.; Liu, Z.; Jeon, G.; Gao, M.; Chisholm, D. SEDRFuse: A symmetric encoder–decoder with residual block network for infrared and visible image fusion. *IEEE Trans. Instrum. Meas.* **2020**, *70*, 5002215. [CrossRef]

32. Li, J.; Huo, H.; Li, C.; Wang, R.; Feng, Q. AttentionFGAN: Infrared and visible image fusion using attention-based generative adversarial networks. *IEEE Trans. Multimed.* **2020**, *23*, 1383–1396. [CrossRef]
33. Long, Y.; Jia, H.; Zhong, Y.; Jiang, Y.; Jia, Y. RXDNFuse: A aggregated residual dense network for infrared and visible image fusion. *Inf. Fusion* **2021**, *69*, 128–141. [CrossRef]
34. Xie, S.; Girshick, R.; Dollár, P.; Tu, Z.; He, K. Aggregated residual transformations for deep neural networks. In Proceedings of the IEEE conference on computer vision and pattern recognition, Honolulu, HI, USA, 21–26 July 2017; pp. 1492–1500.
35. Li, H.; Cen, Y.; Liu, Y.; Chen, X.; Yu, Z. Different input resolutions and arbitrary output resolution: A meta learning-based deep framework for infrared and visible image fusion. *IEEE Trans. Image Process.* **2021**, *30*, 4070–4083. [CrossRef]
36. Ha, Q.; Watanabe, K.; Karasawa, T.; Ushiku, Y.; Harada, T. MFNet: Towards real-time semantic segmentation for autonomous vehicles with multi-spectral scenes. In Proceedings of the 2017 IEEE/RSJ International Conference on Intelligent Robots and Systems (IROS), Vancouver, BC, Canada, 24–28 September 2017; pp. 5108–5115.
37. Sun, Y.; Zuo, W.; Liu, M. Rtfnet: Rgb-thermal fusion network for semantic segmentation of urban scenes. *IEEE Robot. Autom. Lett.* **2019**, *4*, 2576–2583. [CrossRef]
38. Xu, J.; Lu, K.; Wang, H. Attention fusion network for multi-spectral semantic segmentation. *Pattern Recognit. Lett.* **2021**, *146*, 179–184. [CrossRef]
39. Liu, H.; Chen, F.; Zeng, Z.; Tan, X. AMFuse: Add–Multiply-Based Cross-Modal Fusion Network for Multi-Spectral Semantic Segmentation. *Remote Sens.* **2022**, *14*, 3368. [CrossRef]
40. Toet, A. TNO Image Fusion Dataset. 2014. Available online: https://figshare.com/articles/dataset/TNO_Image_Fusion_Dataset/1008029 (accessed on 31 May 2023).
41. Zhang, H.; Ma, J. SDNet: A versatile squeeze-and-decomposition network for real-time image fusion. *Int. J. Comput. Vis.* **2021**, *129*, 2761–2785. [CrossRef]
42. Peng, C.; Tian, T.; Chen, C.; Guo, X.; Ma, J. Bilateral attention decoder: A lightweight decoder for real-time semantic segmentation. *Neural Netw.* **2021**, *137*, 188–199. [CrossRef] [PubMed]

entropy

MDPI

Article

Improved Thermal Infrared Image Super-Resolution Reconstruction Method Base on Multimodal Sensor Fusion

Yichun Jiang [1,2,†], Yunqing Liu [1,*,†], Weida Zhan [1,2,†] and Depeng Zhu [1,2]

1 The College of Electronic and Information Engineering, Changchun University of Science and Technology, Changchun 130022, China; jiangyichun@mails.cust.edu.cn (Y.J.); zhanweida@cust.edu.cn (W.Z.); zhudepeng@mails.cust.edu.cn (D.Z.)
2 National Demonstration Center for Experimental Electrical, Changchun University of Science and Technology, Changchun 130022, China
* Correspondence: mzlyq@cust.edu.cn; Tel.: +86-138-4316-3761
† These authors contributed equally to this work.

Abstract: When traditional super-resolution reconstruction methods are applied to infrared thermal images, they often ignore the problem of poor image quality caused by the imaging mechanism, which makes it difficult to obtain high-quality reconstruction results even with the training of simulated degraded inverse processes. To address these issues, we proposed a thermal infrared image super-resolution reconstruction method based on multimodal sensor fusion, aiming to enhance the resolution of thermal infrared images and rely on multimodal sensor information to reconstruct high-frequency details in the images, thereby overcoming the limitations of imaging mechanisms. First, we designed a novel super-resolution reconstruction network, which consisted of primary feature encoding, super-resolution reconstruction, and high-frequency detail fusion subnetwork, to enhance the resolution of thermal infrared images and rely on multimodal sensor information to reconstruct high-frequency details in the images, thereby overcoming limitations of imaging mechanisms. We designed hierarchical dilated distillation modules and a cross-attention transformation module to extract and transmit image features, enhancing the network's ability to express complex patterns. Then, we proposed a hybrid loss function to guide the network in extracting salient features from thermal infrared images and reference images while maintaining accurate thermal information. Finally, we proposed a learning strategy to ensure the high-quality super-resolution reconstruction performance of the network, even in the absence of reference images. Extensive experimental results show that the proposed method exhibits superior reconstruction image quality compared to other contrastive methods, demonstrating its effectiveness.

Keywords: thermal infrared imaging; super-resolution reconstruction; multimodal sensors; information fusion

Citation: Jiang, Y.; Liu, Y.; Zhan, W.; Zhu, D. Improved Thermal Infrared Image Super-Resolution Reconstruction Method Base on Multimodal Sensor Fusion. *Entropy* **2023**, *25*, 914. https://doi.org/10.3390/e25060914

Academic Editors: Oleg Sergiyenko, Wendy Flores-Fuentes, Julio Cesar Rodriguez-Quinonez and Jesús Elías Miranda-Vega

Received: 30 April 2023
Revised: 1 June 2023
Accepted: 7 June 2023
Published: 9 June 2023

1. Introduction

Thermal infrared imaging is a passive imaging technology that detects the thermal radiation passively emitted by objects to form an image [1]. It has the advantages of strong anti-interference ability and the capability to distinguish between targets and backgrounds. Therefore, super-resolution reconstruction (SR) has been widely applied in fields such as remote sensing imaging [2–4], target tracking [5–7], and autonomous driving [8,9], etc. However, compared with visible light imaging, infrared imaging equipment usually has limited spatial resolution, resulting in lower imaging quality. Therefore, to overcome this limitation, super-resolution reconstruction technology has become an important research field. The super-resolution technology can restore high-frequency information from low-resolution images, which can improve the resolution of infrared images and enrich image details.

Currently, due to the continuous improvement of computational device performance and the increasing maturity of deep learning technology, deep learning-based SR methods have become the mainstream solution to SR problems. Compared with interpolation-based [10], reconstruction-based [11], and sparse representation-based methods [12], they have significant performance advantages. The focus of the single image super-resolution reconstruction (SISR) network is mainly on the reasonable allocation of network resources to high- and low-frequency information reconstruction. SISR relies on the mapping relationship between high- and low-resolution information (HR&LR) images solidified in the weight parameters through training and does not introduce effective external information

Compared with collecting infrared images, high-quality visible light images are more easily obtained and possess higher spatial resolution. Although they operate in different spectral bands, a significant amount of complementary information exists, making it feasible and effective to guide infrared image super-resolution using complementary information from visible light images. Some research has made progress, but several key issues remain:

(1) In the case of multimodal super-resolution, the large resolution difference between infrared and visible images leads to a significant decrease in the accuracy of reconstructed infrared images. High-performance super-resolution reconstruction networks, especially their feature extraction and information transformation mechanisms, still require further research.

(2) Due to the imaging mechanism of thermal infrared sensors, the quality of infrared images remains poor despite high pixel resolution. Existing methods that use simulated degradation and train their inverse process are limited by the quality of the infrared images used as labels, making it difficult to effectively enhance high-frequency details in infrared images.

(3) The existing multimodal super-resolution reconstruction methods have not fully considered the cases where the reference image is missing or of poor quality, which leads to a sharp degradation in the performance of the network and poor quality of the reconstructed images.

To address these issues, we proposed a thermal infrared image super-resolution reconstruction method based on multimodal sensor fusion. The method consists of a novel neural network architecture, a new hybrid loss function, and corresponding training strategies. The input infrared image is reconstructed through the network, during which multimodal features are continuously extracted and fused to obtain a high-quality, high-resolution thermal infrared image. The proposed loss function is used to constrain the network to ensure that the thermal infrared information in the image is not erroneously altered. Moreover, the proposed training strategy ensures that the network can still correctly reconstruct thermal infrared images even when the reference images are missing or of poor quality.

Our main contributions are as follows:

(1) We proposed a super-resolution reconstruction network that continuously fuses information from different scales of visible light images in the iterative process to reconstruct low-frequency and high-frequency information in infrared images, solving the problem of accuracy decline caused by large resolution difference between infrared and visible light images.

(2) We proposed a hierarchical dilated distillation module that can adaptively extract features of different scales, with strong representation ability and fewer learnable parameters.

(3) We proposed an information transformation module based on attention mechanism, which calculates pixel-level correlation between infrared and visible light features to reduce the interference of redundant and unrelated information on reconstructed images, improve information fusion efficiency, and suppress the blurring phenomenon in the infrared image reconstruction process.

(4) We designed a hybrid loss function for multimodal super-resolution to supervise the network to obtain more high-frequency features from visible light images and ensure

the style of infrared images does not deviate by adversarial loss, retaining richer details and more thermal infrared information.

(5) We proposed a modal switching training strategy to solve the problem of degraded performance in reference-based super-resolution reconstruction of thermal infrared images when the reference image is missing, improving the network's robustness.

2. Related work

2.1. Image Super-Resolution Reconstruction Based on Neural Networks

As an ill-posed problem, super-resolution reconstruction is limited in its reconstruction and generalization capabilities if relying solely on manually designed prior methods. As a result, neural networks, which are powerful implicit function fitters, have been employed due to their effectiveness in fitting complex mappings in image processing. Since the introduction of the first convolutional neural network for image super-resolution, SRCNN [13], the use of neural networks in this field has grown exponentially, with a primary focus on optimizing network structures. Early research works such as VDSR [14], SRResNet [15], and EDSR [16] have significantly improved network feature expression ability and reconstruction quality by deepening the network and incorporating the residual structure concept. However, increasing network depth and width to a certain extent becomes inefficient, resulting in diminishing performance gains. To further enhance reconstruction quality and efficiency, new model structures have been designed specifically for SR tasks, optimizing reconstruction while maintaining low complexity. These improvements include multi-scale feature extraction [17,18], feature reuse [19–21], and attention mechanisms [22,23]. Such modifications introduce prior knowledge into the network structure, enhancing the model's adaptability to SR tasks while reducing the network's dependence on learnable parameters and training data.

In addition to improving network structure, efforts have been made to better train neural networks to generate realistic and detailed texture details. References [24,25] investigated several commonly-used loss functions in image restoration and provided guidance for loss function design in super-resolution reconstruction. Although these loss functions calculate the difference between predicted and real data, they may produce significant blur or aliasing artifacts due to the diversity of mappings. Consequently, the use of generative adversarial learning is being explored to obtain the implicit distribution of real images from the dataset [13,26,27], guiding the network to generate clearer reconstruction results. However, this technique often results in apparent reconstruction errors that are difficult to avoid. Despite the current advancements in network structure, loss functions, and training methods for SR, there remains substantial room for further improvement.

2.2. Multimodal Reference-Based Super-Resolution Reconstruction

Compared to SISR, reference-based SR is a technique that uses additional guiding images to transfer relevant structural information to the target image in order to achieve high-quality super-resolution reconstruction [28,29]. In early research, multimodal reference-based super-resolution (multimodal SR) reconstruction mainly used filtering-based [30,31], optimization-based [32], and sparse representation-based [33] methods. However, these methods faced difficulties in reconstructing HR images, especially when there were large differences in image structure or resolution between modalities. Currently, the main method used is learning-based. By utilizing the powerful fitting ability of deep learning, texture conversion and transmission between modalities can be achieved.

However, in recent research, impressive reconstruction quality has been achieved by studying the correlation between the source image and the reference image. Despite this progress, these methods still adhere to the traditional SR training method, which simulates the downsampling process and then learns its inverse process, producing images similar to the original collected data [34–36]. The method suffers from the limitations of the low resolution and imaging mechanism of the thermal infrared sensor. Compared to reconstructing high-quality visible light or near-infrared images, it is difficult to reconstruct high-quality infrared images using this method and many texture details may be lost.

In order to solve this problem, some studies have designed fusion strategies to synthesize visible light and infrared image information, and introduce visible light texture while performing SR of infrared images [37]. However, although this method supplements some details, the generated image not only produces incorrect texture but also does not conform to the thermal information distribution in the infrared source image due to its imperfect network structure, loss function, and supervision design. Therefore, further research and improvement are still needed to develop effective fusion strategies that can better preserve the thermal information distribution and generate high-quality HR images. Additionally, the use of appropriate evaluation metrics is essential to ensure that the generated images meet the requirements of practical applications.

3. Proposed Method

Thermal infrared radiation can be affected by various factors when reaching imaging sensors, such as motion blur, optical blur, and electronic noise, leading to degradation in the quality of infrared images. Super-resolution reconstruction techniques for thermal infrared images are often considered the inverse process to address these issues. However, the pixel size of thermal infrared sensors is larger, and diffraction and scattering effects are more pronounced. As a result, even when the resolution is the same, thermal infrared images appear blurrier. Traditional super-resolution reconstruction methods obtain HR and LR infrared image pairs through simulated downsampling and training the SR mapping in reverse is not effective in reconstructing ideal HR infrared images. We believe that preserving the original infrared thermal information is necessary, while predicting some high-frequency information reasonably can make the reconstructed image more visually appealing. Therefore, our research focused not only on restoring the information in the original infrared image but also on using visible light images to guide neural networks to predict and reconstruct high-frequency information in thermal infrared images to improve the overall quality of reconstructed images. To achieve our goal, we designed specific network structures, loss functions, and training strategies.

3.1. Network Architecture

The network structure is shown in Figure 1. Our proposed network consists of three parts: the primary feature encoding subnetwork, the super-resolution reconstruction subnetwork, and the high-frequency detail fusion subnetwork. Subsequently, we will explicate the operational principles, design concepts, and particular implementations of each component.

3.1.1. Primary Feature Encoding Subnetwork

The Primary Feature Encoding Subnetwork is used to map the input image to a feature space for further processing. It primarily consists of an infrared feature encoder and multiple visible light feature encoders. The infrared feature encoder is only used before the first stage of super-resolution reconstruction subnetwork to encode the input infrared image I_{LR}^{TIR} into primary feature f_b^{TIR} using a straightforward convolutional layer, which can be represented by the following equation:

$$\begin{aligned} f_{b,1}^{TIR} &= \sigma(W_{enc}^{TIR} * I_{LR}^{TIR} + B_{enc}) \\ f_{b,n}^{TIR} &= f_{o,n-1}(n = 2,3\ldots N) \end{aligned} \tag{1}$$

where W_{enc}^{TIR} represents the filter for encoding thermal infrared images, B_{enc} represents the bias value, $f_{o,n}$ represents the output feature map for the n-th stage of super-resolution reconstruction subnetwork, $\sigma(x) = \max(x, 0)$ represents the rectified linear unit, and $*$ represents the convolution operation. The visible light feature encoder uses multiple convolutional layers with varying specifications, depending on the super-resolution reconstruction multipliers, to encode visible light images I^{VIS} at different scales. These layers construct a feature pyramid to generate visible light image features $f_{b,n}^{VIS}(n = 1,2,\ldots,N)$

corresponding to the various stages of the reconstruction process. Mathematically, the process can be expressed as follows:

$$f_{b,n}^{VIS} = \sigma(W_n^{VIS} * I^{VIS} + B)(n = 1, 2 \ldots N) \quad (2)$$

where W_n^{VIS} represents the encoding filter for visible light images used in the n-th stage and N represents the total number of stages. Unlike the infrared encoding filter, the visible light encoding filter is used in each stage, and the corresponding filter uses different convolution kernel sizes and strides. Specifically, for the filter $W_{b,n}^{VIS}$ in the n-th stage, the stride is set to $2^{N-(n-1)}$ to output the current infrared feature map size, and the convolution kernel size is designed as $2^{N-(n-1)} + 1$ to prevent the loss of pixel information in the image. By using this method, we can mainly introduce the low-frequency features of visible light images in the early stage of reconstruction, and focus more on the high-frequency details of visible light images in the later stage of reconstruction and fusion of details.

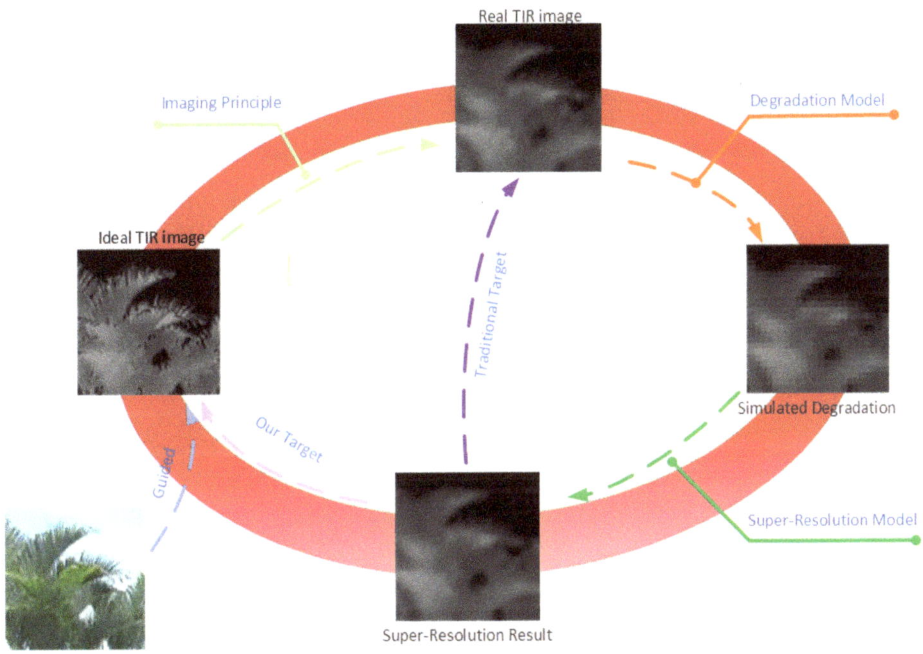

Figure 1. The target of our method.

3.1.2. Super-Resolution Reconstruction Subnetwork

As shown in Figure 2a, the Super-Resolution Reconstruction Subnetwork is the core component of our network, which aims to restore and enhance the resolution and texture details of thermal infrared images. Note that for different stages of super-resolution (SR), we use the same subnetwork for super-resolution reconstruction, with shared weights and identical structure. For this subnetwork, we designed the Feature Extraction Module (FEM), Cross-Attention Transformation Module (CATM), and Upsampling Module (UM) for efficient extraction of structural information from different modal images. By measuring the degree of correlation between multimodal images, we achieved effective texture transfer and super-resolution reconstruction.

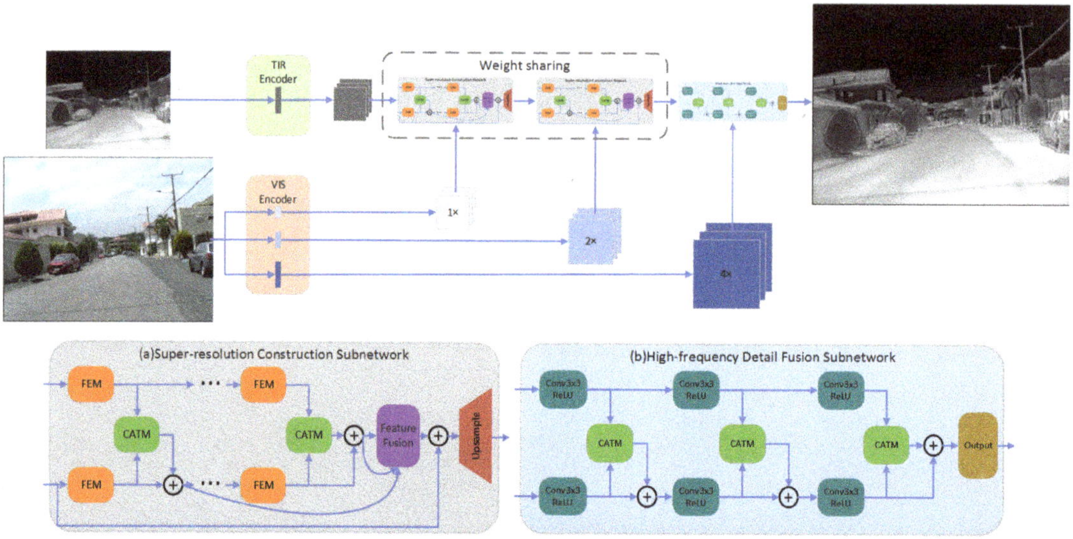

Figure 2. The architecture of our proposed network.

- **Feature Extraction Module (FEM):** We employed the same structure for the FEM used to process both infrared images and visible light features. This approach is based on the fact that visible light features have been previously adjusted to a feature space that matches the infrared features during the primary feature encoding process. Batch normalization layers in the network can destroy the original contrast of images in image reconstruction tasks according to some existing research. Therefore, we specifically removed all batch normalization layers in the network to improve reconstruction performance, reduce redundancy operations, and increase training and inference speed. Our FEM consisted of a series of improved Hierarchical Dilated Distillation Modules (HDDM) as presented in Figure 3a. We designed a multi-scale distillation fusion mechanism for visible or infrared input feature maps f_{in}^{FEM}, which sequentially passes through filters with different dilation rates to separate different frequency components of different image features. This process enhances the representational capacity of the network. After each filter output, the feature map is split into two equal parts along the channel dimension. One part f_{s1}^{FEM} is directly passed on to the subsequent steps for feature fusion, while the other part f_{s2}^{FEM} continues to extract features. This operation can be represented as

$$\begin{aligned} f_{MS,1}^{FEM} &= \sigma(W_{MS,1} * f_{in}^{FEM} + B_{MS,1}) \\ \left[f_{s1,n-1}^{FEM}, f_{s2,n-1}^{FEM}\right] &= f_{MS,n-1}^{FEM} \\ f_{MS,n}^{FEM} &= \sigma(W_{MS,n} * f_{n-1}^{FEM} + B_{MS,n}), n = (2,3,4) \end{aligned} \tag{3}$$

where $W_{MS,n}^{FEM}$ represents the n-th filter in HDDM. Then, we concatenated all the f_{s1}^{FEM} in HDDM into one vector for subsequent operations. This operation can be represented as

$$\begin{aligned} f_{c}^{FEM} &= F_{cat}(f_{s1,1}^{FEM}, f_{s1,2}^{FEM}, \ldots, f_{s1,M-1}^{FEM}, f_{MS,M}^{FEM}) \\ f_{a}^{FEM} &= F_{ca}(f_{c}^{FEM}) \end{aligned} \tag{4}$$

where $F_{ca}(\cdot)$ represents our improved Channel Enhanced Attention Module (CEAM), as shown in Figure 3b. Firstly, in low level visual tasks such as super-resolution, it is more important to focus on the image structure information. Directly using global average pooling to extract information is not appropriate. Therefore, we first

introduced a depthwise separable convolution at the front end of CEAM to process the features of each channel. Then we performed the operation of global average pooling. We also utilized a 1-D convolution to process the compressed channel information, inspired by previous literature [38]. This approach reduces the computational and parameter complexity. We not only avoided the dimensionality reduction operation in channel attention, but also elevated the channel dimension to form multiple subspaces for different aspects of information. By combining different dimension features, we achieved more flexible information interaction between channels.
Inspired by previous research, such as EDSR, we introduced local residual learning into the feature extraction module. This approach can effectively alleviate the potential problem of gradient disappearance in the parameter optimization process, making it possible to construct a deeper super-resolution reconstruction network. To perform point-wise add operation between the output feature map and the input feature map, we set a filter with a convolution kernel size of 1 × 1 at the end. This filter fuses the multi-scale information previously extracted and matches the number of channels with the input feature map. This operation can be represented as follows:

$$f_{out}^{FEM} = \sigma(W_{out}^{FEM} * f_2^{FEM} + B_{out}^{FEM}) + f_{in}^{FEM} \tag{5}$$

- **Cross-Attention Transformation Module (CATM):** In order to guide the process of infrared image SR with visible features, we constructed a Cross-Attention Transformation Module to obtain the attention map of relevant information from the input visible light features and transfer the useful information. The structure of the CATM are shown in Figure 4.
Given the input of infrared and visible feature maps f_{in}^{TIR} and f_{in}^{VIS}, which are obtained by the feature extraction module processing the primary features of infrared and visible light, respectively. After f_{in}^{TIR} and f_{in}^{VIS} were concatenated into a tensor f_{in}^{CATM}, they were input into the attention branch. Unlike previous attention mechanisms, we did not limit the estimation of attention maps to channel or spatial dimensions, but constructed a pixel-level attention mechanism. Firstly, f_{in}^{CATM} was filtered by a 3 × 3 convolutional kernel to extract effective features in the feature map, and the channel number of the feature map was compressed to $1/\beta$ (β was the compression ratio, set to 4 due to performance limitations of server) to improve the computational efficiency of attention map estimation. Then, the number of channels was restored through a 3 × 3 convolutional kernel, and the attention map was reconstructed based on the effective features. This operation can be represented as:

$$\begin{aligned} f_{in}^{CATM} &= F_{cat}(f_{in}^{TIR}, f_{in}^{VIS}) \\ f_{PA1}^{CATM} &= \sigma(W_{PA1} * f_{in}^{CATM} + B_{PA1}) \\ f_{PA2}^{CATM} &= \delta(W_{PA2} * f_{PA1}^{CATM} + B_{PA2}) \end{aligned} \tag{6}$$

where $F_{cat}(\cdot)$ represents the concatenation operation along the channel dimension. $\delta(x) = (1 + e^{-x})^{-1}$ is the Sigmoid function, which is used to restrict the range of the output attention map values to (0,1), ensuring that no error occured during testing and training. Meanwhile, we apply the feature sub-module to the input tensor f_{in}^{CATM}, and obtain the feature map f_{feat}^{CATM} that stores texture information. This operation can be represented as:

$$f_{feat}^{CATM} = \sigma(W_{feat} * f_{in}^{CATM} + B_{feat}) \tag{7}$$

Finally, the feature map f_{feat}^{CATM} and attention map f_{PA2}^{CATM} were multiplied point by point, and added to the infrared feature map f_{in}^{TIR} to introduce the structural features of visible light images and obtain the updated infrared features f_{out}^{TIR}. This operation can be represented by the following formula:

$$f_{out}^{CATM} = f_{in}^{TIR} + f_{PA2}^{CATM} \cdot f_{feat}^{TIR} \tag{8}$$

- **Global Residual Connection and Hierarchical Feature Fusion:** In this task, there is a strong correlation between input features and the output image. Shallow features typically retain a significant amount of low-frequency information. Additionally, as the network goes deeper, an optimization challenge called gradient vanishing occurs. Global residual connection serves as a simple yet effective solution for addressing these issues. It enables the network to concentrate on reconstructing the image's high-frequency information, reduces resource waste, and simultaneously resolves the gradient vanishing problem. However, relying solely on global residual connections during the network inference process cannot fully utilize the abundant image features generated, resulting in information redundancy. As the network depth increases, the spatial representation capacity gradually decreases, while the semantic representation capacity increases. Therefore, fully exploiting these features can enhance the quality of the reconstructed image. To address this issue, we adopted a hierarchical feature fusion mechanism that sent the output of each CATM to the endpoint before up-sampling for processing. Considering the significant amount of redundancy in these features, we added a feature fusion layer, which acts as a bottleneck layer to selectively extract relevant information from the hierarchical features. This layer is crucial for improving network efficiency and performance. The operation can be represented by the following formula:

$$f_c = f_{b,n}^{TIR} + [W_c * F_{cat}(f_{out}^{CATM,1}, f_{out}^{CATM,2}, \dots, f_{out}^{CATM,M}) + B_c] \tag{9}$$

  where $f_{out}^{CATM,m} (m = 1, 2, \dots, M)$ represents the output of the m-th CATM in the reconstructed network, M represents the total number of CATMs. $f_{b,n}^{TIR}$ represents the input infrared feature map of the super-resolution reconstruction network for the n-th stage.
- **Upsamle Module (UM):** Upsampling methods have been extensively studied in super-resolution networks. Some studies process feature maps at low resolutions, and then directly upsample and reconstruct the features to the target scale, which can reduce some computational cost. However, these methods are not conducive to achieving high magnification ratios and convenient interaction of multimodal information. Our proposed network gradually performs feature extraction and information fusion while the feature map is being constantly upsampled by a factor of 2 in each stage, in order to introduce rich texture details of visible light images at different scales. The feature f_c was input into UM and upsampled by 2× through bilinear interpolation. Then, the updated features were filtered using a 3 × 3 convolution kernel to reduce the block effect in the feature maps. This process can be formalized as follows:

$$f_o = W_u * F_{up\uparrow}(f_c) + B_u \tag{10}$$

  where $F_{up\uparrow}$ represents the operation of bilinear interpolation.

3.1.3. High-Frequency Detail Fusion Subnetwork

In order to maximize the utilization of visible light image information, we specially set up a high-frequency detail fusion network to further refine the infrared reconstruction images at the target scale. As it is difficult to control the computational complexity and spatial complexity of the network when operating on HR images, which is not conducive to training and inference, we designed a simple network structure consisting of three pairs of convolutional layers, three CATMs, and one reconstruction layer. The specific structure is shown in Figure 2b.

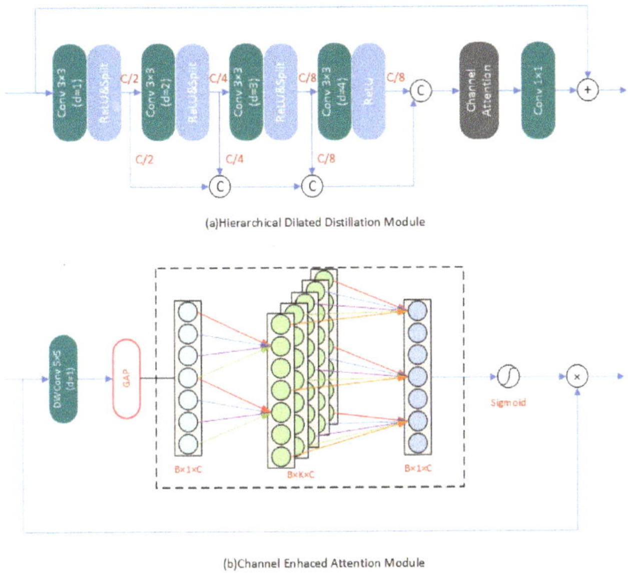

Figure 3. The basic unit of Feature Extraction Module.

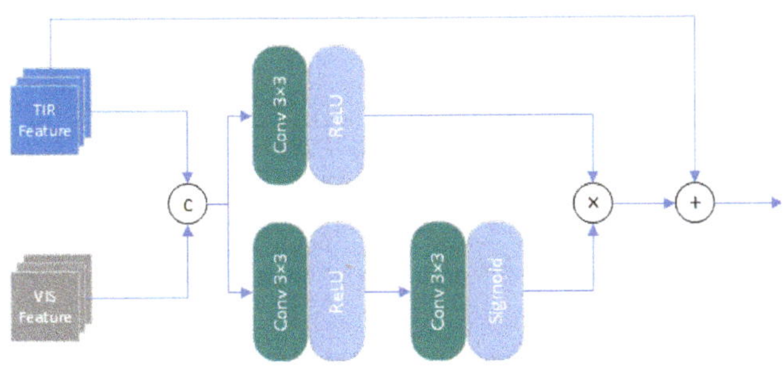

Figure 4. Cross-Attention Transformation Module.

3.2. Loss Function

To train the network proposed in this study, it was necessary to measure the similarity between the network output and Ground Truth(GT) of the infrared image, and restore the thermal information as much as possible. At the start of the third section, we emphasized the need to recover not only the known details in the infrared image, but also texture features that had been lost due to the imaging mechanism with the assistance of visible light images. Therefore, we designed a hybrid loss function, including intensity loss, structure loss, adversarial loss, and perceptual loss, to ensure the real thermal information while retaining valuable multimodal feature. The training process of neural network is shown in Figure 5.

The intensity loss is designed to retain low-frequency information of infrared images, and the main schemes include L1 and L2 loss functions. Many studies have shown that the L1 loss function is superior to the L2 loss function in terms of optimizability and reconstruction quality [24,39], so we adopted the L1 loss as the intensity loss. For the given input training samples $\{x, y, z\}$, in which x, y, and z are, respectively, the LR versions of infrared images, visible light images (Ref) as the reference image, and the HR version of infrared images. The intensity loss can be represented by the following formula:

$$L_i(\theta) = \frac{1}{N}\sum_{n=1}^{N} |G(x,y|\theta) - z| \tag{11}$$

where $G(\cdot,\cdot)$ represents the proposed network model in this article, and θ represents the weight parameters of the network model.

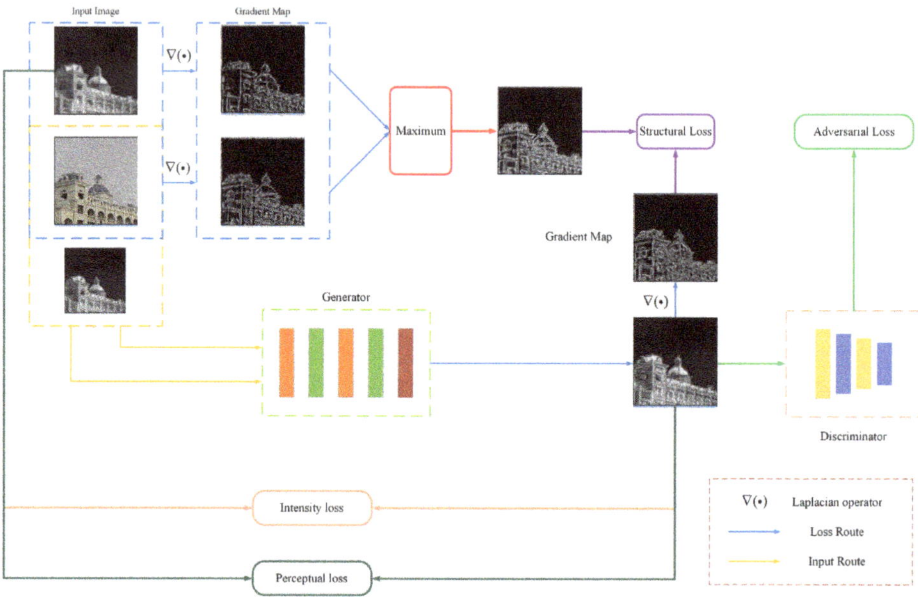

Figure 5. Training Process of the Neural Network. The architecture of the generator is shown in Figure 2. In order to achieve better supervision, we adopted a Markovian discriminator (PatchGAN) [40] as the discriminator to preserve high-resolution details.

The role of the structural loss is to guide the network in obtaining sufficient complementary features from visible light images, which will result in the preservation of high-frequency details of both infrared and visible light images in the reconstructed image. We proposed a gradient-based structural loss to train the network to acquire this ability, employing salient features present in the infrared and reference images as a training target. The following equation represents the loss:

$$L_s(\theta) = \frac{1}{N}\sum_{n=1}^{N} ||\nabla G(x,y|\theta)| - \max(|\nabla y|, |\nabla z|)| \tag{12}$$

where ∇ represents the gradient operator; we used the Laplace operator. Although the utilization of structural loss has the benefit of preserving rich high-frequency details, the ablation study conducted in Section 4.3 indicated that its implementation may lead to serious image distortion. This ultimately results in inaccurate thermal information, especially at the edges and texture details of the image. To address this issue, we added both adversarial loss and perceptual loss into the hybrid loss. These constraints facilitated the generation process of the network and ensured that thermal information in the image was preserved. Furthermore, these additions improved the overall quality of the reconstructed image. Specifically, adversarial and perceptual losses can be represented as follows:

$$L_{adv}(\theta) = \frac{1}{N}\sum_{i=1}^{N} \|1 - D(G(x_i, y_i|\theta))\|^2$$
$$L_p(\theta) = \frac{1}{N}\sum_{i=1}^{N}\sum_{j\in\Omega} \|\varphi_j(G(x_i, y_i|\theta)) - \varphi_j(z_i)\|^2 \quad (13)$$

where $D(\cdot)$ represents the discriminator network and $\varphi_j(\cdot)$ represents the feature map of j-th layer in the VGG19 network. The hybrid loss we proposed can be represented by the following equation:

$$L_{total}(\theta) = L_i(\theta) + L_s(\theta) + \lambda L_{adv}(\theta) + L_p(\theta) \quad (14)$$

where λ is the weight factor of the adversarial loss, which is used to balance the magnitude of other loss function values and adversarial loss. It was set to 0.1 based on experimental settings. Our ultimate goal was to minimize the value of the hybrid loss and obtain the corresponding network parameter weights, as shown in the following equation:

$$\hat{\theta} = \arg\min_{\theta} L_{total}(\theta) \quad (15)$$

3.3. Training Strategies

Although introducing the information of visible light images in reconstruction image can greatly enrich the texture details in the reconstructed image, high-quality visible light images cannot always be obtained under all conditions, often being affected by conditions such as lighting and smoke. In practical application scenarios, infrared images have the characteristics of all-weather and strong anti-interference abilities; the imaging quality is also more stable. Therefore, we hoped that low-resolution infrared images could be used as the main information source in the reconstruction process, with visible light information as supplementary information. To improve the robustness of the proposed method, based on the network structure and loss function we designed, a modal switching training strategy was proposed.

During each single training epoch, we initially input multimodal data of infrared and visible light images to train all weight parameters in the network, which enabled the network to learn the ability to obtain information from input infrared images and visible light images as reference, and reconstruct high-quality images. Subsequently, to prevent significant performance deterioration of the network when no reference images are present, we input infrared images and a black image (filled with zeros) to remove the input visible light images used as references. In this process, only those structure of the network associated with infrared images were updated during training and inference, which enabled the network to attain capabilities similar to single image super-resolution. Therefore, while the reference input was being removeded, we temporarily froze all CATM that were involved in fusion parts of high frequency details and super-resolution reconstruction, as well as the encoder used to process visible light features. During this process, we set their convolutional kernels and biases to zero for a temporary period, so that no updates would be made to these parameters and thereby not impact the infrared branch's training. Finally, The loss function could be trained as normal, without modification in this stage, as the structural loss adropts the maximum value strategy to introduce visible light information.

By using this method, we trained the network proposed by them to reconstruct high-resolution infrared images while reducing reliance on reference images in subsequent trials. During a single round of training, the discriminator was updated only once to prevent mode collapse.

4. Experiment

4.1. Experimental Environment and Dataset Settings

Our proposed network was trained and on a hardware environment with an Intel (R) Core (TM) i9-13900KF CPU, 64.0 GB of RAM and a NVIDIA GeForce RTX 4090 GPU.

We used the PyCharm 2021.3.2 software platform on the Windows 11 operating system, alongside the PyTorch 1.10.1 deep learning framework. The training process took 44.3 h overall, while for each image, the testing speed was 0.62 s.

Deep learning, as a data-driven technology, necessitates a significant amount of well-registered thermal infrared-visible light images for training data. To achieve this objective, we combined three popular multimodal datasets: M3FD [41], FLIR ADAS, and TISR [42]. Sample images from the dataset are exemplified in Figure 6. We partitioned the dataset into three sets, namely, training set, testing set and validation set with the ratio of 8:1:1. This step was conducted to evaluate the generalization capability of our proposed algorithm. Additionally, we trained and tested all other comparative methods using the same dataset.

Figure 6. Visualization of samples from the trainging dataset.

The dataset consisted of a total of 1394 infrared and visible light images of complex scenes, including urban, road, and forest environments, with all images completed at the pixel-level alignment. To achieve data augmentation, all images in the training set were flipped and rotated, and then cropped into image blocks with a size of 256 × 256. Furthermore, we simulated degradation by downsampling the infrared images via bicubic interpolation to obtain the corresponding LR input images.

We trained our model using the ADAM optimizer and set $\beta_1 = 0.9$, $\beta_2 = 0.999$, and $\varepsilon = 10^{-8}$. We set the minibatch size to 16, initial learning rate to 5×10^{-4}, and trained the model for a total of 200 epochs. We reduced the learning rate to 0.1 at the 100-th and 150-th epochs.

4.2. *Comparative Experiments*

In order to demonstrate the effectiveness and superiority of our proposed method, we conducted comparative experiments on multiple classical or state-of-the-art (SOTA) methods in the same test environment. Firstly, we removed the visible light images and information conversion mechanism in the network to test the ability of our proposed method to perform single image super-resolution (SISR) without reference image guidance. In this experiment, we compared RCAN [22], EDSR [16], s-LWSR64 [19], Zou et al. [43] and Wang et al. [37]'s methods. The qualitative analysis, as shown in Figure 7, demonstrates the infrared super-resolution reconstruction results (4×) of three scenarios. Meanwhile, we present the quantitative analysis results of each method on the 8× and 4× test datasets in Table 1, mainly using the peak signal-to-noise ratio (PSNR) and the structural similarity index (SSIM) as the metrics.

In the SISR at the 4× scale, our proposed method performs comparably to EDSR in terms of performance and outperforms all other comparison methods, with slightly lower PSNR but better SSIM. It is worth noting that EDSR, as a rather large model, has about 43M parameters, while our network has only around 700 K trainable parameters (excluding frozen weight parameters), with significant advantages in both computational efficiency and memory usage. At 8× super-resolution reconstruction, our proposed method outperforms other methods, and is more suitable for high-resolution reconstruction than other methods. In terms of visual imaging performance, our proposed method effectively restores the original low-frequency information in infrared images, and is more prominent in reconstructing texture details. Wang et al.'s method, compared to the method proposed in this paper, shows a serious degradation of image quality both in objective metrics and visual perception after masking the reference image, and this is because their training strategy and information transmission mechanism cannot adapt to this situation, while our method effectively avoids this problem.

Table 1. Benchmark test results for SISR.

Methods	Prameters	4×		8×	
		PNSR	SSIM	PNSR	SSIM
RCAN	16 M	31.66	0.8724	28.20	0.7107
s-LWSR64	2.27 M	31.67	0.8894	28.18	0.7098
Zou et al.	3.73 M	31.84	0.8863	28.40	0.7121
EDSR	43.09 M	32.21	0.8913	28.38	0.7166
Wang et al.	573.6 K	21.33	0.6042	-	-
Ours	698.2 K	32.15	0.8921	28.41	0.7283

Overall, in the task of SISR, without introducing external information, the restoration performance of using only single image super-resolution methods to restore high-frequency information was limited, which may be affected by the amount of data and the difficulty of the task. From another perspective, the experiment also verifies that in the super-resolution reconstruction of multimodal information fusion, it is feasible to achieve high-quality single image super-resolution without reference images by using a modal switching strategy for training.

After verifying the infrared super-resolution reconstruction ability of the network, we studied the effect of image super-resolution through multimodal fusion with a reference image. As discussed in Section 3, unlike SISR tasks, we no longer considered the original high-resolution infrared image as the Ground Truth, but rather aimed to restore ideal and high-quality infrared images using multimodal sensor fusion. We selected Real-ESRGAN [27], CMSR [44], and Wang et al. [37] as comparative methods to consider the network's ability to enhance the details of infrared super-resolution reconstructed images with a reference input. The qualitative analysis is shown in Figure 8. From a visual perspective, our method not only obtained clear, high-contrast, and detail-rich infrared images but also avoided generating false textures. There were no visible artifacts or blurs compared to other contrast methods, which benefited from the neural network's feature extraction and information transmission capabilities. To further verify, we used a reference index to analyze the correlation between the reconstructed image and thermal information (i.e., the intensity of the infrared image), including Peak Signal-to-Noise Ratio (PSNR), Structural Similarity (SSIM), Learning-based Image Perceptual Similarity (LPIPS) [45], and Mutual Information (MI). To compare the image quality generated by different methods, we also added non-reference evaluation metrics to evaluate the enhanced-detail infrared images, including Entropy (EN), Average Gradient (AG), Edge Intensity (EI), and Spatial Frequency (SF). The quantitative comparison results are shown in Table 2, where the best and second-best values for each indicator are marked in red and blue, respectively.

In general, our proposed method outperformed other reference-based comparison methods, which indicates that our images have richer details, better contrast, and preserve more infrared thermal information. Although Real-ESRGAN is superior to our algorithm in reference-based metrics, this is due to the fact that our algorithm introduces more additional information to reasonably predict some high-frequency details that are not present in the original infrared image, which would result in a certain degree of decline in reference-based metrics. However, the actual image quality can be significantly improved. The result is consistent with the qualitative analysis results of generated image quality, fully demonstrating the effectiveness of our proposed method.

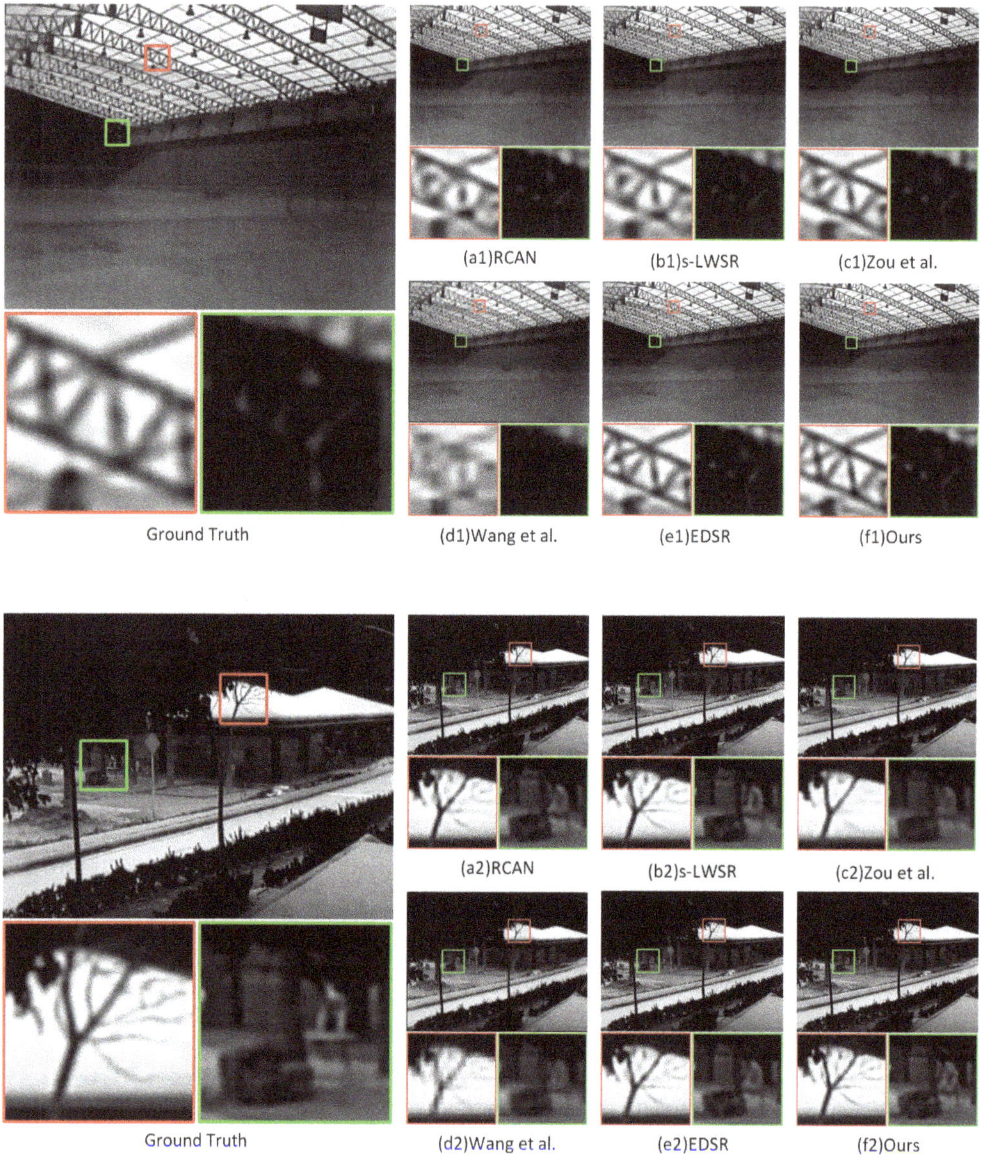

Figure 7. Comparison of SISR results of thermal infrared images under multimodal fusion using different methods. Zoom in for best view.

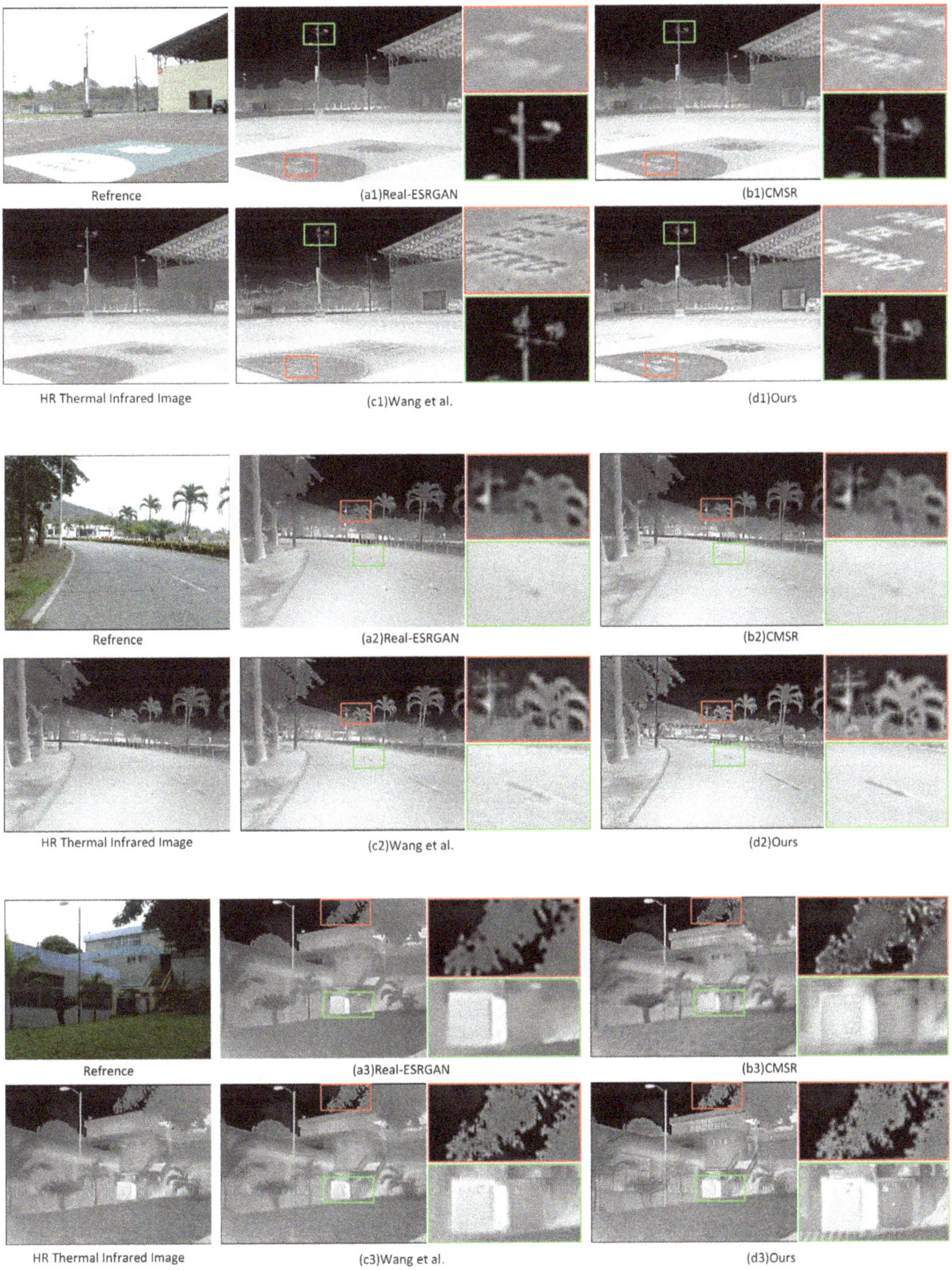

Figure 8. Comparison of multimodal SR results of thermal infrared images under multimodal fusion using different methods. Zoom in for best view.

Table 2. Benchmark test results for multimodal SR.

Methods	Ref.	PSNR	SSIM	LPIPS	MI	EN	AG	EI	SF
Origin TIR	-	-	-	-	-	7.1152	2.3342	30.6074	14.1549
Real-ESRGAN	×	30.23	0.7130	0.1728	3.0812	7.1295	1.6568	20.8512	6.3397
CMSR	✓	29.52	0.6623	0.2213	1.7308	6.8232	3.4218	32.1385	12.2585
Wang et al.	✓	29.38	0.6570	0.2513	1.7603	6.9754	3.0916	37.8797	13.0632
Ours	✓	30.04	0.7041	0.1869	2.8672	7.6473	4.2393	49.9074	18.9905

4.3. Ablation Study

Our approach has been proven superior through the comparative experiments we conducted. Subsequently, to determine the effectiveness of our proposed improvements, we conducted a series of ablation studies.

4.3.1. Ablation Studies of Network Structure

Firstly, we investigated the impact of three mechanisms, Multi-Scale (MS), Information Distillation (ID), and CEAM, in the primary unit HDDM of the FEM on the network reconstruction performance. We observed their performance changes in the SISR task to test their ability to extract features and reconstruct images from infrared images. Table 3 shows the quantitative results. It can be seen that the main improvements, including multi-scale branch, feature distillation, and channel attention, significantly improved the network performance, with all metrics showing improvement. We display the feature maps of different scales in our multi-scale module in Figure 9. It can be seen that this structure can adaptively divide the features into different-frequency components and extract them. Our structure has achieved a good balance between performance and efficiency and can efficiently and effectively extract information from input images.

Figure 9. Visualization of feature maps at different scales in the HDDM.

Table 3. Results of ablation study on the composition of HDDM structure.

MS	ID	CEAM	Param	PSNR	SSIM
✓	×	×	14.5 K	31.84	0.6997
✓	✓	×	14.5 K	32.02	0.7002
✓	✓	✓	15.4 K	32.15	0.7041

We replaced the self-attention module with three different modes: point-wise addition, channel attention, and spatial attention. We conducted experiments to evaluate their impact on performance. This primarily evaluated the performance of visible light image information when generating images using different information fusion mechanisms. The quantitative results are presented in Table 4. The attention mechanism outperforms the point-wise addition calculation mode, as evidenced by the results, which proves the importance of the learnability of information transmission. Compared to the other two attention mechanisms, our proposed CATM generated images with finer details, and had an overall better quality, which was supported by several performance metrics. This validates the effectiveness and rationale of the proposed CATM, which has the ability to extract more relevant information from the reference image.

Table 4. Results of ablation experiments on information transformation methods.

Transformation Type	PSNR	SSIM	EN	AG	EI	SF
point-wise add	23.2113	0.6377	4.6732	2.9369	21.1218	9.5865
Channel Attention	25.8466	0.6902	6.6181	3.6187	31.7487	12.3742
Spatial Attention	28.7214	0.6911	7.2657	4.1258	42.0281	15.8656
Ours	30.0418	0.7041	7.6473	4.2393	49.9074	18.9905

4.3.2. Ablation Study of Loss Function

We conducted ablation experiments to analyze the composition of the loss function and to evaluate the effect of different combinations of loss functions on the quality of reconstructed images. The focus of our research was on gradient loss and adversarial loss as they are the primary approaches for achieving image reconstruction and detail enhancement. We compared the reconstruction effects of the network under three conditions, including only pixel loss, with gradient loss, and complete hyibrid loss. Figure 10 and Table 5 display the specific subjective effects and indicators discerned. After adding the loss functions, the image quality was significantly improved subjectively, with improved details and enhanced contrast and sharpness. The reference metrics showed a significant decrease with the addition of gradient and adversarial loss while the non-reference metrics displayed a significant improvement.

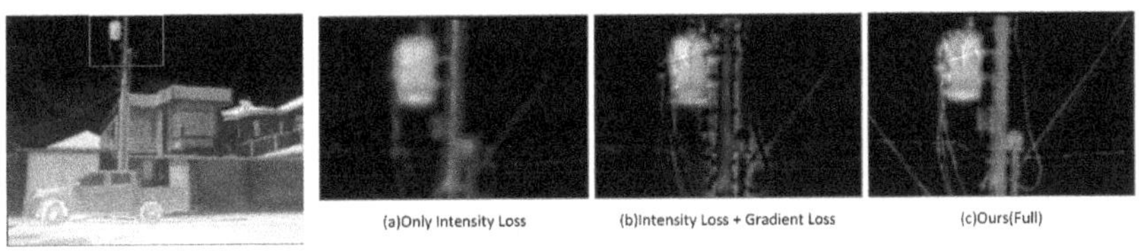

Figure 10. Comparison of reconstruction results using different loss functions. Zoom in for best view.

Table 5. Benchmark test results for multimodal SR of thermal infrared images.

$L_i(\theta)$	$L_s(\theta)$	$L_{adv}(\theta) + L_p(\theta)$	PSNR	SSIM	EN	AG	EI	SF
✓	×	×	33.6149	0.9012	7.0281	2.2627	30.6074	14.1833
✓	✓	×	28.8282	0.6702	7.6657	4.4251	52.0372	16.7221
✓	✓	✓	30.0418	0.7041	7.2657	4.1258	42.0281	15.8656

The purpose of using reference metrics is to measure the difference between the generated images and the GT. However, the thermal images as GT are limited by the imaging mechanism and affected by various factors, resulting in changes to the original signal, such as blurring or noise interference. Therefore, it is necessary to comprehensively judge the reconstruction ability of the network through non-reference metrics and qualitative analysis results. Obviously, the network trained with perfect hybrid loss has the best image reconstruction quality. In contrast, although the non-reference metrics have improved without introducing adversarial loss, a lot of infrared thermal information has been lost. The addition of adversarial loss can effectively solve this problem because the discriminator can prompt the generator to learn the implicit infrared image features. The lack of gradient loss makes it difficult to obtain enough texture details from the reference image, resulting in blurred reconstructed images. Therefore, our proposed hybrid loss can effectively restore the infrared thermal information in the image and obtain enough features from the reference image to enhance the texture details in the SR image.

4.3.3. Ablation Study of Training Strategy

Finally, we examined the effectiveness and necessity of the proposed training strategy through experiments. Figure 11 and Table 6, respectively, show the qualitative and quantitative analysis results of using and not using this training strategy in image reconstruction. Under SISR, if this learning strategy is not used, the quality of the reconstructed infrared images is poor, after masking the reference image and the corresponding network structure. This is mainly because the reconstruction process overly relies on the image information in visible light images. Although some of this information exists in infrared images, ineffective constraints during the training process still lead to serious network performance degradation, as demonstrated by the performance of Wang et al.'s method in the SISR comparative experiment. The network trained using this training strategy performs well in the SISR task and can reconstruct high-quality images without relying on visible light images. In the reference super-resolution task, both methods show quite similar reconstruction quality and achieve good performance, indicating that parallel training or inference of SISR and reference super-resolution tasks is feasible.

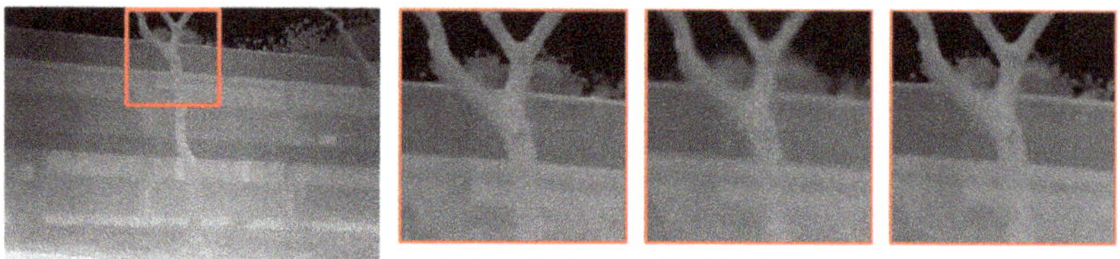

Figure 11. Comparison of reconstruction results of w/ and w/o training strategy.

Table 6. Results of ablation study on the training strategy.

Training Strategy	SISR		Multimodal SR			
	PSNR	SSIM	EN	AG	EI	SF
×	25.64	0.6130	7.1545	4.2258	40.8564	14.9313
✓	32.15	0.7041	7.2657	4.1258	42.0281	15.8656

This experiment proves that the training strategy proposed in this paper can effectively optimize the network, enabling it to maximize the use of effective information in the input infrared image. It is possible to rely solely on the infrared image for reconstruction in situations where visible light reference images are missing or of poor quality, which improves the robustness of our method and provides more options for practical applications.

5. Conclusions

In this paper, we proposed a thermal infrared image super-resolution reconstruction method based on multimodal sensor fusion, which included a multimodal super-resolution reconstruction network, a novel hybrid loss function, and a corresponding training strategy. Our multimodal super-resolution reconstruction network adopted an iterative super-resolution approach to gradually incorporate visible light features of different scales, which could better adapt to large-scale thermal infrared image super-resolution. We designed a hierarchical expansion distillation module to extract features from thermal infrared and visible light images, which was lightweight and high-performance, contributing to generating better reconstruction results. Additionally, we proposed a cross-modal information transformation module with pixel-level attention to achieve more efficient and accurate information fusion between the two modalities. To reasonably supplement lost texture details, a hybrid loss function is proposed, which could fuse and enhance salient details in different modalities while maintaining correct thermal information, improving the imaging quality of generated images. Moreover, we proposed a training strategy for multimodal sensor fusion super-resolution to reduce the network performance degradation caused by missing or low-quality reference images, improve the network's robustness and expand the scope of application in practical scenarios. Through extensive experimentation and comparison with various state-of-the-art methods, our method has demonstrated good performance in both visual quality and quantitative metrics, and improved the reconstruction quality of the images to some extent, validating the potential of our method.

Author Contributions: Conceptualization, Y.J., Y.L. and D.Z.; methodology, Y.J. and Y.L.; data curation, Y.J.; writing—original draft preparation, Y.J. and W.Z.; writing—review and editing, Y.L., W.Z. and D.Z.; supervision, Y.L. and D.Z.; project administration, Y.L.; funding acquisition, Y.L. All authors have read and agreed to the published version of the manuscript.

Funding: This study was supported by the Chongqing Natural Science Foundation (CSTB2022NSCQ-MSX1071).

Institutional Review Board Statement: Not applicable.

Data Availability Statement: Not applicable.

Conflicts of Interest: The authors declare no conflict of interest.

References

1. Zhu, D.; Zhan, W.; Jiang, Y.; Xu, X.; Guo, R. IPLF: A novel image pair learning fusion network for infrared and visible image. *IEEE Sens. J.* **2022**, *22*, 8808–8817. [CrossRef]
2. Yu, Q.; Zhu, M.; Zeng, Q.; Wang, H.; Chen, Q.; Fu, X.; Qing, Z. Weather Radar Super-Resolution Reconstruction Based on Residual Attention Back-Projection Network. *Remote Sens.* **2023**, *15*, 1999. [CrossRef]
3. Zhao, J.; Ma, Y.; Chen, F.; Shang, E.; Yao, W.; Zhang, S.; Yang, J. SA-GAN: A Second Order Attention Generator Adversarial Network with Region Aware Strategy for Real Satellite Images Super Resolution Reconstruction. *Remote Sens.* **2023**, *15*, 1391. [CrossRef]
4. Wang, J.; Shao, Z.; Huang, X.; Lu, T.; Zhang, R.; Li, Y. From artifact removal to super-resolution. *IEEE Trans. Geosci. Remote Sens.* **2022**, *60*, 1–15. [CrossRef]
5. Fang, H.; Ding, L.; Wang, L.; Chang, Y.; Yan, L.; Han, J. Infrared Small UAV Target Detection Based on Depthwise Separable Residual Dense Network and Multiscale Feature Fusion. *IEEE Trans. Instrum. Meas.* **2022**, *71*, 1–20. [CrossRef]
6. Liu, Q.; Yuan, D.; Fan, N.; Gao, P.; Li, X.; He, Z. Learning dual-level deep representation for thermal infrared tracking. *IEEE Trans. Multimed.* **2022**, *25*, 1269–1281. [CrossRef]
7. Yuan, D.; Shu, X.; Liu, Q.; He, Z. Structural target-aware model for thermal infrared tracking. *Neurocomputing* **2022**, *491*, 44–56. [CrossRef]
8. Rivera Velázquez, J.M.; Khoudour, L.; Saint Pierre, G.; Duthon, P.; Liandrat, S.; Bernardin, F.; Fiss, S.; Ivanov, I.; Peleg, R. Analysis of thermal imaging performance under extreme foggy conditions: Applications to autonomous driving. *J. Imaging* **2022**, *8*, 306. [CrossRef]
9. Munir, F.; Azam, S.; Rafique, M.A.; Sheri, A.M.; Jeon, M.; Pedrycz, W. Exploring thermal images for object detection in underexposure regions for autonomous driving. *Appl. Soft Comput.* **2022**, *121*, 108793. [CrossRef]
10. Keys, R. Cubic convolution interpolation for digital image processing. *IEEE Trans. Acoust. Speech Signal Process.* **1981**, *29*, 1153–1160. [CrossRef]
11. Sun, J.; Zhu, J.; Tappen, M.F. Context-constrained hallucination for image super-resolution. In Proceedings of the 2010 IEEE Computer Society Conference on Computer Vision and Pattern Recognition, San Francisco, CA, USA, 13–18 June 2010; pp. 231–238. [CrossRef]
12. Yang, J.; Wright, J.; Huang, T.; Ma, Y. Image super-resolution as sparse representation of raw image patches. In Proceedings of the 2008 IEEE Conference on Computer Vision and Pattern Recognition, Anchorage, AK, USA, 23–28 June 2008; pp. 1–8. [CrossRef]
13. Dong, C.; Loy, C.C.; He, K.; Tang, X. Learning a deep convolutional network for image super-resolution. In *Proceedings of the Computer Vision—ECCV 2014: 13th European Conference, Zurich, Switzerland, 6–12 September 2014*; Proceedings, Part IV 13; Springer: Berlin/Heidelberg, Germany, 2014; pp. 184–199.
14. Kim, J.; Lee, J.K.; Lee, K.M. Accurate image super-resolution using very deep convolutional networks. In Proceedings of the IEEE Conference on Computer Vision and Pattern Recognition, Las Vegas, NV, USA, 27–30 June 2016; pp. 1646–1654.
15. Ledig, C.; Theis, L.; Huszár, F.; Caballero, J.; Cunningham, A.; Acosta, A.; Aitken, A.; Tejani, A.; Totz, J.; Wang, Z.; et al. Photo-realistic single image super-resolution using a generative adversarial network. In Proceedings of the IEEE Conference on Computer Vision and Pattern Recognition, Honolulu, HI, USA, 21–26 July 2017; pp. 4681–4690.
16. Lim, B.; Son, S.; Kim, H.; Nah, S.; Mu Lee, K. Enhanced deep residual networks for single image super-resolution. In Proceedings of the IEEE Conference on Computer Vision and Pattern Recognition Workshops, Honolulu, HI, USA, 21–26 July 2017; pp. 136–144.
17. Li, J.; Fang, F.; Mei, K.; Zhang, G. Multi-scale residual network for image super-resolution. In Proceedings of the European Conference on Computer Vision (ECCV), Munich, Germany, 8–14 September 2018; pp. 517–532.
18. Anwar, S.; Barnes, N. Densely residual laplacian super-resolution. *IEEE Trans. Pattern Anal. Mach. Intell.* **2020**, *44*, 1192–1204. [CrossRef] [PubMed]
19. Li, B.; Wang, B.; Liu, J.; Qi, Z.; Shi, Y. s-lwsr: Super lightweight super-resolution network. *IEEE Trans. Image Process.* **2020**, *29*, 8368–8380. [CrossRef] [PubMed]
20. Zhang, Y.; Tian, Y.; Kong, Y.; Zhong, B.; Fu, Y. Residual dense network for image super-resolution. In Proceedings of the IEEE Conference on Computer Vision and Pattern Recognition, Chengdu, China, 20–22 April 2018; pp. 2472–2481.
21. Mehta, N.; Murala, S. MSAR-Net: Multi-scale attention based light-weight image super-resolution. *Pattern Recognit. Lett.* **2021**, *151*, 215–221. [CrossRef]

22. Zhang, Y.; Li, K.; Li, K.; Wang, L.; Zhong, B.; Fu, Y. Image super-resolution using very deep residual channel attention networks. In Proceedings of the European Conference on Computer Vision (ECCV), Munich, Germany, 8–14 September 2018; pp. 286–301.
23. Zhao, H.; Kong, X.; He, J.; Qiao, Y.; Dong, C. Efficient image super-resolution using pixel attention. In *Proceedings of the Computer Vision—ECCV 2020 Workshops, Glasgow, UK, 23–28 August 2020*; Proceedings, Part III 16; Springer: Berlin/Heidelberg, Germany, 2020; pp. 56–72.
24. Zhao, H.; Gallo, O.; Frosio, I.; Kautz, J. Loss functions for image restoration with neural networks. *IEEE Trans. Comput. Imaging* **2016**, *3*, 47–57. [CrossRef]
25. Liang, J.; Zeng, H.; Zhang, L. Details or artifacts: A locally discriminative learning approach to realistic image super-resolution. In Proceedings of the IEEE/CVF Conference on Computer Vision and Pattern Recognition, New Orleans, LA, USA, 19–20 June 2022; pp. 5657–5666.
26. Wang, X.; Yu, K.; Wu, S.; Gu, J.; Liu, Y.; Dong, C.; Qiao, Y.; Change Loy, C. Esrgan: Enhanced super-resolution generative adversarial networks. In Proceedings of the European Conference on Computer Vision (ECCV) workshops, Munich, Germany, 8–14 September 2018.
27. Wang, X.; Xie, L.; Dong, C.; Shan, Y. Real-esrgan: Training real-world blind super-resolution with pure synthetic data. In Proceedings of the IEEE/CVF International Conference on Computer Vision, Montreal, BC, Canada, 11–17 October 2021; pp. 1905–1914.
28. Zhou, Y.; Wu, G.; Fu, Y.; Li, K.; Liu, Y. Cross-mpi: Cross-scale stereo for image super-resolution using multiplane images. In Proceedings of the IEEE/CVF Conference on Computer Vision and Pattern Recognition, Nashville, TN, USA, 20–25 June 2021; pp. 14842–14851.
29. Zhang, J.; Zhang, W.; Jiang, B.; Tong, X.; Chai, K.; Yin, Y.; Chen, X. Reference-Based Super-Resolution Method for Remote Sensing Images with Feature Compression Module. *Remote Sens.* **2023**, 15, *1103*. [CrossRef]
30. Yang, Q.; Yang, R.; Davis, J.; Nistér, D. Spatial-depth super resolution for range images. In Proceedings of the 2007 IEEE Conference on Computer Vision and Pattern Recognition, Minneapolis, MN, USA, 17–22 June 2007; pp. 1–8.
31. He, K.; Sun, J.; Tang, X. Guided image filtering. *IEEE Trans. Pattern Anal. Mach. Intell.* **2012**, *35*, 1397–1409. [CrossRef]
32. Sun, K.; Tran, T.H.; Krawtschenko, R.; Simon, S. Multi-frame super-resolution reconstruction based on mixed Poisson–Gaussian noise. *Signal Process. Image Commun.* **2020**, *82*, 115736. [CrossRef]
33. Liu, H.c.; Li, S.t.; Yin, H.t. Infrared surveillance image super resolution via group sparse representation. *Opt. Commun.* **2013**, *289*, 45–52. [CrossRef]
34. Dong, X.; Yokoya, N.; Wang, L.; Uezato, T. Learning Mutual Modulation for Self-supervised Cross-Modal Super-Resolution. In *Proceedings of the Computer Vision–ECCV 2022: 17th European Conference, Tel Aviv, Israel, 23–27 October 2022*; Proceedings, Part XIX; Springer: Berlin/Heidelberg, Germany, 2022; pp. 1–18.
35. Zhang, W.; Sui, X.; Gu, G.; Chen, Q.; Cao, H. Infrared thermal imaging super-resolution via multiscale spatio-temporal feature fusion network. *IEEE Sens. J.* **2021**, *21*, 19176–19185. [CrossRef]
36. Du, J.; Zhou, H.; Qian, K.; Tan, W.; Zhang, Z.; Gu, L.; Yu, Y. RGB-IR cross input and sub-pixel upsampling network for infrared image super-resolution. *Sensors* **2020**, *20*, 281. [CrossRef]
37. Wang, B.; Zou, Y.; Zhang, L.; Li, Y.; Chen, Q.; Zuo, C. Multimodal super-resolution reconstruction of infrared and visible images via deep learning. *Opt. Lasers Eng.* **2022**, *156*, 107078. [CrossRef]
38. Wang, Q.; Wu, B.; Zhu, P.; Li, P.; Zuo, W.; Hu, Q. ECA-Net: Efficient channel attention for deep convolutional neural networks. In Proceedings of the IEEE/CVF Conference on Computer Vision and Pattern Recognition, Seattle, WA, USA, 13–19 June 2020; pp. 11534–11542.
39. Ge, L.; Dou, L. G-Loss: A loss function with gradient information for super-resolution. *Optik* **2023**, *280*, 170750. [CrossRef]
40. Isola, P.; Zhu, J.; Zhou, T.; Efros, A.A. Image-to-image translation with conditional adversarial networks. In Proceedings of the IEEE Conference on Computer Vision and Pattern Recognition, Honolulu, HI, USA, 21–26 July 2017; pp. 1125–1134.
41. Ha, Q.; Watanabe, K.; Karasawa, T.; Ushiku, Y.; Harada, T. MFNet: Towards real-time semantic segmentation for autonomous vehicles with multi-spectral scenes. In Proceedings of the 2017 IEEE/RSJ International Conference on Intelligent Robots and Systems (IROS), Vancouver, BC, USA, 24–28 September 2017; pp. 5108–5115.
42. Rivadeneira, R.E.; Sappa, A.D.; Vintimilla, B.X.; Kim, J.; Kim, D.; Li, Z.; Jian, Y.; Yan, B.; Cao, L.; Qi, F.; et al. Thermal image super-resolution challenge results-PBVS 2022. In Proceedings of the IEEE/CVF Conference on Computer Vision and Pattern Recognition, New Orleans, LA, USA, 18–24 June 2022; pp. 418–426.
43. Zou, Y.; Zhang, L.; Liu, C.; Wang, B.; Hu, Y.; Chen, Q. Super-resolution reconstruction of infrared images based on a convolutional neural network with skip connections. *Opt. Lasers Eng.* **2021**, *146*, 106717. [CrossRef]
44. Shacht, G.; Danon, D.; Fogel, S.; Cohen-Or, D. Single pair cross-modality super resolution. In Proceedings of the IEEE/CVF Conference on Computer Vision and Pattern Recognition, Nashville, TN, USA, 20–25 June 2021; pp. 6378–6387.
45. Zhang, R.; Isola, P.; Efros, A.A.; Shechtman, E.; Wang, O. The unreasonable effectiveness of deep features as a perceptual metric. In Proceedings of the IEEE Conference on Computer Vision and Pattern Recognition, Salt Lake City, UT, USA, 18–23 June 2018; pp. 586–595.

entropy

MDPI

Article

Structured Cluster Detection from Local Feature Learning for Text Region Extraction

Huei-Yung Lin [1,*] and Chin-Yu Hsu [2]

1 Department of Computer Science and Information Engineering, National Taipei University of Technology, Taipei 106, Taiwan
2 Department of Electrical Engineering, National Chung Cheng University, Chiayi 621, Taiwan
* Correspondence: lin@ntut.edu.tw

Abstract: The detection of regions of interest is commonly considered as an early stage of information extraction from images. It is used to provide the contents meaningful to human perception for machine vision applications. In this work, a new technique for structured region detection based on the distillation of local image features with clustering analysis is proposed. Different from the existing methods, our approach takes the application-specific reference images for feature learning and extraction. It is able to identify text clusters under the sparsity of feature points derived from the characters. For the localization of structured regions, the cluster with high feature density is calculated and serves as a candidate for region expansion. An iterative adjustment is then performed to enlarge the ROI for complete text coverage. The experiments carried out for text region detection of invoice and banknote demonstrate the effectiveness of the proposed technique.

Keywords: machine vision; structure pattern analysis; text region detection

Citation: Lin, H.-Y.; Hsu, C.-Y. Structured Cluster Detection from Local Feature Learning for Text Region Extraction. *Entropy* **2023**, *25*, 658. https://doi.org/10.3390/e25040658

Academic Editors: Oleg Sergiyenko, Wendy Flores-Fuentes, Julio Cesar Rodriguez-Quinonez and Jesús Elías Miranda-Vega

Received: 15 March 2023
Revised: 11 April 2023
Accepted: 13 April 2023
Published: 14 April 2023

1. Introduction

Due to the prevalence and availability of imaging devices, the use of computer vision techniques is becoming popular in our daily lives. Many objects of image recognition, such as human face, fingerprint and traffic sign, have been extensively investigated as research topics over the past few decades. The detection of regions of interest (ROIs) in images is thus a very important preprocessing stage [1,2]. It is commonly adopted to identify the image region that is meaningful to human perception. Since further analysis can then be carried out for scene-understanding tasks, ROI detection is usually considered as an early stage for extraction of information from acquired images [3]. In general, the overall performance of a machine perception system is highly reliant on the correctness of the ROI detection results.

From the perspective of visual perception, ROI is a fairly general term, and the definition is rather diverse depending on the application scenario [4]. It could represent a variety of pattern classes ranging from natural beings to man-made structures. As an example, the characteristics of images features utilized for the detection of human face and traffic symbols are very different [5,6]. Thus, the methodologies for the extraction of ROIs usually take the identification of some specific pattern structures into consideration. The features for pattern analysis might be manually extracted, using low-level image properties, or derived from learning-based techniques. However, depending on the amount of available training data, encoding high-level features via learning is generally not a simple task. In addition, more computational resources will be required for model training and testing.

To identify the regions of interest based on low-level image properties, local feature analysis is commonly used to obtain the correspondences between the reference and target images [7]. The histogram of oriented gradients (HOG) is a common feature to compute the gradient distributions of the objects. The features can then be adopted by a linear support vector machine (SVM) classifier for pedestrian detection [8]. To deal with the different scales of the targets, Marques et al. adopted size-invariant local features for marine vessel detection [9]. Based on saliency analysis, Achanta et al. presented a frequency-tuned approach to compute salient regions in images using low-level features of color and luminance [10]. It was capable of deriving full-resolution saliency maps with meaningful boundaries. However, the proposed method was mainly utilized to analyze natural scene images. Some good detection results can be obtained using the above approaches, and they have implicitly assumed that the target region is continuous and smooth in the image. The algorithms have not been directly adopted for general cases, where regions contain isolated internal structures.

In this work, we are interested in the detection of structured regions with isolated internal patterns. More specifically, this could be a text region with arbitrary orientation in different scenarios. Due to the wide availability of text descriptions in our living environment, text detection is usually the first step toward scene understanding. In early works, Epshtein et al. [11] converted edge gradients to the width of handwriting texts and used the distribution to localize the text region. A visual attention model was adopted to investigated the feasibility to video applications for salient object detection [12]. The strong signals associated with texts can then be accordingly extracted. To detect the texts in natural scenes, Yin et al. proposed a technique to extract maximally stable extremal regions as the character candidates for grouping [13]. The text classification is performed based on the posterior probability of the text candidates estimated by a character classifier. Most current developments consider the text region as an integrated part for detection, and the algorithms focus on extracting the regional features while minimizing the text localization error for identification.

This paper presents a structured region detection approach based on the distillation of local image features with clustering analysis. We are focused on the extraction of structured clusters from local feature learning. Thus, the objective is not for a general character recognition task. Figure 1 depicts the system flow of the proposed technique. The features in the target image corresponding to the similar structures in the reference images are first extracted, followed by region detection from analysis of the clustering characteristics of the ordered feature points. In our proposed method, the images with multiple characters in the database are used for feature matching. It is able to detect the text clusters under the sparsity of feature points derived from some characters. The location with high feature density is selected as a candidate, and an iterative process is carried out to increase the ROIs for the derivation of some suitable region with structured content. Since fast detection using limited computational resources is the key to the success of real-world applications, it is desirable to reduce the costs of model training and online inference. Different from existing deep neural network approaches, our technique can be easily implemented with hardware-oriented acceleration [14,15]. In the experiments, the text detection and recognition of invoice and banknote have demonstrated the effectiveness of the proposed technique.

The main contributions of this paper are as follows.

- A new approach based on correspondence extraction and clustering analysis of local features is proposed for structured region detection.
- A multi-stage algorithm with robust receptor descriptor is presented for character recognition.
- The proposed technique is capable of fast region detection with limited computational resources and can be easily implemented with hardware acceleration.

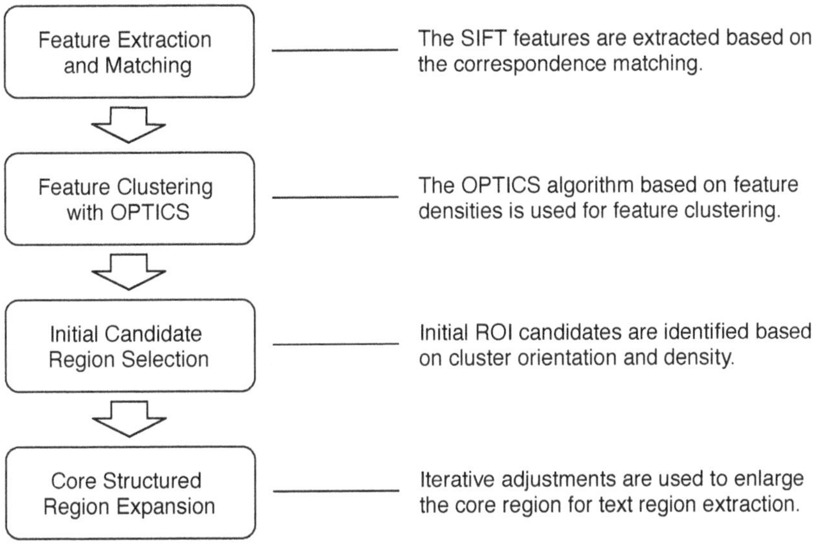

Figure 1. The four stages of our proposed structured region detection technique. It consists of feature extraction and matching, feature clustering, initial region selection and core structured region extraction.

2. Related Works

The investigation of text detection and recognition has been conducted over the decades. It is important due to the necessity of visual communication in the human-centered environment. The techniques are commonly developed based on the application scenarios and can be divided into text region extraction for documents (such as banknotes and invoices) and general scenes. For outdoor environments, the regions of interest usually cover a variety of scene texts, which includes signboards, license plates and digital traffic boards, etc. There also exist some other issues such as language and orientation. These might require the detection of more general structured patterns instead of performing template matching using prior knowledge. Recently, subspace clustering has been developed for various image analysis tasks, including sparse clustering applied to hyperspectral images [16]. It is also used to deal with multi-view data clustering [17] and multivariate time series data [18], and promising results have been reported.

The proper extraction of texts and numbers is the key to automatic document processing and analysis. Different from text detection in general scenes or handwritten documents [19], the pattern structures are usually more constrained in terms of size and format. For banknote recognition applications, Dittimi et al. presented a technique based on multi-class SVM [20]. The classification is carried out via the principal component analysis of HOG features. In [21], Pham et al. proposed a method based on discriminative region selection using the masks derived from a similarity map. The genetic algorithm was then applied out to optimize the banknote regions. More recently, a machine learning-based approach for simultaneous ROI extraction and character classification was presented [22]. Based on the use of knowledge distillation, the complexity can be reduced with a simple model for fast computation.

The technical process for identification of invoice information shares similarities with that of banknote recognition in pattern structure detection. However, the extraction of invoice numbers is usually more complicated due to the variation of background texture. To identify invoice information, Sun et al. proposed a template-based method for region detection [23]. Optical character recognition is then carried out to retrieve the text information. Tian et al. developed an iterative self-learning framework for intelligent financial ticket recognition [24]. A network model was constructed based on Faster R-CNN to recognize multiple ticket formats. In their work [25], Jiang et al. proposed a unified framework to process and recognize VAT invoices. The end-to-end model was trained to handle challenging cases with multi-oriented texts.

Among the text detection and recognition application scenarios, extracting the information for general outdoor scenes is the most challenging task. They contain a wide variety of text arrangements, sizes, styles, etc. In their early works, Coates et al. developed a text detection and recognition system based on a scalable learning algorithm [26]. A band of image features is learned from unlabeled data, followed by a linear classifier used for scene text extraction. Wang et al. [27] presented a texture-based approach for text detection, where a scale-insensitive adaptive region proposal network is first used to create text proposals, followed by a local orthogonal texture-aware method to represent the text. Zhang et al. [28] emphasized application in urban scenes and proposed a deep neural network approach for intelligent transportation systems. A keyword search tool was combined with a GIS system for street scene textual indexing. In [29], Yao et al. presented a unified framework for detecting multi-oriented scene text. A dictionary search-based method was proposed to correct character recognition errors. To deal with multi-lingual scene texts, the ICDAR reading challenge was conducted on the image datasets containing 10 languages [30]. The text structure feature extractor was used to simulate the Chinese text human cognition model [31]. In the recent adversarial learning method, Zhan et al. proposed a geometry aware domain adaption network [32]. It is able to synthesize multiple adapted images with different viewpoints for scene text detection.

In the existing literature, there are not many works focused on the detection of general structured patterns. Compared to the semantic information used for specific applications, low-level features are better suited for structured region detection. Based on the idea of saliency detection, Li et al. proposed a method to measure the 'characterness' of a region [33]. It was constructed using a Bayesian framework to integrate the text region by exploiting the dependencies among the characters. Zhu et al. presented a low-level detector based on MSER and region proposal for text detection [34]. The heuristic features are then used to group the characters into text lines. To ensure that structured pattern detection can be adopted to different high-level image-understanding tasks, it should be able to provide local clustering with a globally consistent scale.

3. Feature Selection and OPTICS Clustering

In the proposed structured region detection pipeline, the first step is to extract the feature points in the target image. These points should possess properties similar to those of the reference images in the database. As in the examples illustrated in Figure 2, the objective is to find the candidate feature locations based on pre-established structures of interest. To perform correspondence matching, the commonly used SIFT descriptor is adopted in this work for feature extraction. In most applications, it is used for object detection or scene matching from different viewpoints. The correspondences between the reference and target image features are established to derive the homography transformation. For our use of structured pattern extraction, the text regions for detection in the images are relatively small. The number of feature points for correspondence matching is very limited. Thus, it is not feasible to use to the distribution of feature points for region extraction, because a large amount of data is generally needed to increase the features for pattern identification.

One important property of the structured region of interest is the spatial proximity of individual building elements. Thus, our idea is to aggregate the few matching correspondences of each element to form the rough region clusters for detection. To maintain the stability of region detection via the aggregation of feature points, it is expected that as many feature correspondences as possible are extracted for each element. However, the increasing number of images for feature matching implies a higher computational cost, which is usually not preferable for the development o real-time systems. Therefore, in addition to the mismatching rate and the storage for reference images, one also needs to consider the computation time when performing the feature matching task.

(a)

(b)

Figure 2. The results of feature correspondence matching between the target image and a database template. Our objective is to find the candidate feature locations based on the pre-established structures of interest. (**a**) An invoice correspondence matching result. (**b**) A banknote correspondence matching result.

To achieve robust feature matching between the target image and reference data, internal feature correspondence extraction among the training images is first conducted. This serves as a training stage for the application-specific feature selection. A correspondence matching is carried out for the SIFT features in all reference images, and the points with good pairing are considered as important features. To be more specific, suppose there are N images in the training dataset, and $\mathbf{p}_i$ is a feature point which belongs to the ROI of the ith image. Let

$$S_i = \{(\mathbf{p}_i, \mathbf{p}_j) \mid 1 \leq j \leq N, j \neq i\} \tag{1}$$

for $1 \leq i \leq N$, where $(\mathbf{p}_i, \mathbf{p}_j)$ denotes the correspondence between image i and j if it exists. Then the point $\mathbf{p}_i$ is defined as a *prominent feature* if the number of correspondence pairs is greater than a preset threshold, or $|S_i| \geq T$.

It should be noted that, depending on the feature extraction criteria, there could be zero to many prominent feature points for a reference image in the training data. The images without any prominent features can then be removed from the dataset. Only the prominent features in the reference images are used to match the features in the target images for region detection. To improve the matching efficiency, one-to-many feature correspondences between the target and reference images are also allowed. This strategy to increase the number of feature points is not applicable to most applications utilizing feature correspondence matching. The feasibility of one-to-many mapping is built upon the use of the certified features derived from the database images. Figure 3a,b show the feature extraction results using the conventional method and the proposed technique, respectively. It can be seen that our method is able to reject the undesired points while allowing the important features for region detection to remain intact.

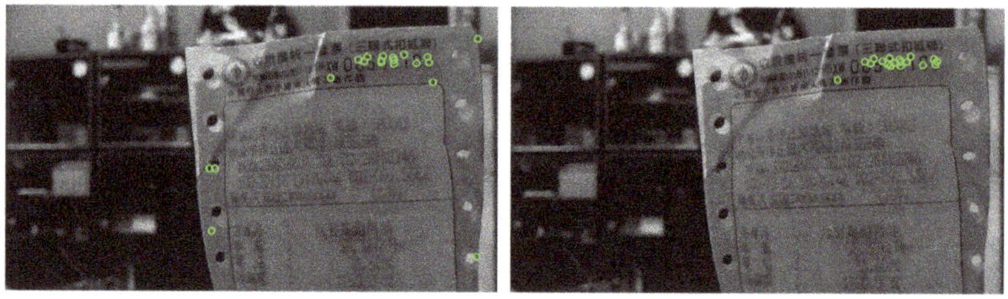

(**a**) Conventional method. (**b**) Our technique.

Figure 3. The feature extraction results using (**a**) the conventional method and (**b**) our technique. The algorithm can reject undesired points while keeping the important features for region detection. It can be seen that our method is able to reject the undesired points while allowing the important features for region detection to remain intact.

After feature extraction and matching, the next stage is to identify the feature points scattered within the structured image region. It is necessary to analyze the spatial relationship among these feature points, and generate dense clusters as candidate regions for ROI detection. However, due to the possible outliers in the set of image features, it is mandatory to perform effective clustering to localize the core feature points to form an initial detection region. The development of a robust clustering method is very important, since the results of candidate detection will be significantly affected by outlier features.

In general, it is required that good parameter settings be provided for clustering analysis algorithms [35]. However, it is not generally the case that there is an internal data structure that can be described clearly using a set of global parameters. The proper parameters are not only difficult to derive but also sensitive to the clustering results. In this work, we present a hierarchical clustering technique for feature aggregation based on the OPTICS (ordering points to identify the clustering structure) algorithm [36]. It does not directly perform the clustering but provides a feature sorting scheme to represent the data. Based on the ordering of reachability distances associated with the feature density, a reachability plot is generated and used for cluster identification. Through the analysis of cluster densities, it is possible to maintain stable feature structures using a wide range of parameters. Figure 4 illustrates a typical example of OPTICS clustering. Two sets of dense feature points located on the corners are shown in Figure 4a. The reachability plot as depicted in Figure 4b indicates the two flat regions in the intervals $(1, 30)$ and $(31, 70)$ are associated with two clusters in Figure 4a. In other words, the peak at around 31 indicates the density-based distance is large and can be adopted for the cluster derivation.

When clustering is performed using a large amount of data, the loss of clusters due to improper parameter settings is less likely. In case there are only a small number of features in the target images for correspondence searching, the final results will be more sensitive to the clustering approaches. Therefore, the OPTICS algorithm is further modified to accommodate extra constraints on the feature distribution to make it more robust under the image scale change. If the density of feature points is less than a preset threshold, the image will be normalized for further hierarchical clustering. The clustering results with two different scales are illustrated in Figure 5, where the core detection regions are indicated by the connected segments in red.

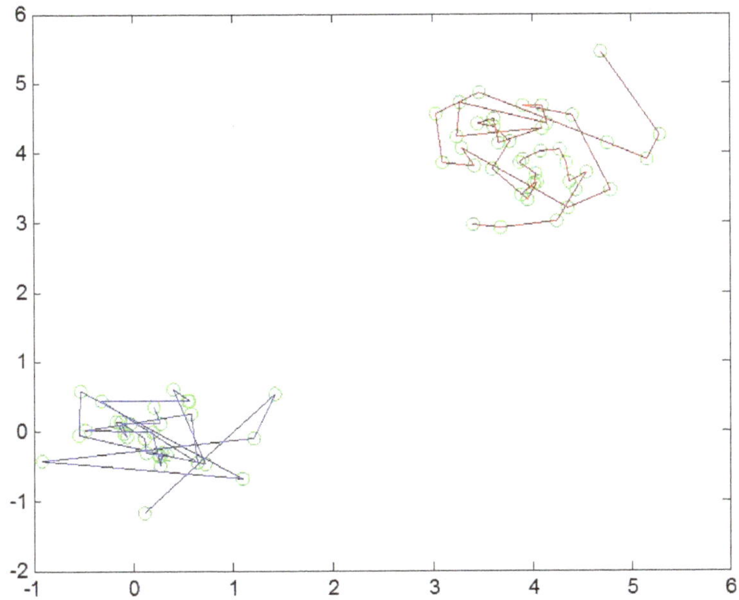

(a) The feature clustering results based on OPTICS.

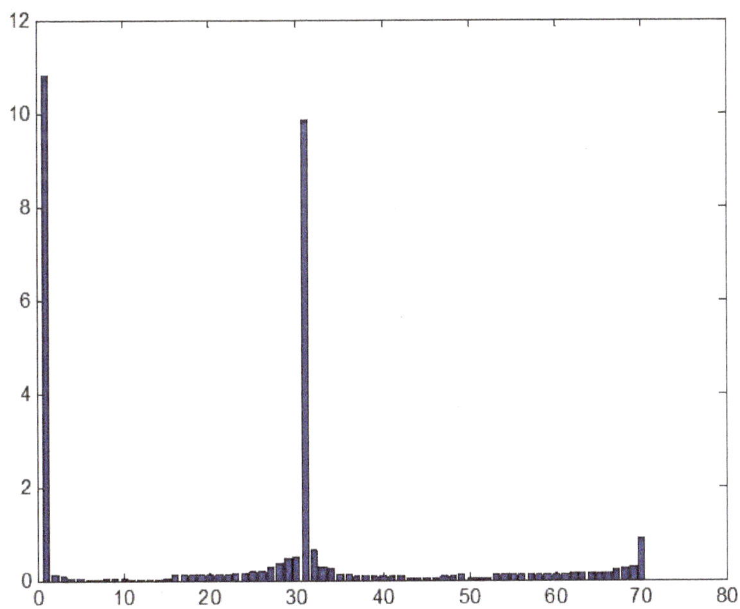

(b) The reachability plot analyzed by OPTICS.

Figure 4. An illustration of the hierarchical feature clustering technique based on the OPTICS algorithm. (**a**) Two sets of dense feature points are located on the corners. (**b**) The two flat regions in the intervals $(1, 30)$ and $(31, 70)$ associated with two clusters are indicated.

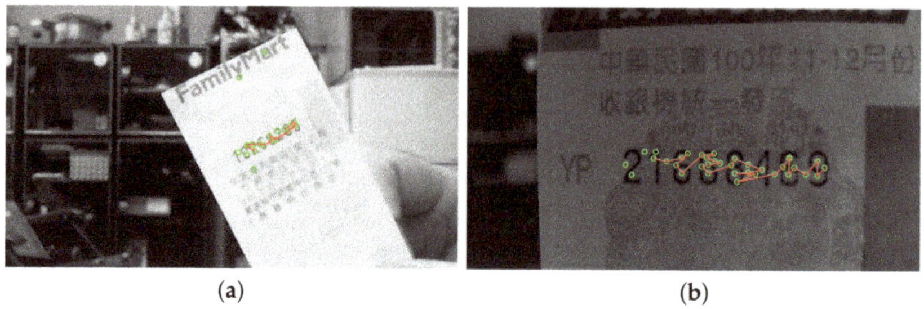

Figure 5. The OPTICS feature clustering results of the input images with two different scales. The core detection regions are indicated by the connected segments in red for (a) small-scale image input; (b) large-scale image input.

4. Structured Region Extraction

Given the results of feature extraction and clustering, a rough candidate region can be obtained. The next stage is to identify the precise ROI for a specific application, including translation, rotation and scaling, according to its pattern structure. To detect a structured region with certain characteristics, it is necessary to use the known templates for learning and pattern analysis. The bounding box with the best structural fitting is then derived by adjusting the candidate region from matching and validation iteratively. Thus, the application-specific ROI templates are extracted from the training images, followed by analyzing the structural characteristics for region identification and validation on the testing data.

In scene text detection, the ROI to be identified usually possesses a similar type of structure in the image. Therefore, it is possible to perform the region extraction using the strong relations among the elements. As an example, a serial number of an invoice or a banknote is composed of several digits and characters. The fixed structural properties include the number of elements in the ROI, the space between the elements, and the aspect ratio of ROI and individual elements. This provides the important information for the extraction of proper regions of interest. Figure 6 illustrates a typical example of an invoice containing 8 digits. In addition to the structural properties, the alternation of character and letter-spacing is also adopted for pattern identification. Since the width ratio between character and letter-spacing is both scale- and space-invariant, it can be used as a stable feature for matching.

In the implementation, the vertical projection of the region of interest is used to derive the histogram of character pixels as shown in Figure 6. Let the widths of character and letter-spacing be denoted by c_{2i-1} and s_{2j}, for $i = 1, 2, \cdots, N$ and $j = 1, 2, \cdots, N-1$, respectively. N is the total number of characters. To increase the robustness of pattern matching, a series of width ratios derived from the neighboring character and letter-spacing is used. That is, a feature vector coded by

$$\left(\frac{c_1}{s_2}, \frac{c_3}{s_4}, \cdots, \frac{c_{2N-3}}{s_{2N-2}}, \frac{c_{2N-1}}{s_{2N-2}}\right)^\top$$

is used for the template matching. Since the degree of letter-spacing is one less than character, the denominator of the last entry is repeated.

The feature selection and clustering analysis have allowed the identification of a cluster in a specific region. This is required to extract the candidate region as close to the true location as possible based on the clustering result. This will facilitate the adjustment of region estimation in the next stage. In serial number extraction, a series of elements are arranged along a straight line, and the feature point should be found in the fixed orientation. Consequently, applying a line-fitting algorithm carried to the feature cluster will allow identifying the features scattered along the text direction. An initial ROI derived from the rectangular region containing the image features can then be constructed. This will serve as the core region for enlargement along the text direction to include all characters in a later stage. An example of core region detection is illustrated in Figure 7a, where the bounding box might contain some outliers and the orientation is not correctly obtained.

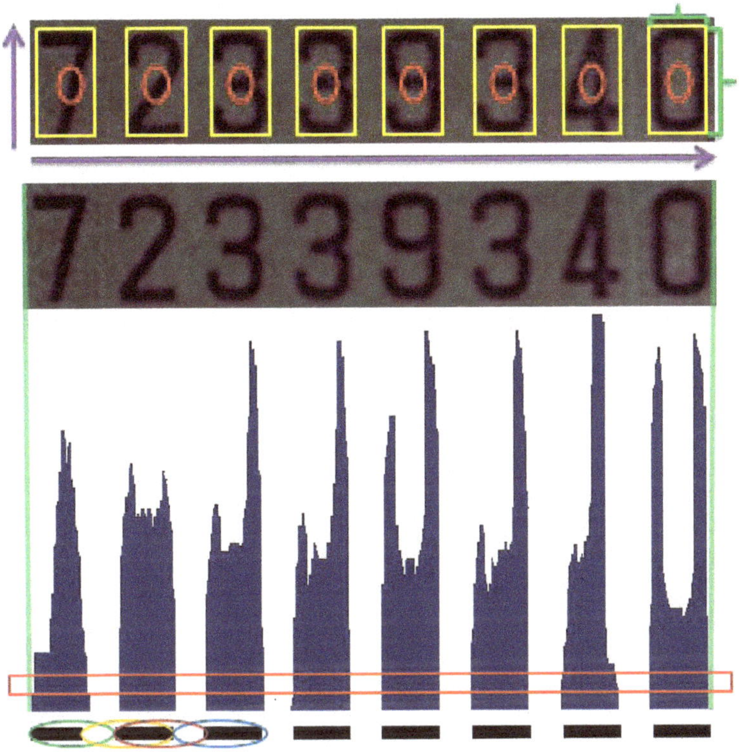

Figure 6. A typical case of an invoice containing eight digits. The structural properties, including the alternation of character and letter-spacing, are adopted for pattern identification. In addition to the structural properties, the alternation of character and letter-spacing is also adopted for pattern identification.

To make the core region extraction more robust, two constraints are adopted to remove the outliers of a cluster. First, the feature points further away from the initial ROI obtained based on the line model are eliminated using RANSAC. Figure 7b depicts the outlier removal results for Figure 7a. Nevertheless, the outlier features along the text direction will still be preserved under this condition, as in the example shown in Figure 8a. Another criterion for outlier rejection is based on the density correlation of ordered feature points obtained from the OPTICS algorithm. Due to the way in which the ordered feature strings are constructed, the outliers commonly appear at the two endpoints. Furthermore, the distances to the connecting feature points are significantly larger than the rest. Thus, a thresholding process is performed on the distance distribution of feature points to reject the outliers based on the variation. Figure 8b shows the filtering result of the outliers in the text direction of Figure 8a.

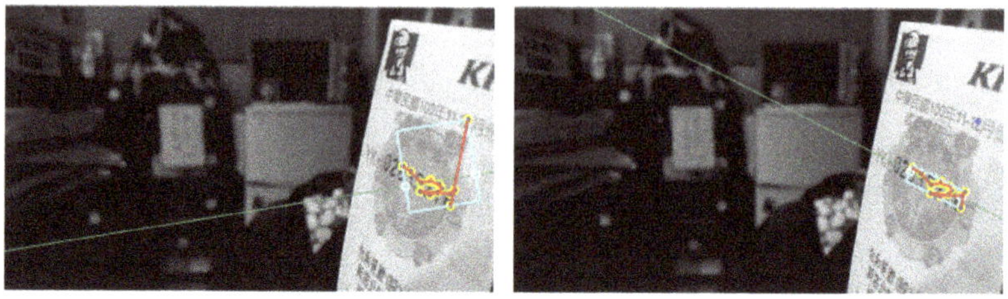

(a) Line fitting only. (b) With RANSAC.

Figure 7. The core region detection based on the clustering results with (**a**) only line fitting and (**b**) additional RANSAC for outlier removal.

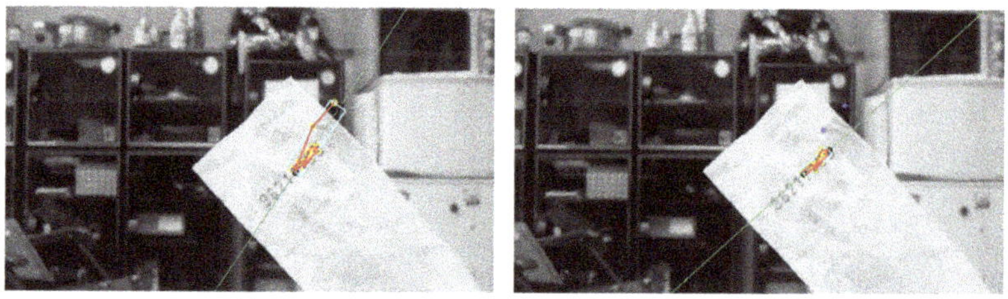

(a) (b)

Figure 8. Due to the way in which the ordered feature strings are constructed, the outliers commonly appear at the two endpoints. The outlier rejection for the feature points along the text direction based on the distance distribution obtained from OPTICS clustering: (**a**) OPTICS clustering; (**b**) with outlier rejection.

The next stage is to adjust the region candidate to correctly match the text region of interest. Our objective is to make the adjustment feasible whenever the candidate location is inside the text region, regardless of the orientation accuracy and the size difference. The procedure will be iteratively carried out, until the extracted region is satisfactory for text detection. First, the image is rotated with respect to the image scanline according to the orientation obtained from the feature point distribution of the region candidate. The purpose of this step is to initialize a rectangular bounding box to enclose the candidate region and use it as the core for expansion. Based on prior knowledge of the specific structures of interest, it is possible to enlarge or shift the bounding box if the projections of connected components in the horizontal or vertical direction are substantially different from the expected region specification. If the border of the current candidate region comes across some characters, this implies that the text is not completely covered by the ROI. Consequently, a region expansion process will be iteratively performed until the characteristics of the text region are satisfied.

To determine the shift direction, we consider the distribution of connected components and the regions formed by the horizontal and vertical projections of the ROI. The direction involves less blank areas and indicates a high possibility of more components to be identified. Since there could be erroneous results due to an inaccurate initial region assessment, timely adjustment is required to avoid the accumulation of improper expansions of the detected region. The rotation for the ROI is determined according to the accumulation error derived from the connected components and refer to an element in the region with an average size. Due to the presence of noise, the height difference between the reference and the remaining elements is used for the assessment of the accumulation error. The ROI will be rotated using the angle derived based on the error if it exceeds the threshold. In addition, the selected reference element is also used for severe noise filtering, since the object sizes are generally similar in the text region. The algorithm for ROI extraction, including the iterative process and region rotation, is depicted in Algorithm 1.

Algorithm 1 ROI Extraction Algorithm

```
Require: The initial region candidate.
Ensure: The text region of interest.
ROI ← RotateByFirstOrientation(candidate ROI)
iterative ← True
WHILE iterative
    iterative ← False
    Binarization(ROI)
    (midAreaCC, quantity) ←
        ConnectedComponentAnalysis(ROI)
    IF heightAccError > heightRatio
        Rotate(ROI)
        Truncate(ROI)
        iterative ← True
        continue
    IF quantity < totalAmount
        IF margin is confused
            trend ← ComputePointDistributed()
            ExpandByTrend(ROI) or
            ShiftByTrend(ROI)
        ELSE
            ExpandByMargin(ROI) or
            ShiftByMargin(ROI)
        iterative ← True
END WHILE
return ROI
```

5. Character Recognition

In this work, we propose a multi-stage approach for character recognition using a neural network. Most of current algorithms require significant training time to iteratively optimize the performance. However, the training time generally grows exponentially in proportion to the amount of training data. One simple method to cope with this problem is to divide the training dataset to a number of smaller subsets. Consequently, the overall training time can be derived by the largest training subset. Based on a similar concept, a preprocessing stage is applied to group characters with the same properties. A fast training network is then developed for the recognition of diverse characters.

The training data are divided into groups in our recognition framework based on the character symmetry properties and the Euler number. Multiple neural networks are used for fast learning and inference. When performing the recognition, no preprocessing is carried out on the input characters. It is not required that a specific network be used according to the associated group. Each character is taken as an input to all networks for processing, and the best three recognition results are selected as the candidates for verification in the second stage.

In the proposed multi-stage character recognition scheme, a similarity evaluation is performed in the second stage using pixelwise comparison. The input characters are compared with the first-stage results from all groups. Let s_i denote the similarity metric defined by the pixelwise region intersection with the ith group. The final result is then determined by the score $\alpha_i s_i$, where α_i is a weight factor. If the input character possesses the same Euler number with the ith group, then set $\alpha_i > 1$ to provide a high similarity weight. Otherwise, we let $\alpha_i = 1$. The first-stage output that has the highest weighted score is taken as the final recognition result.

The character recognition system using conventional neural networks generally takes the pixel values of the images for processing. It is sensitive to character deformation and image noise due to the spatial relations among the characters in the input layer. In this paper, we present a method to extract more robust descriptors from the character using random *receptors*. This is designed to reduce the influence of image normalization for the neural network. The basic idea is to drop some random line segments generated with various orientations and lengths on the images and record the status of intersection between the character and different receptors. Figure 9 illustrates two examples of the descriptors derived with 10 receptors applied on the characters. The values 1 and 0 in the table indicate the presence of intersection with the character.

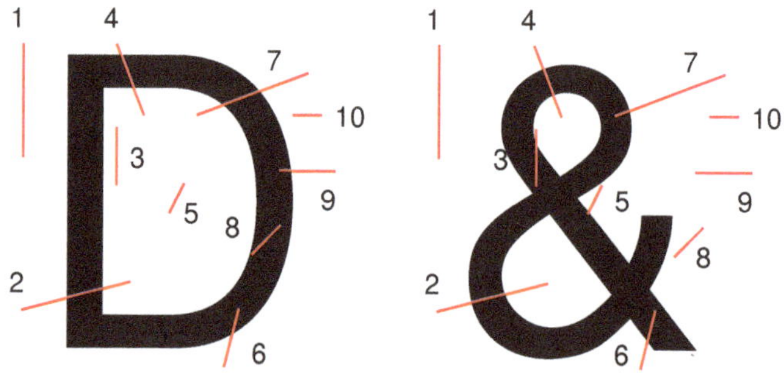

Receptor	1	2	3	4	5	6	7	8	9	10
Descriptor (D)	0	1	0	1	0	1	1	1	1	0
Descriptor (&)	0	1	1	1	1	1	1	0	0	0

Figure 9. Two examples of the descriptors derived with 10 receptors applied on the characters 'D' and '&'. The values 1 and 0 in the table indicate the presence of intersection with the character.

Although the receptors can be manually constructed using a set of parameters, they are usually randomly generated with only the number specified. Since it is not possible to guarantee the uniqueness of feature descriptors derived from characters, they are treated as as additional nodes in the input layer of the network for recognition. This network structure design is able to improve the stability of character recognition results under the influence of image deformation between the training and testing data. In our current implementation, the receptors are adopted for the first-stage recognition network. The same idea might be applied to the second-stage network to transform the character recognition problem to the similarity evaluation of binary codes. This approach can be further investigated, albeit its advantage to the recognition system is not clear.

6. Experiments and Evaluation

In the experiments, the proposed method is carried out on real-world images for structured region detection. Three application scenarios, including text detection for invoices and serial number identification for banknotes, are adopted for performance evaluation. Investigated in the tests are feature matching, feature clustering, region selection, and character recognition. Since all of these stages are highly correlated, we also tabulate the intermediate results for analysis. Compared to the region extraction algorithms based on deep neural networks, the implementation of our technique is simple and easy to use. It does not require a large image dataset for training or high computational power. To ensure the approach is suitable for practical applications, the testing samples are collected in a cluttered environment. The number of images and the contained regions of interest for the different applications are shown in the first and second rows of Table 1.

Since the number of regions for detection is not constrained, there might exist multiple regions of interest, as indicated in the table. The region detection results are then used for performance evaluation.

Table 1. The statistics of experimental results for invoice and banknote applications. The first and second rows indicate the total numbers of testing images and regions of interest, respectively. The numbers of correctly identified clusters and correct regions are shown in the third and fourth rows, respectively.

	Invoice	Banknote
Number of images	113	109
Number of ROIs	116	114
Number of detected clusters	107	106
Number of correct regions	100	102
Accuracy of detection	93%	76%

The regions of interest in the testing images are recorded with different scales, orientations, illuminations and backgrounds. In the first stage, the SIFT features are extracted based on the correspondence matching with the reference dataset images. The green circles marked in Figure 10a,b illustrate the feature extraction results of the invoice and banknote, respectively. It can be seen that the majority of feature points aggregate around the text regions and with only a small number of outliers. This greatly facilitates feature clustering in the following stage. Figure 11 shows the results of OPTICS clustering based on the feature densities, with each individual cluster represented using connected line segments. From the experiment, the clustering efficiency is demonstrated by the perfect match between the features and text region of interest. In the third stage, the initial ROI candidate is identified based on the orientation and density of the cluster. As illustrated in Figure 12, the detected bounding box serves as a core region for further expansion. The text region detection results in the last stage using iterative adjustments are shown in Figure 13. Since the proposed technique does not take the text boundary into consideration, the enclosing region is set as a rectangular bounding box. Consequently, the image captured with severe perspective distortion might result in a slightly larger region.

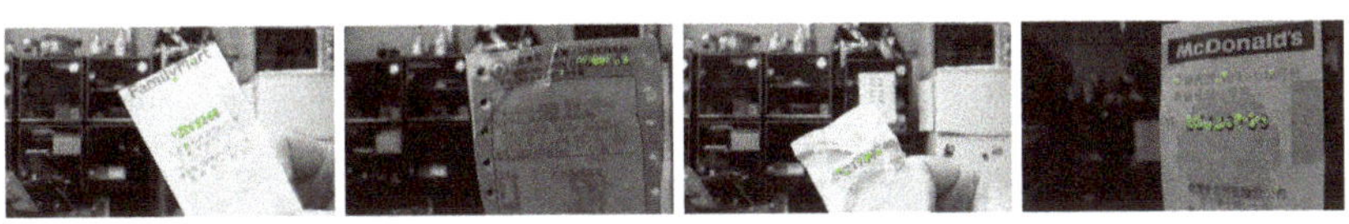

(**a**)

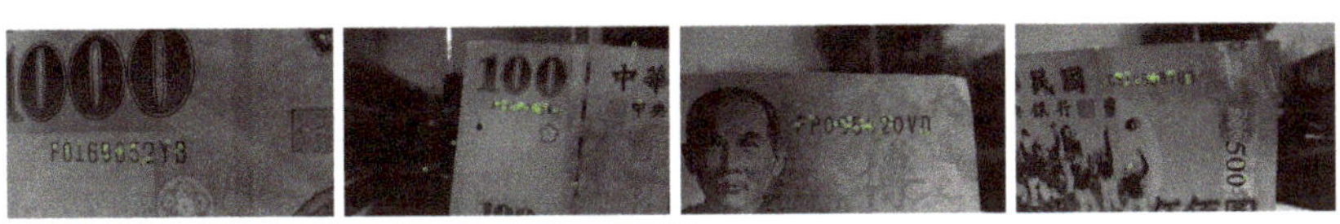

(**b**)

Figure 10. In the first-stage results, the SIFT features are extracted based on the correspondence matching with the reference dataset images. The green circles marked in the images indicate the feature point extraction for (**a**) an invoice and (**b**) a banknote.

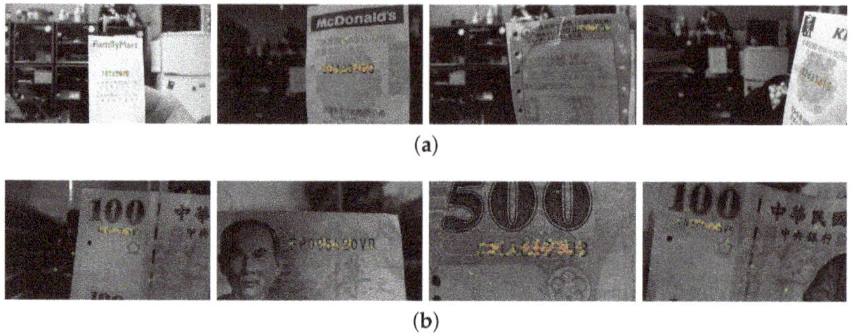

Figure 11. The second-stage results of the OPTICS clustering based on the feature densities, with each individual cluster represented using connected line segments for (**a**) an invoice and (**b**) a banknote.

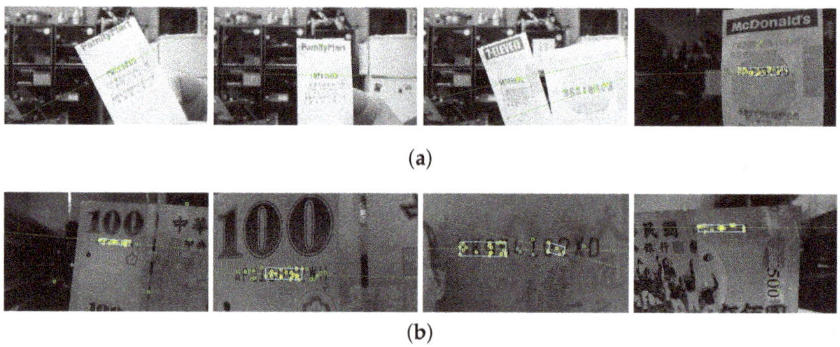

Figure 12. In the third stage, the initial ROI candidate is identified based on the orientation and density of the cluster. The detected bounding box serves as a core region for further expansion in (**a**) an invoice and (**b**) a banknote.

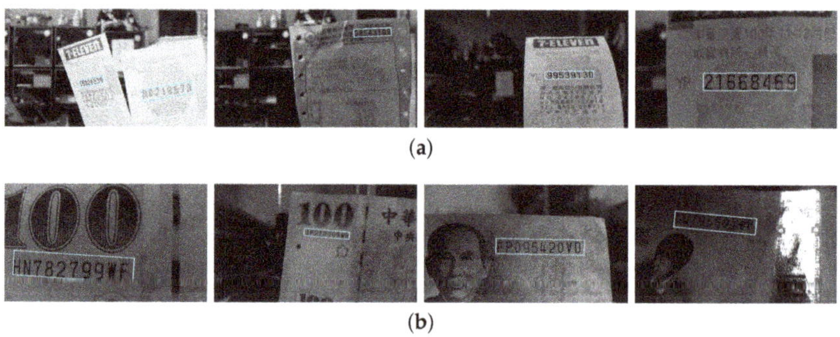

Figure 13. The text region detection results in the last stage, the iterative adjustments are carried out to enlarge the bounding box for the final ROI extraction in (**a**) an invoice and (**b**) a banknote.

Table 2 tabulates the important parameters used in the text region detection. The clustering results for invoice and banknote are tabulated in the third row of Table 1. This indicates that some good results are achieved for invoice and banknote. The evaluation of structured region detection is based on the derivation of regions of interest. Only the detected ROIs that fully cover the text content are considered as a correct result. Since our primary objective is the identification of text regions rather than intermediate feature clustering, we are more interested in the success rates of the ROI detection results. The last two rows of Table 1 tabulate the number of correctly detected regions and the detection accuracy. This shows over 90% of ROI detection rates in the indoor scenes (invoice and banknote). Finally, optical character recognition using the receptors is carried out on the text regions. The performance evaluation for different applications is depicted in Table 3. In general, it takes 2 ms for character recognition using the fast training network.

Table 2. The important parameters used for text region detection in our experiments.

Parameter	Description	Value
SIFT distance	The threshold for feature correspondence matching.	0.8
MinPts1	The minimum number of points to form a cluster in OPTICS. (invoice)	#pt/10
MinPts2	The minimum number of points to form a cluster in OPTICS. (banknote and plate)	#pt/6
ϵ	The control parameter for clustering analysis in OPTICS.	0.05
NoRefFeat	The number of reference feature points for corresponding matching.	300
NoHidNode	The number of hidden nodes in the neural network.	100
NoReceptor	The dimension of receptors for the neural network input.	300

Table 3. The statistics of character recognition results for the experiments on invoice and banknote.

	Invoice	Banknote
Number of characters	808	949
Correct recognition	673	722
Recognition rate	83 %	76 %

7. Conclusions

In this work, we present a new approach for structured region detection based on correspondence extraction and clustering analysis of local features. The proposed technique is designed for diverse application scenarios. It is capable of dealing with the cases where the target region in different orientations, with size changes, or under perspective distortion. The OPTICS algorithm with clustering density analysis is utilized to derive the characteristics of feature correspondences. Based on the initial ROI candidates identified with cluster orientations, the iterative adjustments are performed to enlarge for text region extraction. The experiments carried out on invoice and banknote have demonstrated the feasibility of the proposed method. Nevertheless, one major limitation of the proposed approach is the detection capability of man-made structures as illustrated in the implementation. In future work, the investigation will be conducted for natural scenes to reveal structural patterns for agriculture applications. The code is available at https://github.com/faketifosi/SCFlow.

Author Contributions: Conceptualization, H.-Y.L.; Methodology, H.-Y.L.; Software, C.-Y.H.; Validation, C.-Y.H.; Investigation, H.-Y.L. and C.-Y.H.; Resources, H.-Y.L.; Data curation, C.-Y.H.; Writing—original draft, H.-Y.L. and C.-Y.H.; Writing—review & editing, H.-Y.L.; Visualization, C.-Y.H.; Supervision, H.-Y.L.; Project administration, H.-Y.L.; Funding acquisition, H.-Y.L. All authors have read and agreed to the published version of the manuscript.

Funding: This research received no external funding.

Institutional Review Board Statement: Not applicable.

Data Availability Statement: Not applicable.

Conflicts of Interest: The authors declare no conflict of interest.

References

1. Real-Moreno, O.; Rodríguez-Quiñonez, J.C.; Sergiyenko, O.; Flores-Fuentes, W.; Mercorelli, P.; Ramírez-Hernández, L.R. Obtaining object information from stereo vision system for autonomous vehicles. In Proceedings of the 2021 IEEE 30th International Symposium on Industrial Electronics (ISIE), Kyoto, Japan, 20–23 June 2021; pp. 1–6.
2. Sergiyenko, O.; Flores-Fuentes, W.; Mercorelli, P. *Machine Vision and Navigation*; Springer: Berlin/Heidelberg, Germany, 2020.
3. Huang, N.C.; Lin, H.Y. A multi-stage processing technique for character recognition. In Proceedings of the Advanced Intelligent Mechatronics (AIM), 2012 IEEE/ASME International Conference, Kaohsiung, Taiwan, 11–14 July 2012; pp. 1081–1085. [CrossRef]
4. Alaniz-Plata, R.; Sergiyenko, O.; Flores-Fuentes, W.; Tyrsa, V.V.; Rodríguez-Quiñonez, J.C.; Sepúlveda-Valdez, C.A.; Andrade-Collazo, H.; Mercorelli, P.; Lindner, L. ROS and Stereovision Collaborative System. In *Optoelectronic Devices in Robotic Systems*; Springer: Berlin/Heidelberg, Germany, 2022; pp. 71–113.
5. Hong, F.; Lu, C.; Jiang, W.; Ju, W.; Wang, T. RDNet: Regression Dense and Attention for Object Detection in Traffic Symbols. *IEEE Sens. J.* **2021**, *21*, 25372–25378. [CrossRef]
6. Maheswari, V.U.; Varaprasad, G.; Raju, S.V. Local directional maximum edge patterns for facial expression recognition. *J. Ambient Intell. Humaniz. Comput.* **2021**, *12*, 4775–4783. [CrossRef]
7. Lin, H.Y.; Fan-Chiang, W.C. Reconstruction of shredded document based on image feature matching. *Expert Syst. Appl.* **2012**, *39*, 3324–3332. [CrossRef]
8. Bilal, M.; Hanif, M.S. Benchmark revision for HOG-SVM pedestrian detector through reinvigorated training and evaluation methodologies. *IEEE Trans. Intell. Transp. Syst.* **2019**, *21*, 1277–1287. [CrossRef]
9. Marques, T.P.; Albu, A.B.; O'Hara, P.; Serra, N.; Morrow, B.; McWhinnie, L.; Canessa, R. Size-invariant detection of marine vessels from visual time series. In Proceedings of the IEEE/CVF Winter Conference on Applications of Computer Vision, Montreal, QC, Canada, 11–17 October 2021; pp. 443–453.
10. Achanta, R.; Hemami, S.; Estrada, F.; Susstrunk, S. Frequency-tuned salient region detection. In Proceedings of the 2009 IEEE Conference on Computer Vision and Pattern Recognition, Miami, FL, USA, 20–25 June 2009; pp. 1597–1604. [CrossRef]
11. Epshtein, B.; Ofek, E.; Wexler, Y. Detecting text in natural scenes with stroke width transform. In Proceedings of the 2010 IEEE Conference on Computer Vision and Pattern Recognition, San Francisco, CA, USA, 13–18 June 2010; pp. 2963–2970. [CrossRef]
12. Fan, D.P.; Wang, W.; Cheng, M.M.; Shen, J. Shifting more attention to video salient object detection. In Proceedings of the IEEE/CVF Conference on Computer Vision and Pattern Recognition, Long Beach, CA, USA, 16–17 June 2019; pp. 8554–8564.
13. Yin, X.C.; Yin, X.; Huang, K.; Hao, H.W. Robust Text Detection in Natural Scene Images. *IEEE Trans. Pattern Anal. Mach. Intell.* **2014**, *36*, 970–983. [CrossRef] [PubMed]
14. Long, S.; He, X.; Yao, C. Scene text detection and recognition: The deep learning era. *Int. J. Comput. Vis.* **2021**, *129*, 161–184. [CrossRef]
15. Zhang, S.X.; Zhu, X.; Hou, J.B.; Liu, C.; Yang, C.; Wang, H.; Yin, X.C. Deep relational reasoning graph network for arbitrary shape text detection. In Proceedings of the IEEE/CVF Conference on Computer Vision and Pattern Recognition, Seattle, WA, USA, 13–19 June 2020; pp. 9699–9708.
16. Huang, S.; Zhang, H.; Pižurica, A. Subspace Clustering for Hyperspectral Images via Dictionary Learning With Adaptive Regularization. *IEEE Trans. Geosci. Remote Sens.* **2022**, *60*, 5524017. [CrossRef]
17. Qin, Y.; Feng, G.; Ren, Y.; Zhang, X. Consistency-Induced Multiview Subspace Clustering. *IEEE Trans. Cybern.* **2022**, *53*, 832–844. [CrossRef] [PubMed]
18. He, G.; Jiang, W.; Peng, R.; Yin, M.; Han, M.; IEEE. Soft Subspace Based Ensemble Clustering for Multivariate Time Series Data. *IEEE Trans. Neural Netw. Learn. Syst.* **2022**, 1–14. [CrossRef] [PubMed]
19. Ahamed, P.; Kundu, S.; Khan, T.; Bhateja, V.; Sarkar, R.; Mollah, A.F. Handwritten Arabic numerals recognition using convolutional neural network. *J. Ambient Intell. Humaniz. Comput.* **2020**, *11*, 5445–5457. [CrossRef]
20. Dittimi, T.V.; Hmood, A.K.; Suen, C.Y. Multi-class SVM based gradient feature for banknote recognition. In Proceedings of the 2017 IEEE International Conference on Industrial Technology (ICIT), Toronto, ON, Canada, 22–25 March 2017; pp. 1030–1035. [CrossRef]
21. Pham, T.D.; Kim, K.W.; Kang, J.S.; Park, K.R. Banknote recognition based on optimization of discriminative regions by genetic algorithm with one-dimensional visible-light line sensor. *Pattern Recognit.* **2017**, *72*, 27–43. [CrossRef]
22. Choi, E.; Chae, S.; Kim, J. Machine Learning-Based Fast Banknote Serial Number Recognition Using Knowledge Distillation and Bayesian Optimization. *Sensors* **2019**, *19*, 4218. [CrossRef] [PubMed]
23. Sun, Y.; Mao, X.; Hong, S.; Xu, W.; Gui, G. Template Matching-Based Method for Intelligent Invoice Information Identification. *IEEE Access* **2019**, *7*, 28392–28401. [CrossRef]
24. Zhang, H.; Zheng, Q.; Dong, B.; Feng, B. A financial ticket image intelligent recognition system based on deep learning. *Knowl.-Based Syst.* **2021**, *222*, 106955. [CrossRef]
25. Jiang, F.; Chen, H.; Zhang, L.J. FCN-biLSTM Based VAT Invoice Recognition and Processing. In Proceedings of the International Conference on Edge Computing, Seattle, WA, USA, 25–30 June 2018; pp. 135–143.
26. Coates, A.; Carpenter, B.; Case, C.; Satheesh, S.; Suresh, B.; Wang, T.; Wu, D.J.; Ng, A.Y. Text Detection and Character Recognition in Scene Images with Unsupervised Feature Learning. In Proceedings of the 2011 International Conference on Document Analysis and Recognition, Beijing, China, 18–21 September 2011; pp. 440–445. [CrossRef]

27. Wang, Y.; Xie, H.; Zha, Z.J.; Xing, M.; Fu, Z.; Zhang, Y. Contournet: Taking a further step toward accurate arbitrary-shaped scene text detection. In Proceedings of the IEEE/CVF Conference on Computer Vision and Pattern Recognition, Seattle, WA, USA, 13–19 June 2020; pp. 11753–11762.
28. Zhang, C.; Ding, W.; Peng, G.; Fu, F.; Wang, W. Street view text recognition with deep learning for urban scene understanding in intelligent transportation systems. *IEEE Trans. Intell. Transp. Syst.* **2020**, *22*, 4727–4743. [CrossRef]
29. Yao, C.; Bai, X.; Liu, W. A Unified Framework for Multioriented Text Detection and Recognition. *IEEE Trans. Image Process.* **2014**, *23*, 4737–4749. [CrossRef] [PubMed]
30. Nayef, N.; Patel, Y.; Busta, M.; Chowdhury, P.N.; Karatzas, D.; Khlif, W.; Matas, J.; Pal, U.; Burie, J.C.; Liu, C.L.; et al. ICDAR2019 Robust Reading Challenge on Multi-lingual Scene Text Detection and Recognition—RRC-MLT-2019. In Proceedings of the 2019 International Conference on Document Analysis and Recognition (ICDAR), Sydney, Australia, 20–25 September 2019; pp. 1582–1587. [CrossRef]
31. Ren, X.; Zhou, Y.; Huang, Z.; Sun, J.; Yang, X.; Chen, K. A Novel Text Structure Feature Extractor for Chinese Scene Text Detection and Recognition. *IEEE Access* **2017**, *5*, 3193–3204. [CrossRef]
32. Zhan, F.; Xue, C.; Lu, S. GA-DAN: Geometry-Aware Domain Adaptation Network for Scene Text Detection and Recognition. In Proceedings of the IEEE/CVF International Conference on Computer Vision (ICCV), Seoul, Republic of Korea, 27 October–2 November 2019.
33. Li, Y.; Jia, W.; Shen, C.; van den Hengel, A. Characterness: An Indicator of Text in the Wild. *IEEE Trans. Image Process.* **2014**, *23*, 1666–1677. [CrossRef] [PubMed]
34. Zhu, W.; Lou, J.; Chen, L.; Xia, Q.; Ren, M. Scene text detection via extremal region based double threshold convolutional network classification. *PLoS ONE* **2017**, *12*, e0182227. [CrossRef] [PubMed]
35. Jain, A.K.; Murty, M.N.; Flynn, P.J. Data clustering: A review. *ACM Comput. Surv.* **1999**, *31*, 264–323. [CrossRef]
36. Ankerst, M.; Breunig, M.M.; Kriegel, H.P.; Sander, J. OPTICS: Ordering points to identify the clustering structure. In Proceedings of the 1999 ACM SIGMOD International Conference on Management of Data, Philadelphia, PA, USA, 1–3 June 1999; pp. 49–60. [CrossRef]

entropy

MDPI

Article

Hybrid Multi-Dimensional Attention U-Net for Hyperspectral Snapshot Compressive Imaging Reconstruction

Siming Zheng [1,2,*], Mingyu Zhu [3] and Mingliang Chen [4]

1 Computer Network Information Center, Chinese Academy of Sciences, Beijing 100190, China
2 University of Chinese Academy of Sciences, Beijing 100049, China
3 School of Engineering, Westlake University, Hangzhou 310024, China; zhumingyu@westlake.edu.cn
4 Shanghai Institute of Optics and Fine Mechanics, Chinese Academy of Sciences, Shanghai 201800, China; cml2008@siom.ac.cn
* Correspondence: zhengsiming@cnic.cn

Abstract: In order to capture the spatial-spectral (x, y, λ) information of the scene, various techniques have been proposed. Different from the widely used scanning-based methods, spectral snapshot compressive imaging (SCI) utilizes the idea of compressive sensing to compressively capture the 3D spatial-spectral data-cube in a single-shot 2D measurement and thus it is efficient, enjoying the advantages of high-speed and low bandwidth. However, *the reconstruction process*, i.e., to retrieve the 3D cube from the 2D measurement, is an ill-posed problem and it is challenging to reconstruct high quality images. Previous works usually use 2D convolutions and preliminary attention to address this challenge. However, these networks and attention do not exactly extract spectral features. On the other hand, 3D convolutions can extract more features in a 3D cube, but increase computational cost significantly. To balance this trade-off, in this paper, we propose a hybrid multi-dimensional attention U-Net (HMDAU-Net) to reconstruct hyperspectral images from the 2D measurement in an end-to-end manner. HMDAU-Net integrates 3D and 2D convolutions in an encoder–decoder structure to fully utilize the abundant spectral information of hyperspectral images with a trade-off between performance and computational cost. Furthermore, *attention gates* are employed to highlight salient features and suppress the noise carried by the skip connections. Our proposed HMDAU-Net achieves superior performance over previous state-of-the-art reconstruction algorithms.

Keywords: hyperspectral; snapshot compressive imaging; CASSI; compressive sensing

Citation: Zheng, S.; Zhu, M.; Chen, M. Hybrid Multi-Dimensional Attention U-Net for Hyperspectral Snapshot Compressive Imaging Reconstruction. *Entropy* **2023**, *25*, 649. https://doi.org/10.3390/e25040649

Academic Editors: Oleg Sergiyenko, Wendy Flores-Fuentes, Julio Cesar Rodriguez-Quinonez and Jesús Elías Miranda-Vega

Received: 21 March 2023
Revised: 10 April 2023
Accepted: 10 April 2023
Published: 12 April 2023

1. Introduction

Hyperspectral images contain richer information than common RGB images and are thus widely used in various types of equipment like endoscopic system and remote sensing. To capture the rich spectral information, widely used spectrometers are mostly based on scanning to capture the three-dimensional (3D) spatial-spectral data-cube, i.e., to capture one 2D spatial frame at one wavelength in one shot and then move the next wavelength. The information captured in a 3D data-cube differs from conventional spatial coordinates [1,2], as it includes spectral information in the third dimension. Though high quality hyperspectral images can be obtained, scanning-based techniques are inefficient with respect to capturing dynamic scenes because of accuracy limitations imposed by moving objects or moving devices [3]. Thanks to compressive sensing (CS) [4,5], instead of sampling the spectral data-cube directly, the snapshot compressive-spectral imaging (SCI) [6] system samples the high dimensional data in an indirect manner. In particular, the first designed spectral SCI system, named coded aperture snapshot spectral imaging (CASSI) [7], uses a physical mask (coded aperture) and a disperser to modulate different channels (each channel corresponding to one wavelength) of the hyperspectral image and then captures the modulated data-cube in a snapshot 2D measurement by integrating across the wavelengths.

In this way, a 3D hyperspectral image can be compressed as a 2D measurement (please refer to the left part of Figure 1) and captured by an optical sensor in a short time, thus paving the way for high-speed hyperspectral image sampling [8]. With this high-speed imaging, the data storage and transmission efficiency will be extremely prompted and thus SCI has its promising prospect. After a 2D measurement is acquired, the reconstruction algorithms are employed to recover the 3D spectral data-cube (please refer to the right part of Figure 1).

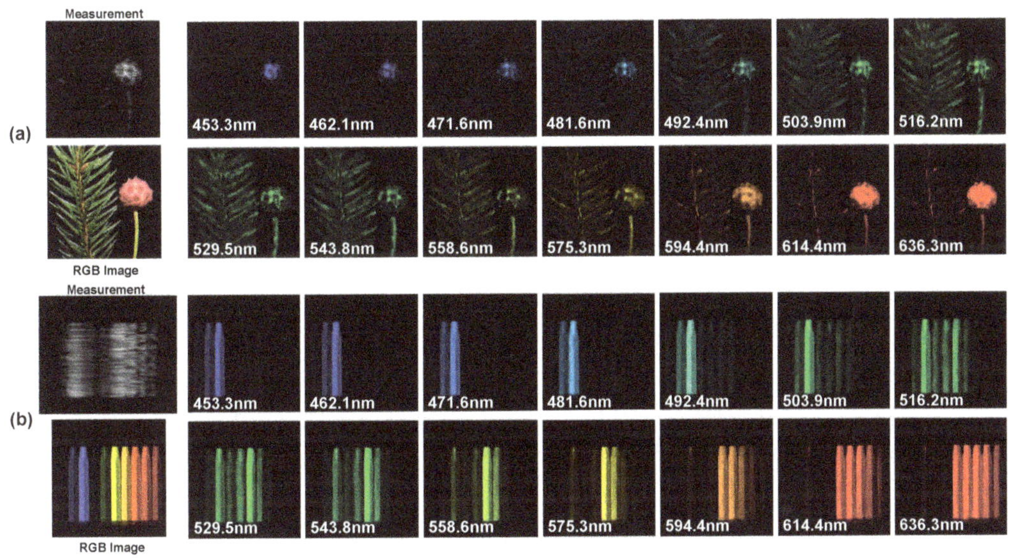

Figure 1. (**a**) The RGB references and reconstructed spectral images of a real measurement captured by [9] with 28 spectral bands (14 are shown here) using our HMDAU-Net. (**b**) One simulated data result (scene 9 in Table 1). The RGB images are shown as a reference.

Table 1. PSNR in dB (left entry in each cell) and SSIM [10] (right entry in each cell) by different algorithms on 10 scenes in simulation. Best results are shown in **bold**.

Algorithm	TwIST	GAP-TV	DeSCI	AE	U-Net	HSSP	λ-Net	TSA-Net	Ours
Scene1	24.81, 0.730	25.13, 0.724	27.15, 0.794	27.45, 0.813	28.28, 0.822	31.07, 0.852	30.82, 0.880	31.26, 0.887	**32.00, 0.898**
Scene2	19.99, 0.632	20.67, 0.630	22.26, 0.694	22.40, 0.709	24.06, 0.777	26.30, 0.798	26.30, 0.846	26.88, 0.855	**28.00, 0.889**
Scene3	21.14, 0.764	23.19, 0.757	26.56, 0.877	26.47, 0.861	26.02, 0.857	29.00, 0.875	29.42, 0.916	30.03, 0.921	**31.37, 0.939**
Scene4	30.30, 0.874	35.13, 0.870	39.00, 0.965	36.96, 0.950	36.33, 0.877	38.24, 0.926	37.37, 0.962	39.90, 0.964	**40.75, 0.971**
Scene5	21.68, 0.688	22.31, 0.674	24.80, 0.778	24.37, 0.797	25.51, 0.795	27.98, 0.827	27.84, 0.866	28.89, 0.878	**29.08, 0.893**
Scene6	22.16, 0.660	22.90, 0.635	23.55, 0.753	24.64, 0.776	27.97, 0.794	29.16, 0.823	30.69, 0.886	31.30, 0.895	**31.41, 0.919**
Scene7	17.71, 0.694	17.98, 0.670	20.03, 0.772	20.04, 0.786	21.15, 0.799	24.11, 0.851	24.20, 0.875	25.16, 0.887	**25.71, 0.901**
Scene8	22.39, 0.682	23.00, 0.624	20.29, 0.740	24.33, 0.783	26.83, 0.796	27.94, 0.831	28.86, 0.880	**29.69**, 0.887	29.49, **0.900**
Scene9	21.43, 0.729	23.36, 0.717	23.98, 0.818	25.10, 0.793	26.13, 0.804	29.14, 0.822	29.32, 0.902	30.03, 0.903	**31.38, 0.920**
Scene10	22.87, 0.595	23.70, 0.551	25.94, 0.666	24.55, 0.701	25.07, 0.710	26.44, 0.740	27.66, 0.843	**28.32**, 0.848	28.31, **0.859**
Average	22.44, 0.703	23.73, 0.683	25.86, 0.785	25.63, 0.797	26.80, 0.803	28.93, 0.834	29.25, 0.886	30.15, 0.893	**30.75, 0.909**

It has been over 14 years since the first CASSI was built; though different variants of the hardware have been developed [11–13], the reconstruction algorithm has been the long-term bottleneck that precludes the wide applications of spectral SCI. Conventionally, the iterative algorithms developed for CS have been used [14–17], but have been limited by the speed [18] or performance. Fortunately, recent advances in deep learning (DL) open a new window for the inverse problem in imaging [19]. Motivated by this, different DL-based algorithms have been proposed for spectral SCI [9,20–25]. However, most existing DL methods borrow the idea from other image restoration problems; for example, both λ-Net [20] and TSA-Net [9] are based on U-Net [26]. These networks usually use 2D convolutional neural networks (CNNs) that ignore the strong correlation among different spectral channels in the data-cube, though some preliminary attention modules are employed. On the other hand, the 3D CNN is able to extract high-dimensional features but suffers from low speed during training and testing.

Bearing these in mind, in this paper, we propose a hybrid multi-dimensional attention U-Net (HMDAU-Net) to reconstruct hyperspectral images from the 2D measurement in an end-to-end manner. HMDAU-Net integrates 3D and 2D convolutions in an encoder–decoder structure to fully utilize the abundant spectral information of hyperspectral images with a trade-off between performance and computational cost. Furthermore, *attention gates* [27] are employed to highlight salient features and suppress the noise carried through the skip connections.

Note that while reconstructing hyperspectral images, we not only need to focus on the spatial resolution but also need to take the spectral resolution into consideration. Though 2D convolutions can capture spatial features well, they lack the ability to effectively investigate the spectral correlation across the third dimension. Hence, we introduce 3D CNN for reconstruction. Due to the greater computational cost of 3D CNN which will increase the inference time, we integrate 3D and 2D CNN for the trade-off of reconstruction fidelity and speed. The utilization of attention gates helps the model to suppress irrelevant regions during training which makes the model pay more attention to the reconstruction details.

1.1. Review of the CASSI System

As mentioned above, the key idea of CASSI is to modulate different wavelengths in the spectral data-cube by different weights and then integrate the light to the sensor. The first version of CASSI used a fixed mask and two dispersers to modulate the spatial information over all wavelengths in the spectral cube, termed DD-CASSI [28]; here DD means dual disperser. Following this, the single-disperser (SD) CASSI [7] was developed, which achieves modulation by removing a disperser. Below, we mathematically model the SD-CASSI sensing process.

Let $\boldsymbol{X} \in \mathbb{R}^{W\times H\times B}$ denote the to-be-captured spectral data-cube at the top-left of Figure 2 and $\boldsymbol{M} \in \mathbb{R}^{W\times H}$ denote the fixed physical mask, where W, H and B denote the width, height and number of spectral channels, respectively. The spectral data-cube modulated by the coded aperture is $\mathbf{X}'(:,:,b) = \mathbf{X}(:,:,b) \odot \mathbf{M}$, where $\mathbf{X}'$ is the same size as $\mathbf{X}$, $b = 1,2,\ldots,B$ and $\odot$ represents the element-wise multiplication. After propagation through the disperser, each channel of $\mathbf{X}'$ is shifted along the $\boldsymbol{H}$-*axis* according to a liner dispersion $\boldsymbol{d}$ and the respective wavelength. We then use $\mathbf{X}'' \in \mathbb{R}^{W\times \tilde{H}\times B}$, where $\tilde{H} = H + d\times(B-1)$, to denote the shifted cube and assume λ_c to be the center wavelength which is not shifted when passing through the disperser. We can obtain $\mathbf{X}''(i,j,b) = \mathbf{X}'(i,j+d\times(\lambda_b-\lambda_c),b)$, where (i,j) represents the coordinate system on the plane of the sensor and λ_b is the wavelength at the b-th channel; $d\times(\lambda_b-\lambda_c)$ indicates the spatial shifting of the b-th channel. Thus, the 2D SCI measurement $\mathbf{Y} \in \mathbb{R}^{W\times \tilde{H}}$ we obtain on the detector is a sum over the wavelength dimension of a mask-modulated and later shifted data-cube. It can be modeled as

$$\mathbf{Y} = \sum_{b=1}^{B} \mathbf{X}''(:,:,b) + \mathbf{N}, \tag{1}$$

where $\mathbf{N} \in \mathbb{R}^{W\times H}$ denotes the measurement noise. To facilitate the description of the model, the coding process could be considered as modulating with a shifted mask $\tilde{\mathbf{M}} \in \mathbb{R}^{W\times \tilde{H}\times B}$

corresponding to different wavelengths and the liner dispersion d, i.e., $\tilde{M}(i,j,b) = M(w, h + d \times (\lambda_b - \lambda_c))$. Correspondingly, the shifted version $\mathbf{X} \in \mathbb{R}^{W \times \tilde{H} \times B}$ of the original data-cube is $\tilde{\mathbf{X}}(i,j,b) = \mathbf{X}(w, h + d \times (\lambda_b - \lambda_c), b)$. According to this, the 2D measurement $\mathbf{Y}$ can be modeled as

$$\mathbf{Y} = \sum_{b=1}^{B} \tilde{\mathbf{X}}(:,:,b) \odot \tilde{\mathbf{M}}(:,:,b) + \mathbf{N}. \tag{2}$$

By vectorizing the spectral data-cube and measurement, that is $x = \mathrm{vec}(\tilde{\mathbf{X}}) \in \mathbb{R}^{W\tilde{H}B}$ and $y = \mathrm{vec}(\mathbf{Y}) \in \mathbb{R}^{W\tilde{H}}$, this model can be rewritten as

$$y = \mathbf{A}x + n, \tag{3}$$

where $\mathbf{A} \in \mathbb{R}^{W\tilde{H} \times W\tilde{H}B}$ denotes the sensing matrix (coded aperture) which is a concatenation of diagonal matrices, that is $\mathbf{A} = [\mathbf{D}_1, \ldots, \mathbf{D}_B]$, where $\mathbf{D}_b = Diag(vec(\tilde{\mathbf{M}}(:,:,b)))$ is the diagonal matrix with $vec(\tilde{\mathbf{M}}(:,:,b))$ as the diagonal elements. Note that $\mathbf{A}$ is a very sparse matrix and the theoretical bounds have been developed in [29,30].

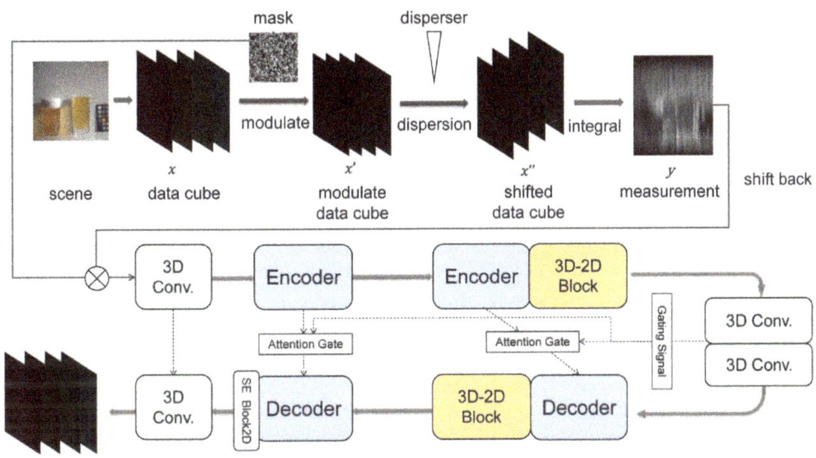

Figure 2. The proposed Hybrid Multi-dimensional Attention U-Net (HMDAU-Net) for CASSI reconstruction. The upper part is an SD-CASSI forward process and the measurement and mask are used as inputs of HMDAU-Net. The network structure shown in the lower part uses the backbone of a two layer U-net, composed of an encoder and a decoder including 3D CNN, 3D Res2Net [31] and 3D maxpooling/transpose 3D CNN. Attention gates [32] and SE (Squeeze-and-Excitation) blocks [33] are employed to extract important correlation information.

After obtaining the measurement y, we will focus on recovering 3D or multi-dimensional information from the 2D measurements, specifically using a novel deep learning network.

1.2. Contributions of This Work

In this paper, we propose a new end-to-end deep learning algorithm to reconstruct high quality images for the SD-CASSI system. Our contributions are summarized as follows:

- **Hybrid 3D/2D CNN network**: To balance the performance and computational cost, a hybrid 3D/2D block is employed to reduce parameters. Higher performance is achieved than existing 2D CNN-based algorithms; In addition, the proposed hybrid 3D/2D network shows superiority compared to the pure 3D and 2D counterparts.
- **Wider rather than deeper**: We evaluate that a two layer U-Net has similar performance to a four layer one in CASSI reconstructions.

- Effects of **attention gate** and SE (Squeeze-and-Excitation) block [33] in CASSI are evaluated. Attention gate is implemented to filter the noisy information from U-Net bottleneck and former layers. A simple 2D-CNN SE block is used to focus on important channels.

1.3. Related Work

After the first CASSI system [28] was designed, many revised CASSI were proposed. A single disperser CASSI (SD-CASSI) system was designed [7] the following year. Wang et al. [12] designed a Dual-camera CASSI system. Zhang et al. [34] proposed a novel snapshot spectral imaging system that can dynamically capture the spectral images with low computational burden and high light efficiency.

For CASSI reconstruction, the early algorithms are based on iterative optimization algorithms like TwIST [14], GAP-based [15,35] and other algorithms [16–18,36–39]. To promote these iterative algorithms, a deep neural network is inserted into an iteration step as a deep denoiser prior named deep plug-and-play algorithm [40]. Deep unfolding and deep unrolling methods [23,41–45] unfold an iterative algorithm and insert a deep learning network with better performance than common iterative algorithms and maintain their interpretation. The recent work [43] introduced a data-driven prior to exploit both the local and non-local correlations among the spectral image adaptively.

On the other hand, end-to-end deep learning-based algorithms enjoy its high reconstruction speed and excellent performance [46–49]. Researchers [22,50] proposed a CNN-based method to learn the deep prior externally (dataset) and internally (spatial-spectral constraint of inputs). Meng et al. proposed a TSA-Net [9] to exploit the self-attention mechanism to reconstruct the HSI images by capturing the information across dimensions. A generative adversarial network (GAN) [20] was also introduced in reconstruction.

Real CASSI systems always include noise and thus influence the reconstruction. Zhang et al. [51] modeled the real noise with non-zero mean that generalizes the traditional zero mean noise to characterize the optical imaging principle and boost the reconstruction quality of CASSI. The work [9] found that the shot noise is more suitable for real data training than Gaussian noise as well.

2. Proposed Network for CASSI Reconstruction

In this section, we first overview the hybrid multi-dimensional attention U-Net (HMDAU-Net). Following this, different modules of the proposed network are described in detail.

2.1. Overall Network Structure

As shown in Figure 2 (lower part), our network consists of a two layer U-Net [26] backbone, 3D–2D hybrid blocks, SE blocks and attention gates. The backbone is a two layer U-Net which is a trimmed version of TSA-Net backbone but without the attention module [9]. The encoder includes 3D CNN, 3D Res2Net and 3D maxpooling and the decoder includes 3D transpose CNN, 3D Res2Net and 3D CNN. The ReLU follows each CNN operation without batch normalization. We remove two layers from the original TSA backbone and change it into a 3D CNN with one initial 3D CNN and one end 3D CNN to match channels. A 2D SE block is employed to set the weight of the feature map and enhance the weight of important ones. Due to the large increase in parameters using cascade 3D CNN like DenseNet [52], we employ a hybrid 2D/3D CNN block named E-HCM [53] to solve our CASSI reconstruction problem. Furthermore, Attention gates [27,32] are implemented in our network to *reduce inessential information among each layer*.

2.2. Hybrid 2D/3D CNNs

Hyperspectral images contain abundant information across spectral channels; thus, the reconstruction needs to fully explore this information. Two-dimensional CNN extracts feature maps in each channel but lacks the content and relationship among spectral channels.

To address this challenge, 3D CNN for hypterspecral image reconstruction is employed in our network. It has been observed in previous work [54] that a 3D full CNN (3D-FCNN) exploring both spatial context and spectral correlation can achieve excellent results on other applications. Different from 2D convolution, a regular 3D convolution is implemented via 3D kernels and feature maps and thus is capable of investigating correlations across spectral channels. However, 3D CNN generates a large amount of parameters during computing. Some methods use split 3D convolution to reduce parameters (i.e., splitting the filter $k \times k \times k$ as $k \times 1 \times 1$ and $1 \times k \times k$) [55] to mitigate this shortcoming. However, redundant information along the spectral dimension will be generated due to the high spatial similarity among spectral channels. This also reduces the learning ability of the model in space, which is extremely important for the reconstruction purpose as considered in our work.

To address this challenge, MCNet [56] was proposed to share the context among 3D and 2D units. A split adjacent spatial and spectral convolution (SAEC) was proposed in [53] to tackle this difficulty. It implements 3D convolution along height–width, spectral-height and spectral-width (i.e., filters are $1 \times k \times k$, $k \times 1 \times k$ and $k \times k \times 1$). After reshaping, feature maps go through a few 2D convolution units. This *hybrid 3D/2D CNN module* is dubbed E-HCM. In detail, the 3D unit is employed to analyze the relationship of spectra and either horizontal or vertical direction in space. Since the spectral information is acquired, the feature maps after the 3D unit are reshaped into four dimensions to implement 2D convolution to *further extract the spatial information* in the desired image. Based on the consideration of efficiency and computational cost, we employ this module at the end of encoders and decoders in our HDMAU-Net.

2.3. Attention Gate

Attention Gates (AGs) [32] are initially proposed to *capture a sufficiently large receptive field or semantic contextual information* in medical images. The AGs are incorporated into the standard U-Net architecture to highlight salient features that are passed through the skip connections. Information extracted from the coarse scales is used in gating to disambiguate irrelevant and noisy responses in the skip connections.

As shown in Figure 2, the gating signal $g \in \mathbb{R}^{F_g \times N_g}$ is generated via a 3D CNN block, including batch normalization and ReLU. The input feature in the l-th layer is $x^l \in \mathbb{R}^{F_l \times N_l}$. N_g and N_l are the sizes of a feature map (i.e., *channel* $\times$ *width* $\times$ *height*), $N_g < N_l$, F_g and F_l correspond to the number of feature maps. g and x^l are inputs of the attention gate in each layer, which can be represented by:

$$\phi_g = \text{upsample}(\Omega_g(g)), \qquad \phi_x = \Omega_x(x^l), \tag{4}$$

$$q^l_{att} = \psi(\text{ReLU}(\phi_x + \phi_g)), \qquad \alpha^l_{att} = \text{sigmoid}(q^l_{att}), \tag{5}$$

where $\Omega(\cdot)$ and $\psi(\cdot)$ denote linear transformation (e.g., $\Omega(u) = W_u^T u + b_u$, $b_u \in \mathbb{R}^{F_{int} \times M}$, $W_u^T \in \mathbb{R}^{(F_l \times N_l) \times (F_{int} \times M)}$) conducted by $1 \times 1 \times 1$ 3D convolutions. ϕ_g, ϕ_x and $q^l_{att} \in \mathbb{R}^{F_{int} \times M}$, where F_{int} and M are intermediate numbers of a feature map and sizes of a feature map, respectively. Attention coefficient $\alpha^l_{att} \in [0, 1]$. When the attention is generated, we multiply it with x^l from skip connection and then input into decoder.

Motivated by the attention U-Net [27], the same-scale features from the encoder and decoder can be augmented and combined by attention gates. We firstly use attention gates to boost reconstruction of subtle texture in hyperspectral images and enhance the content of each layer during scale transformation in our HMDAU-Net. The output of AGs is then produced by the decoder with scaling conducted by Res2net and upsampling.

3. Experimental Results

We now verify the performance of our proposed HMDAU-Net for CASSI reconstruction, firstly on simulation data and then real data captured via the CASSI system [9]. More results are shown in Appendix A.

3.1. Simulated Data

We train our model for simulated data (256 $\times$ 256 measurement on the CAVE [57], 31 spectral images of 256 $\times$ 256 $\times$ 31) and test it on 10 scenes cropped from the KAIST [58] dataset provided by the TSA-Net [9], which adopts spectral interpolation on the simulation data to acquire an image of the 28 channels (ranging from 450 nm to 650 nm) as ground truth. Similar to TSA-Net, we randomly crop the hyperspectral data-cube into a spacial size of 256 $\times$ 256 with 28 channels and then use real mask and shift the data-cube via a 2 pixel step to generate a 256 $\times$ 310 measurement. After shifting it back to a 256 $\times$ 256 $\times$ 28 data-cube, we put it into our network. Three-dimensional CNN need five dimensions to input and thus we unsqueeze it into a $batchsize \times 1 \times 28 \times 256 \times 256$ data. The number of 3D feature maps after the first 3D CNN is 32 (the second dimension). After it leaves the last block, we squeeze the data into four dimensions.

3.1.1. Comparison with State-of-the-Art Methods

We compare our proposed reconstruction method with several state-of-the-art (SOTA) methods, including three optimization methods (TwIST [14], GAP-TV [15] and DeSCI [18]), a convolutional autoencoder-based method (AE [58]), a deep unfolding method (HSSP [23]), a GAN-based method (λ-Net [20]) and two end-to-end deep learning methods (U-Net [26] and TSA-Net [9]). AE does not perform as well as in the DD-CASSI system shown in Ref [58] because we use their pre-trained model which differs from our SD-CASSI data scenes, wavelenth distributions and spacial sizes. Other experimental results are from [9]. We use the same training dataset in TSA-Net and 10 scenes for test. We can see that the deep learning-based methods achieve better results and our proposed method is better than the past SOTA algorithm TSA-Net. Specifically, as shown in Table 1, our method outperforms the second best method TSA-Net by 0.6dB in average PSNR and 0.016 in average SSIM.

Figure 3 plots selected channels (4 out of 28) and spectral curves of the reconstructed images using the methods above. We can obverse that the images reconstructed via the proposed method have clearer texts and stripes. Please notice the letters on the cup and the sharp edges of the color checker. In addition, our method has more accurate spectral density than the other methods.

As depicted in Figure 3, the top-left panel showcases two designated boxes labeled "a" and "b", accompanied by corresponding reconstructed outcomes and numerical assessments. The assessment procedure involved computing the mean values of boxes "a" and "b" across all wavelengths (each red dot represents an average value of a specific wavelength), followed by correlation analysis of the spectra based on the reference parameter. Our spectral-wise quantitative metrics are shown in the figure and clearly higher than other methods.

3.1.2. Ablation Study

We design several ablation studies to evaluate the effect of different modules in the proposed network. The comparison includes numbers of layers of U-Net backbone, attention gates and hybrid dimensional convolution modules.

To save training time, the experiments in this subsection in simulated data are trained with 16 channels when input into encoder. As shown in Table 2 left, we can observe that a two layer 3D U-Net (using the backbone in TSA-Net and replacing all convolutions by 3D-CNN) has performance similar to a four layer one in CASSI reconstruction. It even achieves 0.22 dB higher PSNR. However, by doubling the feature maps initially input into the encoder, we can see a raise of 0.44 dB. This shows that the assistance of a deeper network is not so distinct and even not beneficial to our SCI reconstruction. Instead, the wider one has much more influence. We find that this may due to the fact that too many downsamplings and upsamplings in spatial and spectral dimensions will cause information loss.

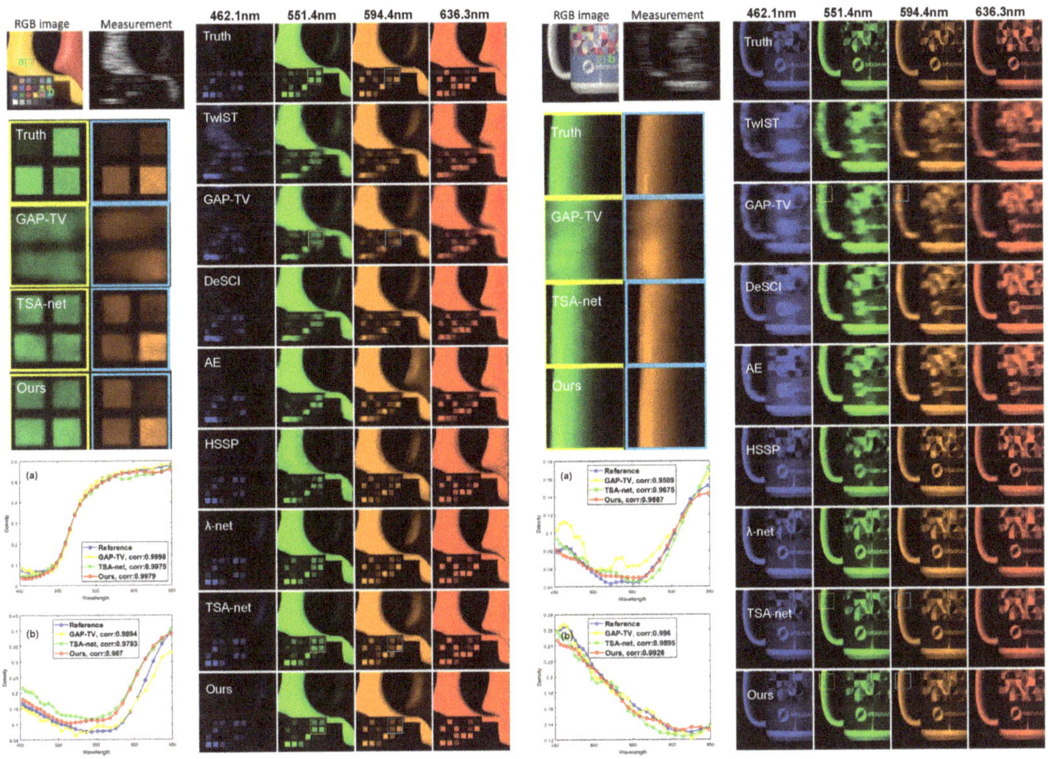

Figure 3. Two reconstructed scenes with four spectral channels using seven methods. We compare the recovered spectra of the selected region (shown with a, b on the RGB images) and spatial details. box "a" and box "b" have been chosen to perform correlation analysis.

Table 2. Left: The comparisons of using different numbers of layers in the 3D U-net backbone showing average PSNR in dB, SSIM on the 10 scenes. Right: The comparisons of using different modules in our proposed algorithm.

Method	PSNR	SSIM
4 layers-16	29.38	0.892
2 layers-16	29.51	0.886
2 layers-32	**29.95**	**0.898**
Our Backbone	29.77	0.888
+SE Block	30.02	0.893
+AGs	29.82	0.887
+AGs and SE Block	**30.14**	**0.899**

In Table 2 right, we evaluate different modules in our proposed method. Both SE block and attention gates improved our reconstruction results. In particular, SE Block can improve them more (0.25 dB in PSNR) while AGs just edge up a little bit (0.05 dB in PSNR). As we put them together, the promotion is expanded, leading to a 0.27 dB gain in PSNR. This presents the consistency of the two attentions in our reconstruction, without excessively filtering necessary spatial-spectral information.

In Table 3, we implemented different types of convolution in our U-net backbone. Our hybrid backbone uses E-HCM on the second encoder and the first decoder is a two layer U-Net backbone. E-HCM includes three 3D convolution operations and four 2D convolution operations. For the full 3D convolution, we replace the E-HCM by the same number of layer residual blocks (seven layers per module). For full 2D convolution, we replace all 3D convolution operations by 2D and keep the number of layers unchanged. For instance, taking a 2D convolution layer with kernel size K, input and output channels C_{in}, C_{out} as an example, the number of MACs is $K \times K \times C_{in} \times H_{out} \times W_{out} \times C_{out}$, where H_{out} and W_{out} denote the height and width of the output feature map, respectively. Compared to 2D, 3D improves the PSNR value but significantly increases the computational workload as well. By using our hybrid backbone, we can decrease parameters and computational load to a large extent (40%) in contrast to 3D and even have achieved higher performance than pure 3D and 2D ones. This observation suggests that the pure 3D CNN is not as practical as 2D ones because of the soaring of computational load. However, we can mix it with 2D CNN to make a balance.

Table 3. The comparisons of using 2D, 3D and hybrid convolution as U-net backbone in our proposed algorithm showing average PSNR, SSIM, model parameters and computational loads on the 10 scenes.

Method	PSNR	SSIM	Parameters	MACs
Our backbone	29.77	0.888	0.446 M	114.9 G
Full 3D Convolution	29.55	0.885	0.700 M	152.9 G
Full 2D Convolution	29.09	0.884	0.270 M	4.6 G

3.2. Real Data

For the real data captured by the system built in [9], we again borrow the experimental results of other methods. The real data is a 660 × 714 measurement with 28 wavelengths ranging from 450 nm to 650 nm. It was shifted 54 pixels with respect to dispersion in the column. We train our model again using the real data mask, i.e., 660 × 660 coded mask and cropped training set. This model is much larger than the simulated one and it takes a huge increase in GPU memory usage (even more than 45 GB for batch size = 1 per batch) and time cost. Thus, we take the advantages of the Automatic Mixed Precision (AMP) module provided by Pytorch to train our model by mixed precision (half precision and single precision real numbers).

The reconstruction results of two scenes, `Lego` and `Strawberry` are shown in Figure 4, where we plot four reconstructed frames at different wavelengths and spectral density curves to demonstrate the performance of our proposed method. We observe that our result contains more detail in Legoman's face area because our model produces sharper edges than other models. In the Strawberry testcase, our result has higher spatial resolution in all selected wavelengths. Similar to Figure 3, we attached a visualization of numerical assessment in Figure 4 and the method to obtain such assessment is the same as described above. We observe that our curve (red) is closest to the reference curve (blue) among all other curves.wo more real data results of `Plants` and `Legoplants` are shown in Figures 1 and 5 with 14 and 7 selected reconstructed channels, respectively. We selected 7 spectral channels out of 28 as shown in Figure 5. Our model achieves superior reconstruction results in terms of clarity and aesthetics compared to TSA-net. Specifically, our model produces more pointed edges that elevate the overall reconstruction quality.s shown in the plots, our method provides sharper edges and more spacial details such as the hands and clothes of the `Lego` man. The spectral density curves reveal our method is closer to the ground truth as well.

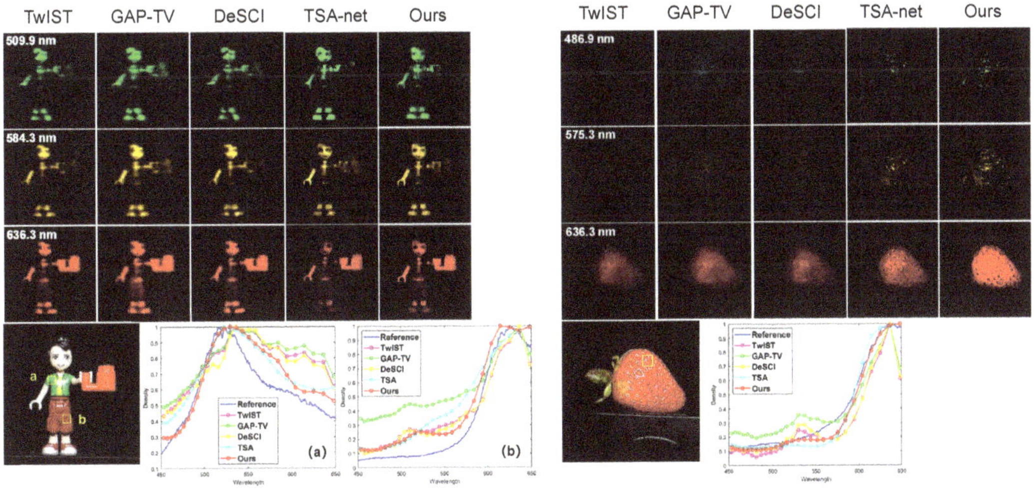

Figure 4. Real data: the reconstructed images of `Lego` man (**left**) and `Strawberry` (**right**) for 4 out of 28 spectral channels. area "a" and "b" have been chosen to perform correlation analysis. The spectral curves are shown at the lower part of the figure, the reference curves and RGB images are from [9].

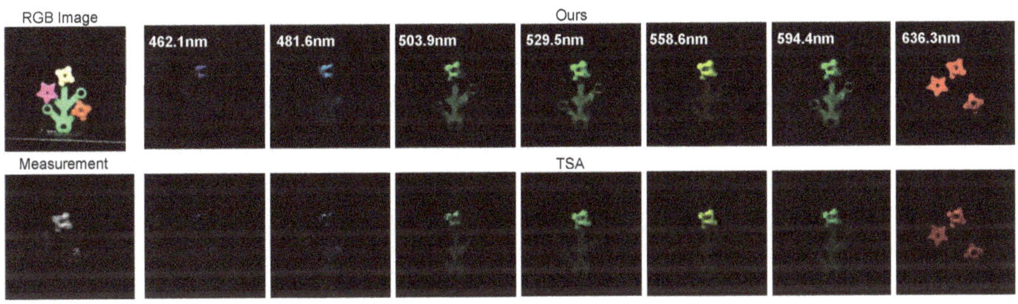

Figure 5. Real data: The RGB references and result images of a real measurement for 7 out of 28 spectral channels reconstructed via TSA and our proposed method.

4. Conclusions

We proposed an end-to-end hybrid multi-dimensional attention U-net for hyperspectral snapshot compressive imaging reconstruction. The algorithm employed hybrid 3D/2D convolutions instead of using one of them alone to balance the trade-off of computational cost and performance. Our proposed network achieved superior results over previous end-to-end CNN based algorithms.

One important observation from our experiments is that for SCI reconstruction tasks, it is not necessary that the backbone network (e.g., U-Net) be deep, but it needs to be wider (more kernels in each layer) to provide good results. This may due to the task difference between image reconstruction (to recover details) and image classification (to extract features). We further used the attention gate to extract essential correlations in the spectral data-cube to improve the reconstruction performance in our network.

In addition to spectral SCI reconstruction as shown in this work, we do believe our network can be used in medical images [59], image compression [60], temporal compressive coherent diffraction imaging [61], and video compressive sensing [62–66].

Author Contributions: Conceptualization, S.Z. and M.C.; Methodology, S.Z. and M.Z.; Software, S.Z.; Validation, S.Z.; Writing—original draft, S.Z.; Writing—review & editing, M.Z. and M.C. All authors have read and agreed to the published version of the manuscript.

Funding: This work was supported by the National Natural Science Foundation of China under Grant No. 62271414, Zhejiang Provincial Natural Science Foundation of China under Grant No. LR23F010001. We would like to thank Research Center for Industries of the Future (RCIF) at Westlake University for supporting this work and the funding from Lochn Optics.

Institutional Review Board Statement: Not applicable.

Data Availability Statement: The data that support the plots within these paper and other findings of this study are available from the corresponding authors upon reasonable request.

Conflicts of Interest: The authors declare no conflict of interest.

Appendix A. Experimental Results

Appendix A.1. Simulated Data Results

Figures A1–A10 show the simulation results with 28 spectral channels for 10 scenes from KAIST. Truth, measurements and RGB images are shown for reference. We compare our proposed method with the TSA-net and λ-net algorithms and list the corresponding PSNR and SSIM.

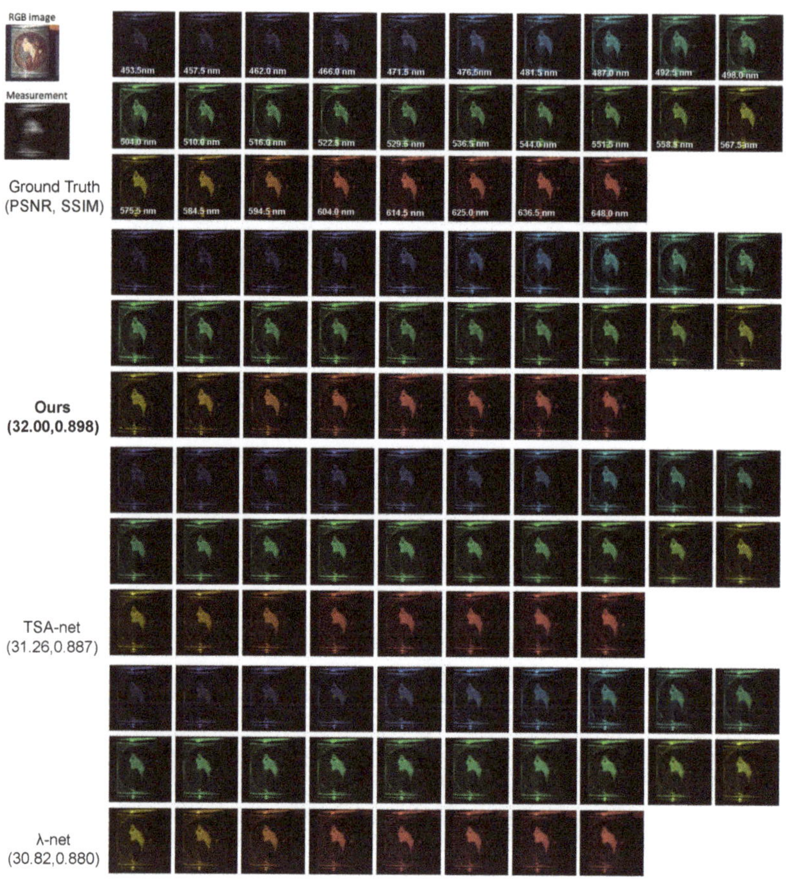

Figure A1. Simulation: RGB image, measurement, ground truth and reconstructed results by Tour proposed method with the TSA-net and λ-net Scene 1. The PSNR in dB and SSIM for the result images are shown in the parenthesis.

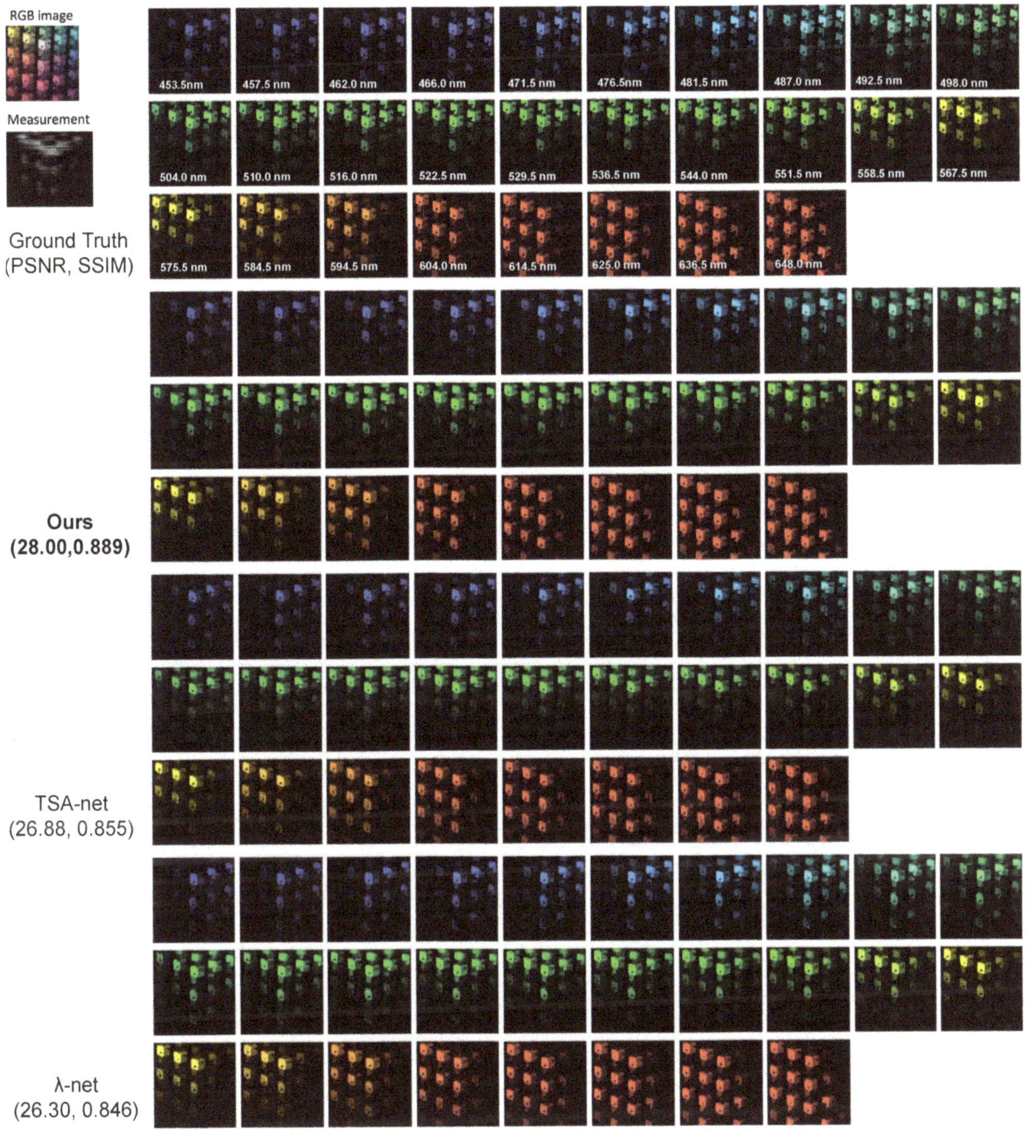

Figure A2. Simulation: RGB image, measurement, ground truth and reconstructed results by Tour proposed method with the TSA-net and λ-net Scene 2. The PSNR in dB and SSIM for the result images are shown in the parenthesis.

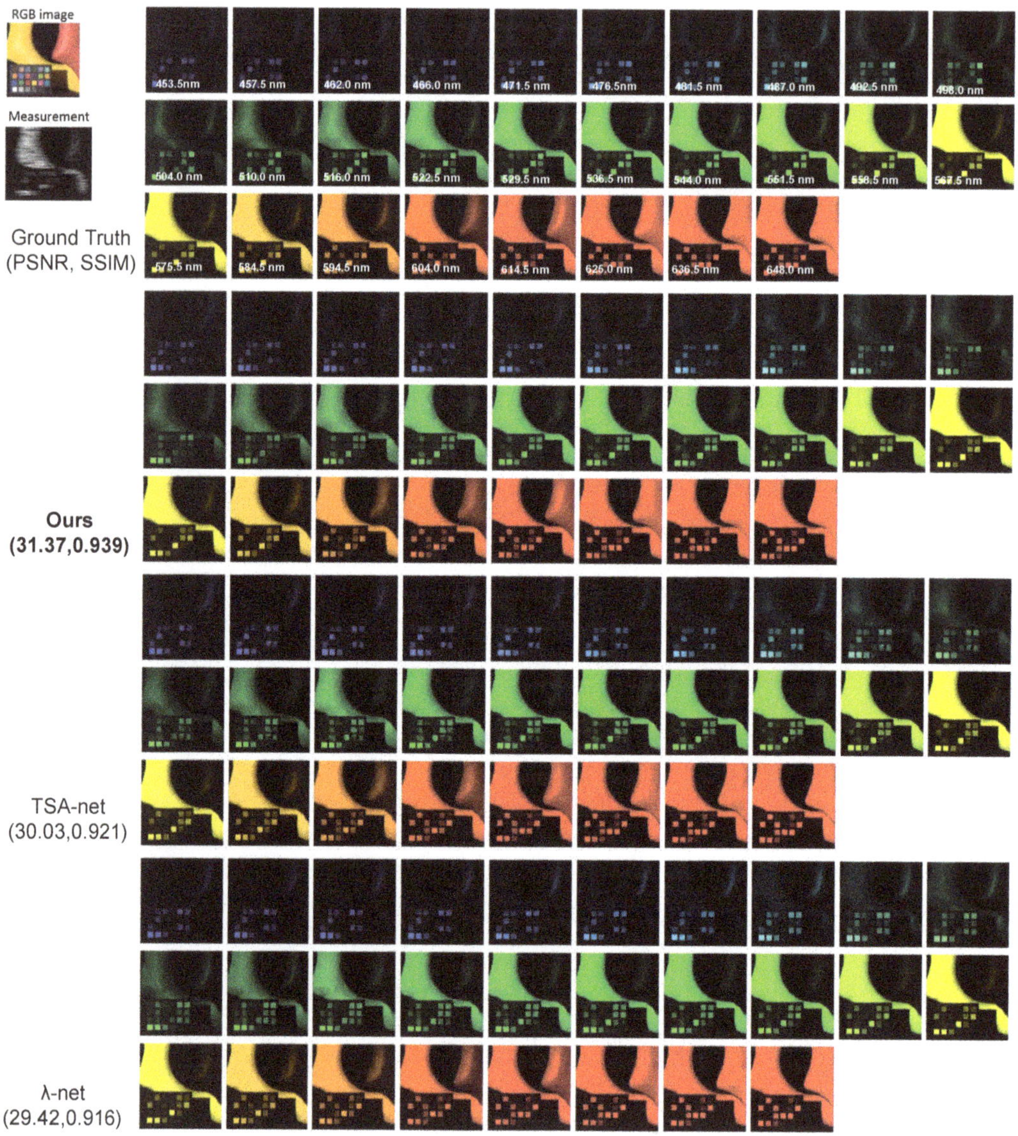

Figure A3. Simulation: RGB image, measurement, ground truth and reconstructed results by Tour proposed method with the TSA-net and λ-net Scene 3. The PSNR in dB and SSIM for the result images are shown in the parenthesis.

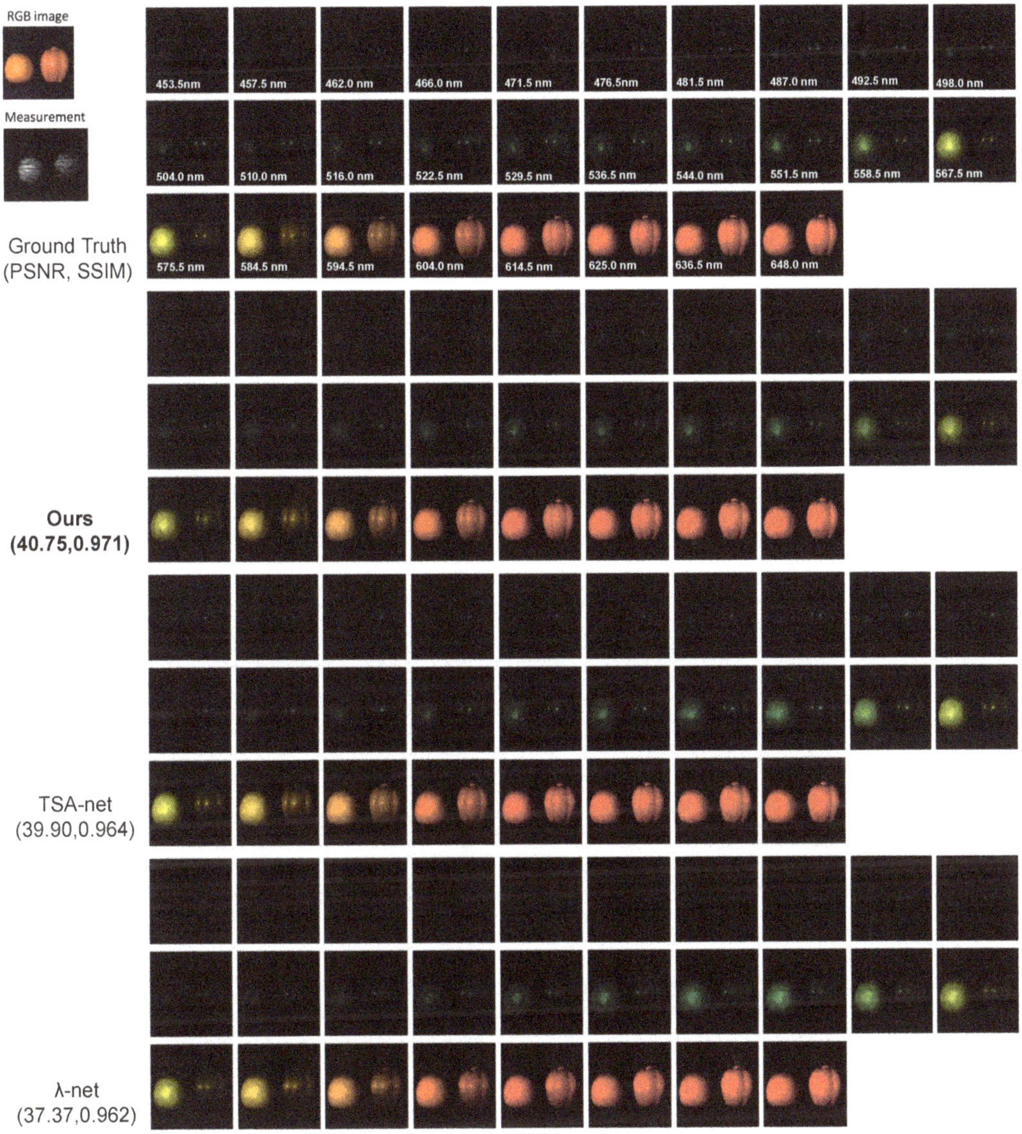

Figure A4. Simulation: RGB image, measurement, ground truth and reconstructed results by Tour proposed method with the TSA-net and λ-net Scene 4. The PSNR in dB and SSIM for the result images are shown in the parenthesis.

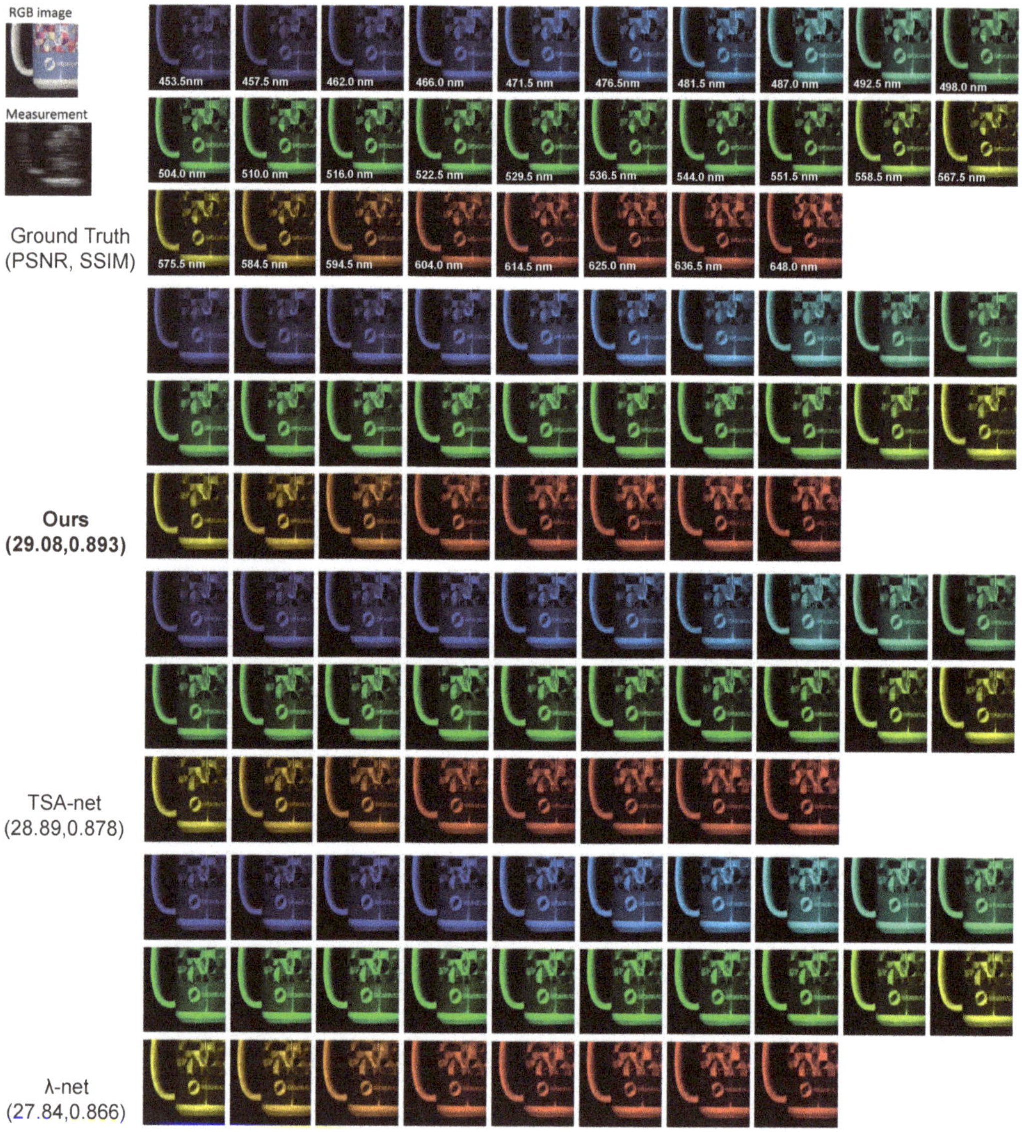

Figure A5. Simulation: RGB image, measurement, ground truth and reconstructed results by Tour proposed method with the TSA-net and λ-net Scene 5. The PSNR in dB and SSIM for the result images are shown in the parenthesis.

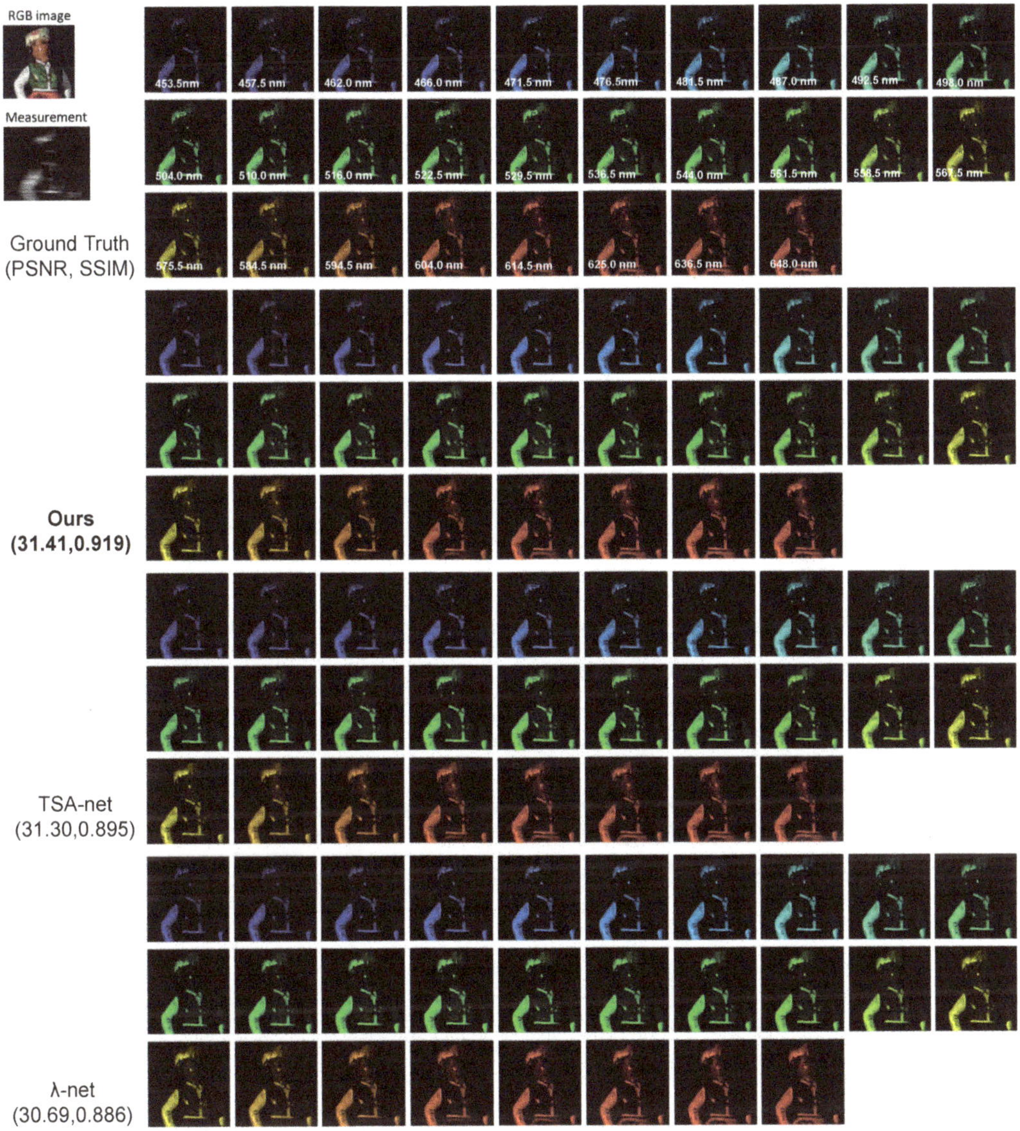

Figure A6. Simulation: RGB image, measurement, ground truth and reconstructed results by Tour proposed method with the TSA-net and λ-net Scene 6. The PSNR in dB and SSIM for the result images are shown in the parenthesis.

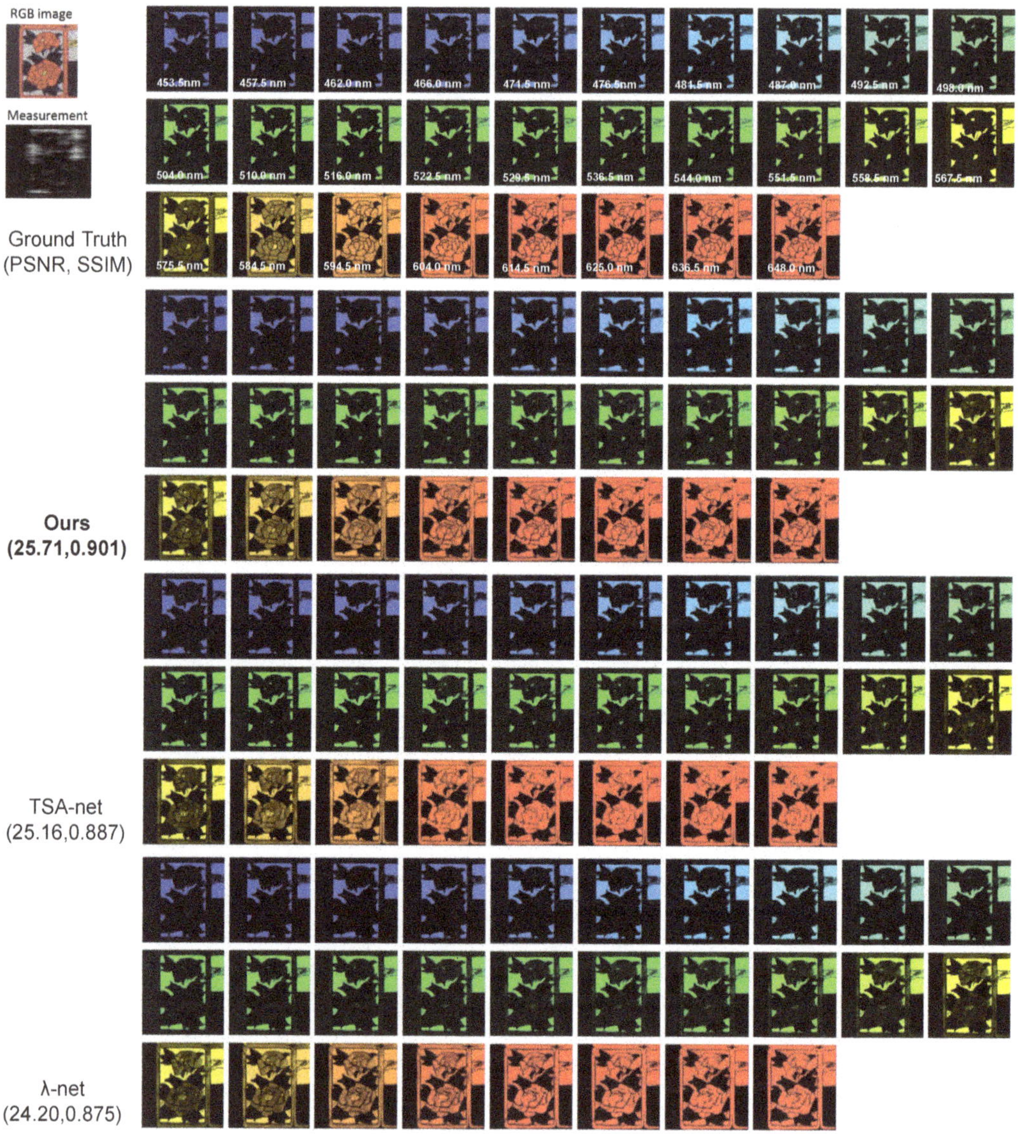

Figure A7. Simulation: RGB image, measurement, ground truth and reconstructed results by Tour proposed method with the TSA-net and λ-net Scene 7. The PSNR in dB and SSIM for the result images are shown in the parenthesis.

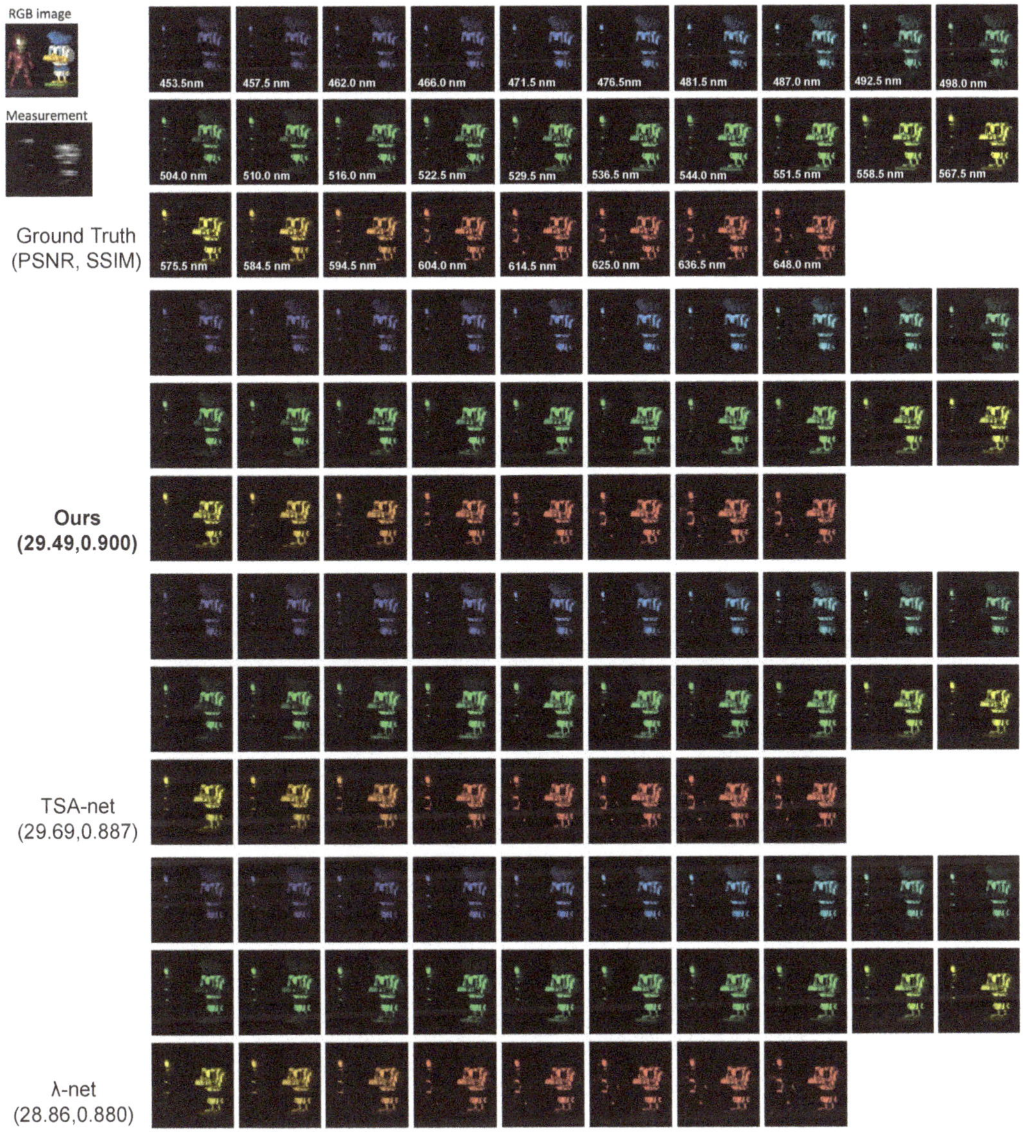

Figure A8. Simulation: RGB image, measurement, ground truth and reconstructed results by Tour proposed method with the TSA-net and λ-net Scene 8. The PSNR in dB and SSIM for the result images are shown in the parenthesis.

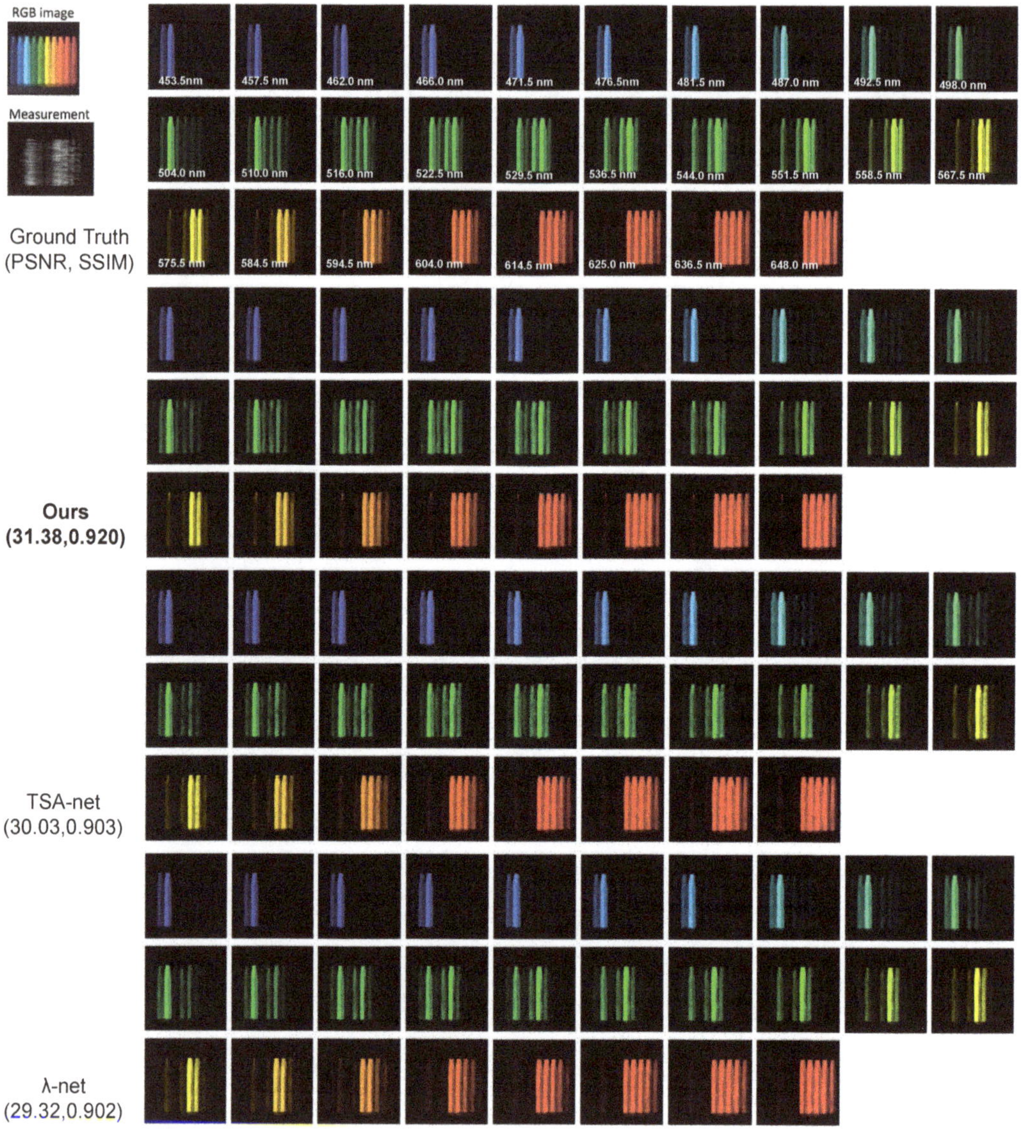

Figure A9. Simulation: RGB image, measurement, ground truth and reconstructed results by Tour proposed method with the TSA-net and λ-net Scene 9. The PSNR in dB and SSIM for the result images are shown in the parenthesis.

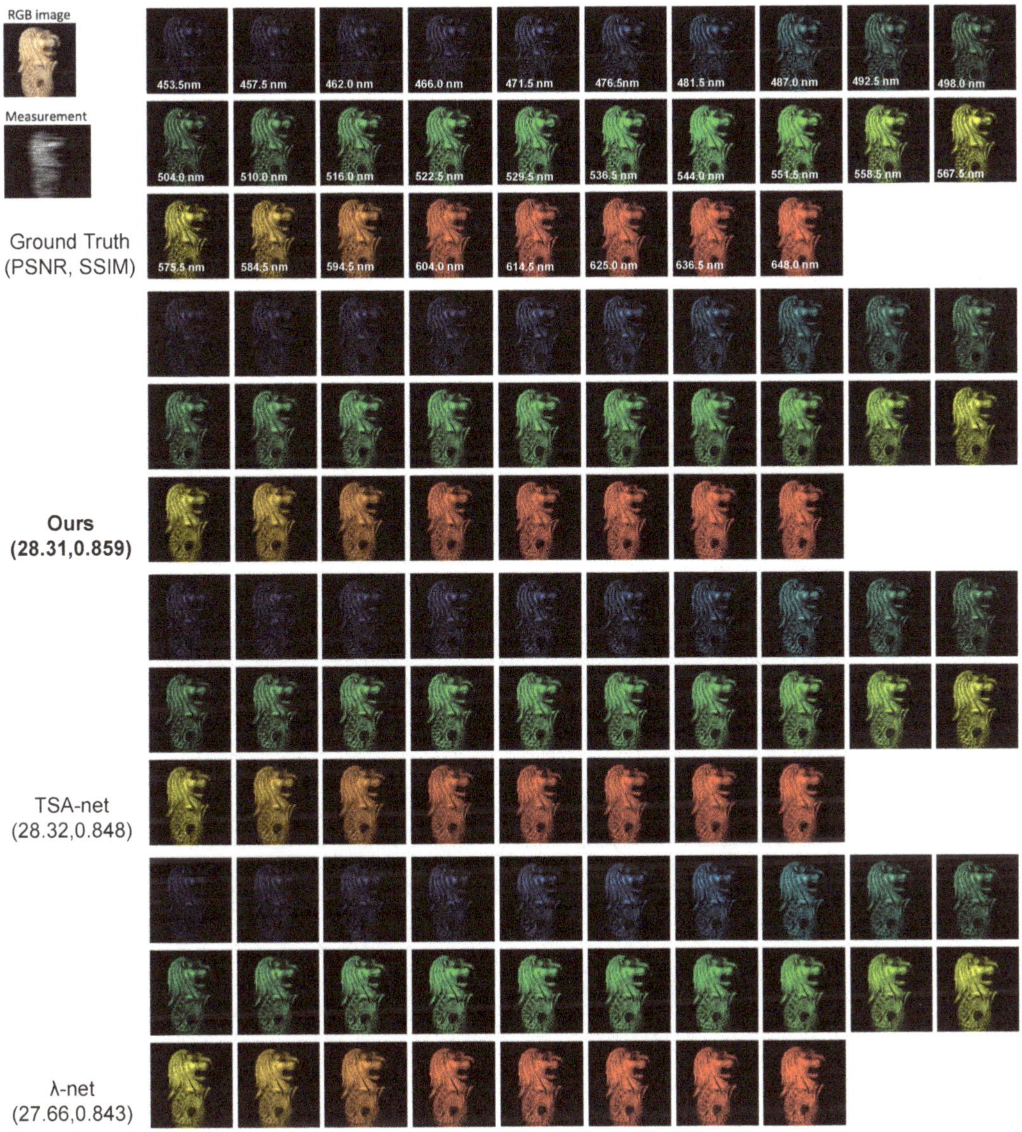

Figure A10. Simulation: RGB image, measurement, ground truth and reconstructed results by Tour proposed method with the TSA-net and λ-net Scene 10. The PSNR in dB and SSIM for the result images are shown in the parenthesis.

Appendix A.2. Real Data Results

Figures A11–A14 show the RGB images, measurements and the reconstructed 28 spectral channels for four real scenes with a size of 660 × 660 pixels captured by real CASSI system.

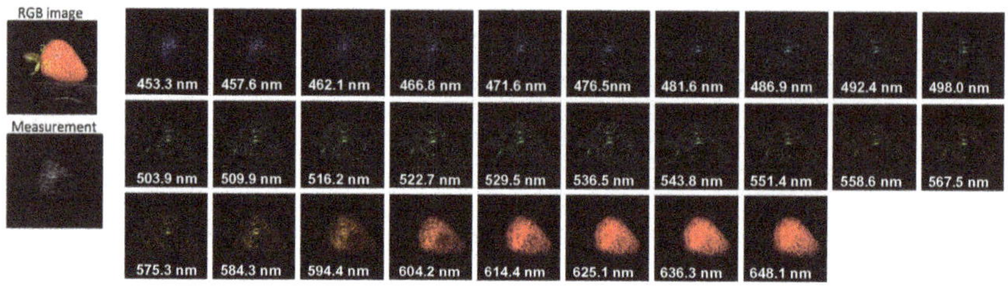

Figure A11. Real data: RGB image, measurement and reconstructed results by our proposed method for scene 1.

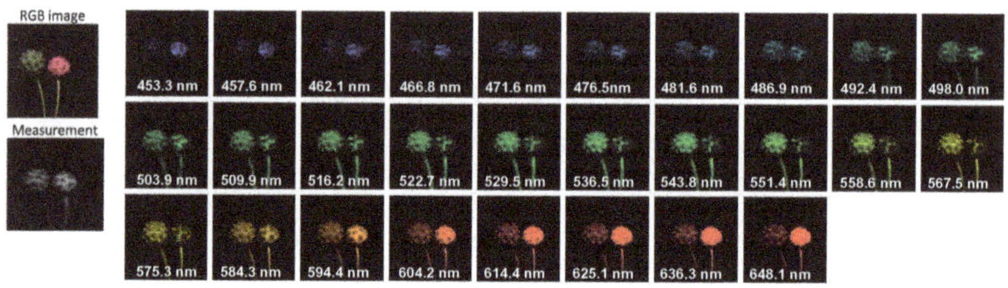

Figure A12. Real data: RGB image, measurement and reconstructed results by our proposed method for scene 2.

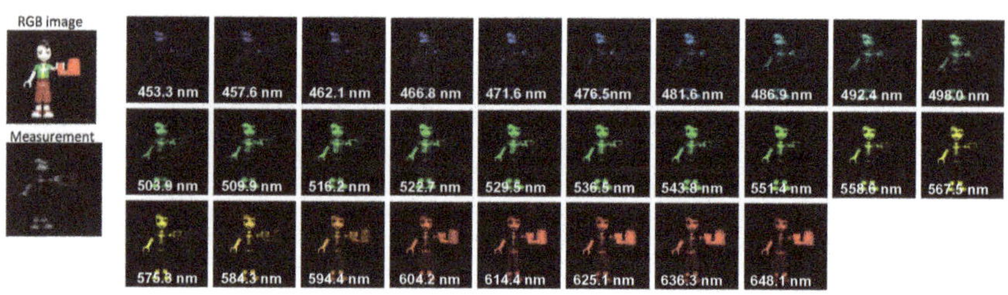

Figure A13. Real data: RGB image, measurement and reconstructed results by our proposed method for scene 3.

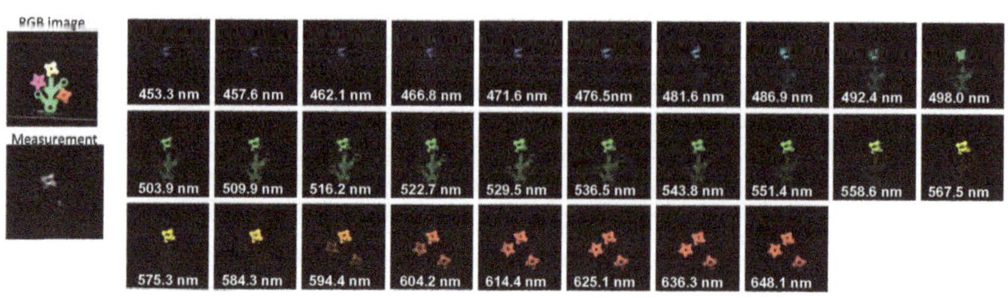

Figure A14. Real data: RGB image, measurement and reconstructed results by our proposed method for scene 4.

References

1. Flores-Fuentes, W.; Trujillo-Hernández, G.; Alba-Corpus, I.Y.; Rodríguez-Quiñonez, J.C.; Mirada-Vega, J.E.; Hernández-Balbuena, D.; Murrieta-Rico, F.N.; Sergiyenko, O. 3D spatial measurement for model reconstruction: A review. *Measurement* **2023** *207*, 112321. [CrossRef]
2. Rodríguez-Quiñonez, J.C.; Sergiyenko, O.; Flores-Fuentes, W.; Lopez, M.; Hernez-Balbuena, D.; Rascon, R.; Mercorelli, P. Improve a 3D distance measurement accuracy in stereo vision systems using optimization methods' approach. *Opto-Electron. Rev.* **2017**, *25*, 24–32. [CrossRef]
3. Rodríguez-Quiñonez, J.C.; Sergiyenko, O.; Hernandez-Balbuena, D.; Rivas-Lopez, M.; Flores-Fuentes, W.; Basaca-Preciado, L.C. Improve 3D laser scanner measurements accuracy using a FFBP neural network with Widrow-Hoff weight/bias learning function. *Opto-Electron* **2014**, *22*, 224–235. [CrossRef]
4. Emannuel, C.; Romberg, J.; Tao, T. Robust uncertainty principles: Exact signal reconstruction from highly incomplete frequency information. *IEEE Trans. Inf. Theory* **2006**, *52*, 489–509
5. Donoho, D.L. Compressed sensing. *IEEE Trans. Inf. Theory* **2006**, *52*, 1289–1306. [CrossRef]
6. Yuan, X.; Brady, D.J.; Katsaggelos, A.K. Snapshot Compressive Imaging: Theory, Algorithms and Applications. *IEEE Signal Process. Mag.* **2021**, *38*, 65–88. [CrossRef]
7. Wagadarikar, A.; John, R.; Willett, R.; Brady, D. Single disperser design for coded aperture snapshot spectral imaging. *Appl. Opt.* **2008**, *47*, B44–B51. [CrossRef]
8. Tsai, T.H.; Llull, P.; Yuan, X.; Carin, L.; Brady, D.J. Spectral-temporal compressive imaging. *Opt. Lett.* **2015**, *40*, 4054–4057. [CrossRef]
9. Meng, Z.; Ma, J.; Yuan, X. End-to-End Low Cost Compressive Spectral Imaging with Spatial-Spectral Self-Attention. In Proceedings of the European Conference on Computer Vision (ECCV), Glasgow, UK, 23–28 August 2020.
10. Wang, Z.; Bovik, A.C.; Sheikh, H.R.; Simoncelli, E.P. Image quality assessment: From error visibility to structural similarity. *IEEE Trans. Image Process.* **2004**, *13*, 600–612. [CrossRef]
11. Yuan, X.; Tsai, T.H.; Zhu, R.; Llull, P.; Brady, D.; Carin, L. Compressive Hyperspectral Imaging with Side Information. *IEEE J. Sel. Top. Signal Process.* **2015**, *9*, 964–976. [CrossRef]
12. Wang, L.; Xiong, Z.; Gao, D.; Shi, G.; Wu, F. Dual-camera design for coded aperture snapshot spectral imaging. *Appl. Opt.* **2015**, *54*, 848–858. [CrossRef] [PubMed]
13. Arguello, H.; Rueda, H.; Wu, Y.; Prather, D.W.; Arce, G.R. Higher-order computational model for coded aperture spectral imaging. *Appl. Opt.* **2013**, *52*, D12–D21. [CrossRef] [PubMed]
14. Bioucas-Dias, J.; Figueiredo, M. A New TwIST: Two-Step Iterative Shrinkage/Thresholding Algorithms for Image Restoration. *IEEE Trans. Image Process.* **2007**, *16*, 2992–3004. [CrossRef]
15. Yuan, X. Generalized alternating projection based total variation minimization for compressive sensing. In Proceedings of the 2016 IEEE International Conference on Image Processing (ICIP), Pheonix, AZ, USA, 25–28 September 2016.
16. Yang, J.; Liao, X.; Yuan, X.; Llull, P.; Brady, D.J.; Sapiro, G.; Carin, L. Compressive Sensing by Learning a Gaussian Mixture Model from Measurements. *IEEE Trans. Image Process.* **2015**, *24*, 106–119. [CrossRef] [PubMed]
17. Wang, L.; Xiong, Z.; Shi, G.; Wu, F.; Zeng, W. Adaptive Nonlocal Sparse Representation for Dual-Camera Compressive Hyperspectral Imaging. *IEEE Trans. Pattern Anal. Mach. Intell.* **2017**, *39*, 2104–2111. [CrossRef]
18. Liu, Y.; Yuan, X.; Suo, J.; Brady, D.J.; Dai, Q. Rank Minimization for Snapshot Compressive Imaging. *IEEE Trans. Pattern Anal. Mach. Intell.* **2019**, *41*, 2990–3006. [CrossRef]
19. Ongie, G.; Jalal, A.; Metzler, C.A.; Baraniuk, R.G.; Dimakis, A.G.; Willett, R. Deep Learning Techniques for Inverse Problems in Imaging. *IEEE J. Sel. Areas Inf. Theory* **2020**, *1*, 39–56. [CrossRef]
20. Miao, X.; Yuan, X.; Pu, Y.; Athitsos, V. λ-net: Reconstruct Hyperspectral Images from a Snapshot Measurement. In Proceedings of the IEEE/CVF Conference on Computer Vision (ICCV), Seoul, Republic of Korea, 27 October–2 November 2019.
21. Meng, Z.; Jalali, S.; Yuan, X. GAP-net for Snapshot Compressive Imaging. *arXiv* **2020**, arXiv:2012.08364.
22. Fu, Y.; Zhang, T.; Wang, L.; Huang, H. Coded Hyperspectral Image Reconstruction using Deep External and Internal Learning. *IEEE Trans. Pattern Anal. Mach. Intell.* **2021**, *44*, 3404–3420. [CrossRef]
23. Wang,L.; Sun, C.; Fu, Y.; Kim, M.H.; Huang, H. Hyperspectral Image Reconstruction Using a Deep Spatial-Spectral Prior. In Proceedings of the 2019 IEEE/CVF Conference on Computer Vision and Pattern Recognition (CVPR), Long Beach, CA, USA, 16–20 June 2019.
24. Yu, Z.; Liu, D.; Cheng, L.; Meng, Z.; Zhao, Z.; Yuan, X.; Xu, K. Deep learning enabled reflective coded aperture snapshot spectral imaging. *Opt. Express* **2022**, *30*, 46822–46837 [CrossRef]
25. Wang, Y.; Han, Y.; Wang, K.; Zhao, X. Total variation regularized nonlocal low-rank tensor train for spectral compressive imaging. *Singal Process.* **2022**, *195*, 108464 [CrossRef]
26. Ronneberger, O.; Fischer, P.; Brox, T. U-Net: Convolutional Networks for Biomedical Image Segmentation. In Proceedings of the International Conference on Medical Image Computing and Computer-Assisted Intervention (MICCAI), Munich, Germany, 5–9 October 2015.
27. Oktay, O.; Schlemper, J.; Folgoc, L.L.; Lee, M.; Heinrich, M.P.; Misawa, K.; Mori, K.; McDonagh, S.; Hammerla, N.Y.; Kainz, B.; et al. Attention U-Net: Learning Where to Look for the Pancreas. *Med. Image Anal.* **2018**, *51*, 63–76.

28. Gehm, M.E.; John, R.; Brady, D.J.; Willett, R.M.; Schulz, T.J. Single-shot compressive spectral imaging with a dual-disperser architecture. *Opt. Express* **2007**, *15*, 14013–14027. [CrossRef] [PubMed]
29. Jalali, S.; Yuan, X. Compressive imaging via one-shot measurements. In Proceedings of the 2018 IEEE International Symposium on Information Theory (ISIT), Vail, CO, USA, 17–23 June 2018.
30. Jalali, S.; Yuan, X. Snapshot compressed sensing: Performance bounds and algorithms. *IEEE Trans. Inf. Theory* **2019**, *65*, 8005–8024. [CrossRef]
31. Gao, S.H.; Cheng, M.M.; Zhao, K.; Zhang, X.Y.; Yang, M.H.; Torr, P.H. Res2net: A new multi-scale backbone architecture. *IEEE Trans. Pattern Anal. Mach. Intell.* **2019**, *43*, 652–662. [CrossRef]
32. Schlemper, J.; Oktay, O.; Chen, L.; Matthew, J.; Knight, C.; Kainz, B.; Glocker, B.; Rueckert, D. Attention-Gated Networks for Improving Ultrasound Scan Plane Detection. *IEEE Trans. Med. Imaging* **2018**, *38*, 5, 1069–1078.
33. Hu, J.; Shen, L.; Sun, G. Squeeze-and-excitation networks. In Proceedings of the IEEE Conference on Computer Vision and Pattern Recognition (CVPR), Salt Lake City, UT, USA, 19–21 June 2018.
34. Zhang, M.; Wang, L.; Zhang, L.; Huang, H. High light efficiency snapshot spectral imaging via spatial multiplexing and spectral mixing. *Opt. Express* **2020**, *28*, 19837–19850. [CrossRef]
35. Liao, X.; Li, H.; Carin, L. Generalized Alternating Projection for Weighted-$\ell_{2,1}$ Minimization with Applications to Model-based Compressive Sensing. *SIAM J. Imaging Sci.* **2014**, *7*, 797–823. [CrossRef]
36. Zhang, S.; Wang, L.; Fu, Y.; Zhong, X.; Huang, H. Computational hyperspectral imaging based on dimension-discriminative low-rank tensor recovery. In Proceedings of the IEEE/CVF International Conference on Computer Vision (ICCV), Seoul, Republic of Korea, 27 October–2 November 2019.
37. Golbabaee, M.; Vandergheynst, P. Hyperspectral image compressed sensing via low-rank and joint-sparse matrix recovery. In Proceedings of the 2012 IEEE International Conference on Acoustics, Speech and Signal Processing (ICASSP), Kyoto, Japan, 25–30 March 2012.
38. Gelvez, T.; Rueda, H.; Arguello, H. Joint sparse and low rank recovery algorithm for compressive hyperspectral imaging. *Appl. Opt.* **2017**, *56*, 6785–6795. [CrossRef]
39. Fu, Y.; Zheng, Y.; Sato, I.; Sato, Y. Exploiting spectral-spatial correlation for coded hyperspectral image restoration. In Proceedings of the IEEE Conference on Computer Vision and Pattern Recognition (CVPR), Las Vegas, NV, USA, 26 June–1 July 2016.
40. Zheng, S.; Liu, Y.; Meng, Z.; Qiao, M.; Tong, Z.; Yang, X.; Han, S.; Yuan, X. Deep plug-and-play priors for spectral snapshot compressive imaging. *Photonics Res.* **2021**, *9*, B18–B29. [CrossRef]
41. Yang, Y.; Sun, J.; Li, H.; Xu, Z. Deep ADMM-Net for Compressive Sensing MRI. In Proceedings of the Neural Information Processing Systems 29, Barcelona, Spain, 5–10 December 2016.
42. Ma, J.; Liu, X.; Shou, Z.; Yuan, X. Deep Tensor ADMM-Net for Snapshot Compressive Imaging. In Proceedings of the IEEE/CVF Conference on Computer Vision (ICCV), Seoul, Republic of Korea, 27 October–2 November 2019.
43. Wang, L.; Sun, C.; Zhang, M.; Fu, Y.; Huang, H. DNU: Deep Non-Local Unrolling for Computational Spectral Imaging. In Proceedings of the IEEE/CVF Conference on Computer Vision and Pattern Recognition (CVPR), Virtual, 14–19 June 2020.
44. Zhang, J.; Ghanem, B. ISTA-Net: Interpretable optimization-inspired deep network for image compressive sensing. In Proceedings of the IEEE Conference on Computer Vision and Pattern Recognition (CVPR), Salt Lake City, UT, USA, 19–21 June 2018.
45. Sogabe, Y.; Sugimoto, S.; Kurozumi, T.; Kimata, H. ADMM-Inspired Reconstruction Network for Compressive Spectral Imaging. In Proceedings of the 2020 IEEE International Conference on Image Processing (ICIP), Abu Dhabi, United Arab Emirates, 25–28 October 2020.
46. Li, H.; Xiong, Z.; Shi, Z.; Wang, L.; Liu, D.; Wu, F. HSVCNN: CNN-based hyperspectral reconstruction from RGB videos. In Proceedings of the 2018 IEEE International Conference on Image Processing (ICIP), Athens, Greece, 7–10 October 2018.
47. Xiong, Z.; Shi, Z.; Li, H.; Wang, L.; Liu, D.; Wu, F. Hscnn: Cnn-based hyperspectral image recovery from spectrally undersampled projections. In Proceedings of the IEEE International Conference on Computer Vision Workshops, Venice, Italy, 22–29 October 2017.
48. Wang, L.; Zhang, T.; Fu, Y.; Huang, H. Hyperreconnet: Joint coded aperture optimization and image reconstruction for compressive hyperspectral imaging. *IEEE Trans. Image Process.* **2018**, *28*, 2257–2270. [CrossRef] [PubMed]
49. Kohei, Y.; Han, X.H. Deep Residual Attention Network for Hyperspectral Image Reconstruction. In Proceedings of the 2020 25th International Conference on Pattern Recognition (ICPR), Milan, Italy, 10–15 January 2021.
50. Zhang, T.; Fu, Y.; Wang, L.; Huang, H. Hyperspectral image reconstruction using deep external and internal learning. In Proceedings of the IEEE/CVF International Conference on Computer Vision (ICCV), Seoul, Republic of Korea, 27 October–2 November 2019.
51. Zhang, M.; Wang, L.; Zhang, L.; Huang, H. Compressive hyperspectral imaging with non-zero mean noise. *Opt. Express* **2019**, *27*, 17449–17462. [CrossRef] [PubMed]
52. Huang, G.; Liu, Z.; Van Der Maaten, L.; Weinberger, K.Q. Densely connected convolutional networks. In Proceedings of the IEEE Conference on Computer Vision and Pattern Recognition (CVPR), Honolulu, HI, USA, 21–26 July 2017.
53. Li, Q.; Wang, Q.; Li, X. Exploring the Relationship Between 2D/3D Convolution for Hyperspectral Image Super-Resolution. *IEEE Trans. Geosci. Remote Sens.* **2021**, *59*, 5622–5637. [CrossRef]
54. Mei, S.; Yuan, X.; Ji, J.; Zhang, Y.; Wan, S.; Du, Q. Hyperspectral image spatial super-resolution via 3D full convolutional neural network. *Remote Sens.* **2017**, *9*, 1139. [CrossRef]
55. Wang, Q.; Li, Q.; Li, X. Spatial-Spectral Residual Network for Hyperspectral Image Super-Resolution. *arXiv* **2020**, arXiv:2001.04609.

56. Li, Q.; Wang, Q.; Li, X. Mixed 2d/3d convolutional network for hyperspectral image super-resolution. *Remote Sens.* **2020**, *12*, 1660. [CrossRef]
57. Yasuma, F.; Mitsunaga, T.; Iso, D.; Nayar, S.K. Generalized assorted pixel camera: Postcapture control of resolution, dynamic range and spectrum. *IEEE Trans. Image Process.* **2010**, *19*, 2241–2253. [CrossRef]
58. Choi, I.; Jeon, D.S.; Nam, G.; Gutierrez, D.; Kim, M.H. High-Quality Hyperspectral Reconstruction Using a Spectral Prior. In Proceedings of the SIGGRAPH Asia 2017, Bangkok, Thailand, 27–30 November 2017.
59. Meng, Z.; Qiao, M.; Ma, J.; Yu, Z.; Xu, K.; Yuan, X. Snapshot multispectral endomicroscopy. *Opt. Lett.* **2020**, *45*, 3897–3900. [CrossRef]
60. Xue, Y.; Zheng, S.; Tahir, W.; Wang, Z.; Zhang, H.; Meng, Z.; Tian, L.; Yuan, X. Block modulating video compression: An ultra low complexity image compression encoder for resource limited platforms. *arXiv* **2022**, arXiv:2205.03677.
61. Chen, Z.; Zheng, S.; Tong, Z.; Yuan, X. Physics-driven deep learning enables temporal compressive coherent diffraction imaging. *Optica* **2022** *25*, 677–680. [CrossRef]
62. Cheng, Z.; Chen, B.; Liu, G.; Zhang, H.; Lu, R.; Wang, Z.; Yuan, X. Memory-efficient network for large-scale video compressive sensing. In Proceedings of the IEEE/CVF Conference on Computer Vision and Pattern Recognition (CVPR), Virtual, 19–25 June 2021; pp. 14887–14896.
63. Zhao, Y.; Zheng, S.; Yuan, X. Deep equilibrium models for video snapshot compressive imaging. *arXiv* **2022**, arXiv:2201.06931.
64. Zhang, Z.; Zhang, B.; Yuan, X.; Zheng, S.; Su, X.; Suo, J.; Brady, D.J.; Dai, Q. From compressive sampling to compressive tasking: Retrieving semantics in compressed domain with low bandwidth. *PhotoniX* **2022**, *3*, 1–22. [CrossRef]
65. Zheng, S.; Wang, C.; Yuan, X.; Xin, H.L. Super-compression of large electron microscopy time series by deep compressive sensing learning. *Patterns* **2021**, *2*, 100292. [CrossRef]
66. Zheng, S.; Yang, X.; Yuan, X. Two-stage is enough: A concise deep unfolding reconstruction network for flexible video compressive sensing. *arXiv* **2022**, arXiv:2201.05810.

entropy

Article

Multi-Receptive Field Soft Attention Part Learning for Vehicle Re-Identification

Xiyu Pang [1,2], Yilong Yin [1,*] and Yanli Zheng [2]

1 School of Software, Shandong University, No. 1500, Shunhua Road, High-Tech Industrial Development Zone, Jinan 250101, China
2 School of Information Science and Electrical Engineering, Shandong Jiaotong University, No. 5001, Haitang Road, Changqing District, Jinan 250357, China
* Correspondence: ylyin@sdu.edu.cn

Abstract: Vehicle re-identification across multiple cameras is one of the main problems of intelligent transportation systems (ITSs). Since the differences in the appearance between different vehicles of the same model are small and the appearance of the same vehicle changes drastically from different viewpoints, vehicle re-identification is a challenging task. In this paper, we propose a model called multi-receptive field soft attention part learning (MRF-SAPL). The MRF-SAPL model learns semantically diverse vehicle part-level features under different receptive fields through multiple local branches, alleviating the problem of small differences in vehicle appearance. To align vehicle parts from different images, this study uses soft attention to adaptively locate the positions of the parts on the final feature map generated by a local branch and maintain the continuity of the internal semantics of the parts. In addition, to obtain parts with different semantic patterns, we propose a new loss function that punishes overlapping regions, forcing the positions of different parts on the same feature map to not overlap each other as much as possible. Extensive ablation experiments demonstrate the effectiveness of our part-level feature learning method MRF-SAPL, and our model achieves state-of-the-art performance on two benchmark datasets.

Keywords: vehicle re-identification; multi-receptive field; part-level features

Citation: Pang, X.; Yin, Y.; Zheng, Y. Multi-Receptive Field Soft Attention Part Learning for Vehicle Re-Identification. *Entropy* **2023**, *25*, 594. https://doi.org/10.3390/e25040594

Academic Editors: Oleg Sergiyenko, Wendy Flores-Fuentes, Julio Cesar Rodriguez-Quinonez and Jesús Elías Miranda-Vega

Received: 21 February 2023
Revised: 26 March 2023
Accepted: 27 March 2023
Published: 31 March 2023

1. Introduction

The vehicle re-identification (Re-ID) task identifies the same vehicle from multiple nonoverlapping cameras in surveillance systems. This task is particularly useful when a car's license plate is occluded or cannot be seen clearly. In these scenarios, the vehicle Re-ID method can effectively locate the vehicle of interest from the monitoring database, which has important applications in intelligent transportation, public safety, smart cities, and other fields. In recent years, vehicle Re-ID has received increasing attention from the computer vision community.

Due to drastic changes in illumination, occlusion, resolution, viewing angle, and background, vehicle Re-ID is still a very challenging task, particularly when the vehicle images are obtained from a large number of different cameras. To address this Re-ID task, many deep learning models [1–3] for extracting global vehicle information have been proposed in recent years. Although these works have achieved remarkable success in vehicle Re-ID tasks, since global feature learning only captures the most important information representing different identities, the identification ability of global features tends to decline severely when the differences in vehicle appearance are not significant. As shown in Figure 1, different vehicles belonging to the same model may look quite similar. We can distinguish these challenging vehicle image samples by subtle clues, such as the annual inspection signs and decorations marked by the circle shown in Figure 1. Therefore, learning rich fine-grained local features is crucial for vehicle Re-ID tasks.

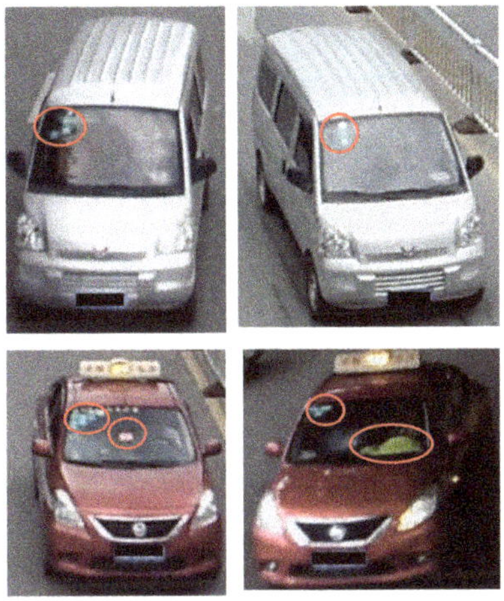

Figure 1. The four cars are divided into two groups: the silver-white cars at the top and the red cabs at the bottom. Each group is comprised by different vehicles of the same model. The detailed information distinguishing each vehicle is marked with red circles; for example, the car in the lower left has rows of stickers including annual inspection signs and a small red screen, while the car in the lower right has three stickers and a yellow item. These details can completely distinguish the two red cars.

Recently, part-based models [4–7] have made great progress in learning effective local feature representations for pedestrian Re-ID and have obtained highly promising results. By horizontally dividing one feature map into multiple parts in space, these models can mine fine-grained discriminative features on each part. Aggregating all part-level features can effectively identify pedestrians. Since person and vehicle Re-ID tasks are conceptually similar and both of them belong to the image retrieval problem, techniques from one task can usually be adapted to the other. To solve the problem of high similarity in Figure 1, Ref. [8] divided the feature maps of vehicle images along various directions to extract rich fine-grained local features. They applied the most advanced methods of pedestrian Re-ID to vehicle Re-ID. However, on the one hand, the change in vehicle appearance from different perspectives is much larger than that of pedestrians. As shown in Figure 2, the texture or color of the clothes worn by a person does not change drastically under different viewing angles, meaning that the images of the same person from different cameras will always have more in common and can be roughly spatially aligned, so that the body can be vertically segmented into several parts to extract part-level features. By contrast, the appearance of the same vehicle can change drastically due to the change in viewpoint, and the misalignment of parts is more severe than that of pedestrians, such that a simple rigid spatial division cannot align vehicle parts well enough to learn the part-level features effectively. On the other hand, the simple rigid division of feature maps breaks the semantic continuity within parts.

To overcome the above-described challenges, some methods [9–12] focus on enabling the networks to identify the vehicle perspective and learn the fine-grained information related to the perspective through vehicle key point detection, parsing networks, and pose estimation. These methods solve the above problems to some extent but increase complexity and rely on additional annotations. In addition, other methods use attention mechanisms to effectively mine identity-related salient information. Ref. [13] enhanced the discriminative power of the features on two branches by using nonlocal spatial attention and channel attention. Although these methods can effectively discover salient information globally, they cannot find rich detailed clues. Analytically, we find that an effective part-level feature learning mechanism for vehicle Re-ID should follow three criteria: (1) the detected parts/regions should be aligned and maintain internal consistency; (2) the detected parts/regions should be semantically diverse to cover as much discriminant information as possible in vehicle images; (3) The detected part semantics should be multilevel because receptive fields of different sizes can capture part information with different semantic levels. To meet these demands, we propose a model called multi-receptive field soft attention part learning (MRF-SAPL) for part-level feature learning. Without the need for additional annotations, MRF-SAPL locates parts under multiple receptive fields and learns rich, multi-semantic-level part-level features associated with vehicle identities.

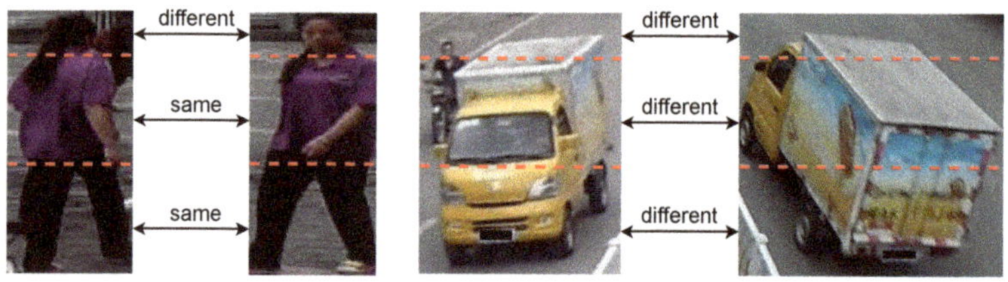

Figure 2. The two images on the left are of the same pedestrian taken from different viewpoints. It can be intuitively seen that the color of the clothes worn by the pedestrian does not change drastically from different viewpoints, and when the pedestrian images are divided vertically into three parts, there are still many commonalities between the corresponding parts of the two images. The two images on the right are of the same vehicle taken from different viewpoints, and it can be seen that the change in viewpoint causes drastic changes in the appearance of the same vehicle, and that the misalignment between the corresponding parts is more severe.

In the MRF-SAPL model, the backbone network is extended to a series of ordered branches, one of which is a global branch for learning global features, and the rest are local branches for learning part-level features. Each local branch mines multiple part-level features with a specific semantic level under a receptive field, so that multiple local branches can obtain enough part-level features with different semantic levels from the entirety of the vehicle image. Within each local branch, we use the soft attention part learning (SAPL) module to learn to locate part positions and extract part features. Specifically, first, the final feature map output by a local branch is adaptively divided into several internal semantically continuous parts/regions using soft attention. The adaptive division of regions can automatically align the corresponding vehicle parts from different images. Second, to ensure that the multiple parts extracted by the same branch are semantically irrelevant, we propose a new loss function called the overlapping region penalty (ORP) to force the corresponding regions of different parts on the feature map to not overlap each other as much as possible in order to obtain parts with different semantic patterns. Finally, after positioning the regions where the parts are located, we use a part feature extractor to extract the corresponding part features from each part region. Our contributions can be summarized as follows:

(1) We propose a multi-receptive soft attention part learning (MRF-SAPL) model for vehicle Re-ID that does not require rigid space partitioning or additional labeling and can flexibly discover enough part-level features with multiple semantic levels;
(2) To align the vehicle part features from different images, we exploit soft attention to adaptively divide the space of the feature map to obtain the locations of parts with internal semantic continuity;
(3) Extensive experimental results show that a higher performance can be obtained compared to that of other state-of-the-art methods on two large datasets, where a new loss function, ORP, is proposed to force each local branch of MRF-SAPL to semantically learn complementary part-level features.

2. Related Work

2.1. Local-Based Re-ID

The design of existing Re-ID methods is mainly based on handcrafted features [14,15], metric learning [16–18] and deep learning networks [5,9,19–27]. Some recent approaches learn features at the part level and achieve state-of-the-art performance in Re-ID tasks. Existing part-based Re-ID methods can be generalized into two categories: methods with external cues and partition-based methods.

Refs. [10,12,28] used external cues utilize human parsing, pose estimation, and object segmentation to precisely align body parts under the supervision of additional semantic labels. Miao et al. [12] learned the visibility of body parts using pose landmarks and extracted useful features for pedestrians using the generated attention masks. Gao et al. [10] learned part features with the help of attention maps guided by pose estimation and trained the visibility of parts through pseudolabels generated by graph matching. He et al. [28] introduced an object detection network to generate the ROI (region of interest) for each vehicle part and then projected the ROIs into the global feature map generated by a global module to capture local information. However, the required external cues limit the usage and robustness of their method in practical deployments. By contrast, our model can align the corresponding parts of different vehicles using only identity labels under the supervision of the overlapping region penalty (ORP) constraint.

Common segmentation-based models mainly align body parts by rigidly segmenting images/feature maps. Sun et al. [5] horizontally partitioned the final feature map output by the network to learn fine-grained part-level features of pedestrians from each region. Chen et al. [8] divided the feature maps of vehicle images in various directions to fully mine fine-grained local features. Although these methods match part features by region partitioning without using external labels and models, they assume that the same part appears at the same location in different images, making it difficult to overcome the serious spatial misalignment problem inherent in vehicle Re-ID. Recently, Li et al. [29] adaptively learned discriminative body part features for occluded person re-identification tasks by enhancing interpart associations from a global perspective through a transformer encoder-decoder architecture. Both our method and the method of Ref. [29] can adaptively align parts to suppress the spatial misalignments. Different from Ref. [29], MRF-SAPL can generate aligned parts without relying on the complex transformer architecture.

2.2. Multiscale Features

Convolutional neural networks extract the features of the target in a layer-by-layer abstract manner through the convolution layer and the pooling layer. The design of the receptive field size has an important impact on the performance of the networks. Small receptive fields can only observe local information; in contrast, large receptive fields can only observe global information. Therefore, researchers have designed various multiscale model architectures to capture features at different semantic levels. He et al. [30] proposed a spatial pyramid pooling network that can obtain fixed-size feature maps and capture information at different scales through different downsampling steps. Zhao et al. [31] proposed the pyramid scene parsing network (PSP Net) that utilizes downsampling and upsampling operations to extract local and global information, making scene recognition more reliable. The Inception module proposed by Szegedy et al. [32] consists of four parallel channels, namely, 1×1 convolution, 3×3 convolution, 5×5 convolution, and 3×3 maximum pooling, which are combined to extract the features of the previous layer of different scales. Tolstikhin et al. [33] proposed a multilayer perceptron Mixer (MLP-Mixer) architecture for computer vision that uses a depthwise separable filter with a maximum receptive field and interchannel parameter sharing to mix tokens to capture global information. Li et al. [34] facilitated visual representation learning via 3×3 convolutional static context and contextual self-attention-based dynamic context. In this paper, we let each local branch focus on capturing discriminative information under a specific receptive field through a downsampling operation.

3. Method

3.1. Network Structure

Figure 3 shows the overall network architecture of the MRF-SAPL model, which includes a ResNet-50-based backbone, a global branch for extracting global information, and three local branches (LB_1, LB_2, LB_3) for extracting part-level information. For the backbone network, we use ResNet-50 [35] as the basis for the construction of feature map extraction. As with previous works [13,36], we further remove the original fully connected layer for multi-loss training and replicate the res_conv4_2 and subsequent blocks to build four independent branches. One branch is the global branch, and the others are local branches.

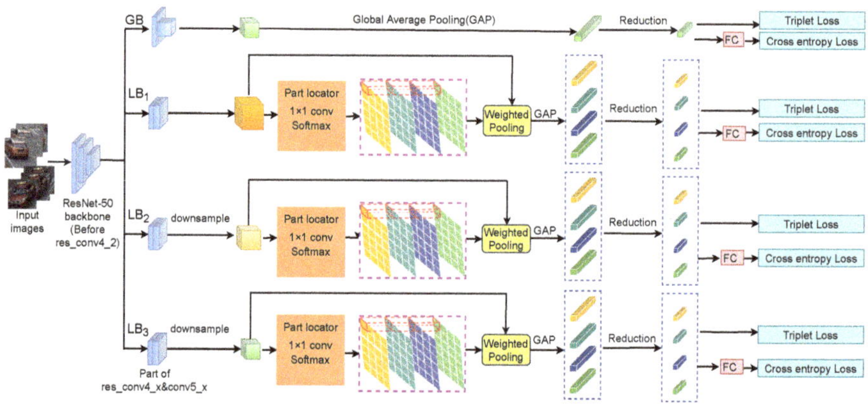

Figure 3. Network architecture of the MRF-SAPL model. It consists of a ResNet-50-based backbone, a global branch for extracting global information, and three local branches (LB_1, LB_2, LB_3). The three local branches extract part features with different semantic levels under different perceptual fields by the soft attention part learning (SAPL) module that consists of a part locator, an overlapping region penalty (ORP) constraint, and a part feature extractor.

The global branch learns a compact global feature representation. In this branch, we use the downsampling convolution layer with a step size of 2 in the res_conv5_1 block and conduct the global average pool (GAP) [37] operation on the final output feature map to obtain a 2048-dimensional feature vector. The dimension of the vector is further reduced to 256 through a dimensionality reduction module that consists of a 1×1 convolution layer, a batch normalization layer and a ReLU layer. We use the subnetwork composed of the backbone network and the global branch as the baseline network (baseline) in our experiments.

Intuitively, with the change in the receptive field size, human beings naturally observe an object from different semantic levels. Integrating discriminative information at different semantic levels can help people better identify objects. Therefore, in our network, we introduce three local branches to capture the semantics at different levels to obtain a large amount of discriminative information related to vehicle identities. To preserve enough detailed information, in all local branches, we do not use the downsampling operation in the res_conv5_1 block to provide appropriate space for the change in the receptive field. For each local branch, we first change the resolution of the final feature map to obtain the feature map under a specific receptive field. On the obtained feature map, a part locator uses soft attention to locate the internal semantic continuous parts and uses the ORP constraint to make the semantic patterns between parts different. Then, a part feature extractor generates the corresponding part features according to the positions of the parts on the feature map. Finally, we use the GAP operation on a part feature to obtain a 2048-dimensional feature vector, and the dimension of the vector is further reduced to 256 by a similar dimension reduction module in the global branch for cross entropy loss and triple loss.

The 256-dimensional feature vectors from the global and the three local branches are combined as the final feature representation for the vehicle Re-ID task. The global branch learns the overall discriminative information of vehicles, and the local branches learn the local information at different semantic levels. The global and local branches complement and cooperate with each other to improve their performance. Combining global features with local features can construct a more robust feature representation.

3.2. *Soft Attention Part Learning Module*

Some methods [10,12,38–40] train detection models with part labels to detect part locations and extract part-level features. However, it is difficult to collect the additional labels required by these methods. Our proposed SAPL module does not require any labels related to parts and can adaptively learn the locations of the parts on the feature map and extract the part features. It consists of a part locator, an overlapping region penalty constraint and a part feature extractor. After the final feature map of a local branch passes through the SAPL module, we obtain the position of each part on the feature map and a constant number of part-level features.

We define the feature map generated by an image via a backbone and a local branch as a three-dimensional tensor T with the size of $h \times w \times c$ (h, w, and c represent the channel height, width, and channel number, respectively). We define the activation vector viewed along the channel dimension as pixel z, which indicates the semantic information of its location. The purpose of the part locator is to locate the spatial positions of the parts on the T and to ensure the continuity and consistency of the internal semantics of the parts. Therefore, according to the semantic similarity between pixels, we use soft attention to assign them to each part. Specifically, the part locator is implemented by a fully connected layer followed by a softmax function, which is given by:

$$P(P_i \mid z) = \text{softmax}\left(W^T z\right) = \frac{\exp\left(W_i^T z\right)}{\sum_{j=1}^{p} \exp\left(W_j^T z\right)}, \tag{1}$$

where $P(P_i \mid z)$ is the prediction probability of part P_i at the z of the feature map T and W is the weight matrix of the fully connected layer. p is the number of vehicle parts.

After applying the part locator on each pixel of T, we obtain a set of attention maps $A = \{A_i \mid i = 1, \ldots, p\}$, where $A_i \in R^{h \times w}$ indicates the position of the i-th part on the feature map T and can be reshaped into a vector with dimension hw. To obtain multiple parts with different semantic patterns in a branch, rather than just focusing on the main discriminant area, the corresponding positions of different semantic parts should have a small overlap in space. Therefore, the overlapping region penalty (ORP) constraint is proposed to measure the area of the overlapping region of A that is defined as:

$$L_{\text{orp}} = \sum_{i \neq j} \frac{A_i^T A_j}{\|A_i\|_2 \cdot \|A_j\|_2}, \tag{2}$$

where $\|\cdot\|_2$ is the $L2$ norm. The ORP constraint adaptively softly divides the semantic space and generates multiple parts with different semantics. The combination of the part locator and ORP constraint has two beneficial effects for part segmentation. On the one hand, semantically similar features from a particular part are encouraged to be grouped together so that a strong part locator can be learned and corresponding parts from different images can be aligned. On the other hand, different semantic patterns between parts are encouraged to obtain multiple semantic complementary parts.

After obtaining the attention map of each part on the feature map, the part feature extractor generates the corresponding features for each part. Given that pixel z on the feature map belongs to the prediction probability of part P_i, the part feature extractor generates feature f_i of the part by weighted pooling that is calculated using the following formula:

$$f_i = \frac{\sum_{z \in T} P(P_i \mid z) \times z}{C_i + \epsilon}, \tag{3}$$

where divisor C_i is the accumulation of $P(P_i \mid z)$ on T and represents the salience of each part on the image. It should be noted that if a vehicle part is not visible in an image, all values of the attention map generated by the part locator for the part are close to 0. Hence, to avoid using 0 as a divisor when C_i is 0, ϵ is a small constant, which is set to 0.05 in our implementation.

3.3. Multi-Receptive-Field Granularity

Humans can capture different levels of semantics of a vehicle (such as vehicle type, lamp shape, and annual inspection sign) under different receptive fields (such as viewing distance or image resolution). Some types of semantics (e.g., with or without annual inspection) may be easier to capture in small receptive fields, while others (e.g., car door style) may be easier to capture in large receptive fields. Inspired by this, we propose a multi-receptive field soft attention part learning (MRF-SAPL) model to capture discriminative information on different semantic levels.

In MRF-SAPL, each local branch corresponds to a receptive field granularity, and we distinguish different receptive field granularities through different resolutions of the final feature maps of the local branches. According to the previous description, when a vehicle image passes through all local branches, we obtain a set of feature maps $T_{all} = \{T_m \mid i = 1,2,3\}$, where $T_i \in R^{h \times w \times c}$ includes $h \times w$ pixels (h, w, c represent the height, width, and number of channels, respectively). For the m_{th} granularity, we perform spatial average pooling with a downsampling factor m on the m_{th} feature map of T_{all} and obtain the downsampled feature map $T_m^{'} \in R^{h_m \times w_m \times c}$ of $h_m \times w_m$ pixels, where $h_m = h - 4(m-1)$ and $w_m = w - 4(m-1)$. The factorized feature map set is $T_{all}^{'} = \left\{T_m^{'} \mid i = 1,2,3\right\}$. We apply the SAPL module separately on all feature maps of $T_{all}^{'}$ to obtain multiple part features on different semantic levels.

3.4. Multitask Training

Multitask learning combines several related subtasks for overall learning and has been shown to be effective in Re-ID problems. We train our network by three types of supervision, i.e., the cross-entropy loss, the triplet loss, and the ORP loss L_{orp} in Equation (2). The cross-entropy loss is expressed as:

$$L_{id} = -\mathbb{I}_{k=y} \log(h), \tag{4}$$

where $\mathbb{I}_{k=y}$ returns 1 only when the predicted class k of a sample is equal to its supervised class y; otherwise, it returns 0. h is the probability that the sample is predicted to be class k.

The triplet loss separates the distance between examples of the same vehicle and the distance between examples of different vehicles by a certain threshold. We adopt the triplet loss with hard mining of Ref. [36]. During model training, P vehicles and K images of each vehicle are randomly sampled for each mini-batch to meet the triplet loss requirement. The triplet loss can be defined as:

$$\begin{aligned} L_{\text{tp}} = \sum_{i=1}^{P} \sum_{a=1}^{K} \Big[& \alpha + \max_{p=1,\ldots,K} \|a_i - p_i\|_2 \\ & - \min_{n=1,\ldots,K, j=1,\ldots,P, j \neq i} \|a_i - n_j\|_2 \Big]_+ , \end{aligned} \tag{5}$$

where α is the margin hyperparameter that controls the differences of intra and inter distances, and a_i, p_i, and n_j are the feature representations extracted from anchor, positive, and negative samples, receptively.

The cross-entropy loss and the triplet loss are used to supervise the network to learn identity-related global and local features. The overall training loss is formulated by:

$$L = L_{orp} + L_{id} + L_{tp}, \tag{6}$$

where L_{orp} prefers that the activated regions of different parts are nonoverlapping, and L_{id} and L_{tp} guide the model MRF-SAPL to activate the image discriminative regions rather than the background.

4. Experiments

4.1. Datasets and Evaluation Metric

We evaluate our proposed model on the VeRi-776 and VehicleID datasets, which are two mainstream datasets used in vehicle re-identification tasks.

VeRi-776 is the benchmark dataset of the vehicle Re-ID task. It consists of 49,357 images of 776 different vehicles captured by 20 nonoverlapping cameras in various directions and lighting conditions. The training and test sets contain 37,781 images of 576 vehicles and 11,579 images of 200 vehicles, respectively. According to the evaluation protocol in Ref. [2], we employ an image-to-trajectory cross-camera search, that is, using a vehicle image of a camera to search the trajectory of the same vehicle in other cameras. We measure the performance of our proposed model using mean average precision (mAP) and the Top-1 and Top-5 accuracies of cumulative matching curves (CMC).

VehicleID is another data-heavy benchmark consisting of 221,567 images from 26,328 different vehicles, of which 113,346 images from 13,164 vehicles are used for training and the rest are used for testing. The test set is further divided into three subsets of different sizes (small, medium, and large). In the inference phase, for each subset, one image is randomly selected from the images of each vehicle to form the gallery set, and the other images are used as query images. The average result of 10 repeated random samplings is regarded as the performance of our model on the VehicleID dataset. The evaluation indices of the VehicleID dataset are the Top-1 and Top-5 accuracies of CMC.

4.2. Implementation Details

Prior to feeding the vehicle images into the MRF-SAPL model, we resize them to 256×256 for more detailed information. The weights of the backbone and branches of MRF-SAPL are initialized with ResNet-50 [35] pretrained on ImageNet. During the training phases, we only randomly flip the input images horizontally for data augmentation. By randomly selecting 16 vehicles with 4 images per vehicle, the batch size is set to 64. We set the margin parameter of the triplet loss to 1.2 in all experiments. We choose stochastic gradient descent (SGD) as the optimizer. The initial learning rate is set to 0.01 and decays to 1×10^{-3} after 300 epochs and 1×10^{-4} after 400 epochs. The total training process lasted for 500 epochs. During testing, we concatenate all dimensionality-reduced feature vectors as a feature representation for each image in the query and gallery sets. The feature representations extracted from the original and horizontally flipped images are summed and normalized as the final vehicle feature embedding for the input image. Our model is implemented on two NVIDIA RTX 2080Ti GPUs using the PyTorch framework.

4.3. Comparison with State-of-the-Art Methods

We compared the proposed model MRF-SAPL in this paper with the current methods on the VeRi-776 and VehicleID datasets with the corresponding evaluation indices.

VeRi-776: Table 1 presents the comparison of previous methods and our model on the VeRi-776 dataset. Among these methods, Siamese+Path [1] relies on the temporal and spatial information of the vehicle images in the VeRi-776 dataset. TCPM [25] divides the final feature map from the horizontal and vertical directions and uses an external memory module to store partial features to model the global feature vector. Dual+SA [41] uses self-attention to generate attention maps about the vehicle model and vehicle ID and inputs the attention map to the part localization module to obtain the fine region features of ROIs Relying only on visual information, our proposed model MRF-SAPL achieves 81.5% mAP, 94.7% Top-1 accuracy, and 98.7% Top-5 accuracy. Our model is superior to these advanced methods in terms of mAP and Top-1 accuracy. A good mAP score shows that MRF-SAPL has a stronger ability to retrieve all corresponding images with the same identity in the gallery set, both for different camera attributes and viewpoint changes.

Table 1. The mAP, Top-1, and Top-5 on VeRi-776.

Method	mAP	Top-1	Top-5
Siames+Path [1]	0.583	0.835	0.900
VAMI [11]	0.501	0.770	0.908
RAM [42]	0.615	0.886	0.940
EALN [43]	0.574	0.844	0.941
AAVER [44]	0.612	0.890	0.947
PRN [28]	0.743	0.943	0.989
VCAM [40]	0.686	0.944	0.969
SPAN [26]	0.689	0.940	0.976
TCPM [25]	0.746	0.940	0.971
VSCR [45]	0.755	0.941	0.979
LCDNet+BRL[46]	0.760	0.946	0.980
Dual+SA [41]	0.786	0.944	0.992
MRF-SAPL (Ours)	0.815	0.947	0.987

VehicleID: We compared the scores of Top-1 and Top-5 on this dataset because each query vehicle has only one corresponding image in the gallery set. The comparison of the results on the Vehicle-ID dataset is shown in Table 2. VAMI [11] utilizes an adversarial training network and vehicle attributes to infer the features of the input vehicle under different viewpoints. PRN [28] utilizes an object detection network to generate the ROI for each vehicle part and extract part features. LRPT+TSAM+CP [47] lets a parameter generator network capable of generating complex image transform regions and a recognizer compete with each other to enhance images. An examination of the results presented in Table 2 shows that our MRF-SAPL outperforms SOTA TCPM by 2.3%, 0.8%, and 1.7% in Top-1 accuracy on small, medium, and large subsets, respectively. Compared with other models, our MRF-SAPL model achieves the best performance. Without resorting to additional labels, object detection, and parsing networks, our proposed model can learn rich fine-grained local features for vehicle Re-ID.

Table 2. The Top-1 and Top-5 on Vehicle ID.

Method	Small		Medium		Large	
	Top-1	Top-5	Top-1	Top-5	Top-1	Top-5
DRDL [48]	0.490	0.735	0.428	0.668	0.382	0.616
OIFE [9]	-	-	-	-	0.670	0.829
VAMI [11]	0.631	0.833	0.529	0.751	0.473	0.703
RAM [42]	0.752	0.915	0.723	0.870	0.677	0.845
AAVER[44]	0.747	0.938	0.686	0.900	0.635	0.856
EALN [43]	0.751	0.881	0.718	0.839	0.693	0.814
PRN [28]	0.784	0.923	0.750	0.883	0.742	0.864
SAVER[17]	0.799	0.952	0.776	0.911	0.753	0.883
TCPM [25]	0.820	0.964	0.788	0.943	0.746	0.907
Dual+SA[41]	-	-	-	-	0.738	0.835
SN++ [49]	0.767	0.870	0.748	0.842	0.739	0.836
LRPT + TSAM + CP[47]	0.779	0.935	0.779	0.907	0.745	0.865
MRF-SAPL (Ours)	0.843	0.977	0.796	0.941	0.763	0.916

4.4. Ablation Study

We conducted extensive experiments on the VeRi-776 dataset and compared the performance of different structures to determine the optimal structure of the proposed model.

Soft attention part learning module. In Table 3, "+" indicates the combination of different branches. Baseline+LB_1, Baseline+LB_2, and Baseline+LB_3 outperform Baseline by 6.9%, 6.8%, and 4.7% in mAP, respectively. Baseline+LB_1 (W/O SAPL) means removing the SAPL module from Baseline+LB_1 and dividing the final feature map of the LB_1 branch evenly into four parts vertically. We observe a 1.1% decrease in mAP with Baseline+LB_1(W/O SAPL) compared to Baseline+LB_1. This demonstrates the effectiveness of our proposed soft-attention part learning module.

Table 3. Performance comparison of MRF-SAPL with different architecture on VeRi-776.

Method	mAP	Top-1	Top-5
Baseline	0.726	0.918	0.973
Baseline+LB_1	0.795	0.932	0.985
Baseline+LB_1(W/O SAPL)	0.784	0.928	0.980
Baseline+LB_2	0.794	0.938	0.982
Baseline+LB_3	0.773	0.924	0.983
Baseline+LB_1+LB_2	0.813	0.935	0.983
Baseline+LB_1+LB_3	0.805	0.935	0.982
Baseline+LB_2+LB_3	0.795	0.945	0.982
LB_1+LB_2+LB_3	0.802	0.938	0.982
Baseline+Single(LB_1+LB_2+LB_3)	0.771	0.923	0.979
MRF-SAPL (Ours)	0.815	0.947	0.987

Multi-receptive field granularity. Our framework contains three local branches with different receptive field granularities, namely fine-grained, medium-grained, and coarse-grained branches, which are responsible for part segmentation and feature extraction under different receptive fields. We investigate the role of multiple receptive field granularities in MRF-SAPL by progressively combining local branches based on the baseline. From Table 3, we can observe that combining two local branches with different receptive field granularities can further improve the performance, and MRF-SAPL using three receptive fields of different sizes to learn part-level features achieves the best performance; this shows that learning part-level features with different semantic-level preferences using different granularities of receptive fields is effective.

Global branch. In Table 3, LB_1+LB_2+LB_3 means that the global branch is removed from MRF-SAPL, and only three local branches are used to train the network. At test time, feature vectors from the three local branches are extracted and concatenated to compute a similarity score. Compared with MRF-SAPL, the accuracy of LB_1+LB_2+LB_3 decreases by 1.3% in mAP. This is because the global branch with a larger receptive field can learn the overall discriminant information of vehicles, complementing the local branches that learn fine-grained local discriminant information.

Multiple local branches. In our method, we use three local branches to learn part features with different semantic levels from vehicle images; therefore, we would like to know whether it is possible to learn part features with different semantic levels using a single branch. To verify this hypothesis, we can perform spatial soft segmentation on the final feature map of the same local branch under multiple receptive fields and apply the corresponding constraints of the method proposed in this paper. From Table 3, we can observe that Baseline+Single(LB_1+LB_2+LB_3) relying on a single local branch has a 4.4% performance drop in mAP compared to MRF-SAPL. This may because using different receptive field granularities to softly divide the space of the same feature map will have different or even the opposite effects on its res_conv5 layer.

Influence of the number of parts. To study the impact of the number of parts on the Re-ID accuracy, we introduce several divisions with different numbers of parts. Specifically, we conduct experiments with the 2, 3, 4, and 5 parts. In each experiment, the feature maps on the three local branches are divided into the same number of regions. The experimental results are summarized in Table 4. With an increasing number of parts, mAP first increases, but does not always increase. When the number of parts is equal to 5, mAP starts to decrease. This is because when the spatial of a feature map is too finely divided, some semantic information that is meaningful for vehicle Re-ID will be decomposed into segments that do not have general discriminative abilities. In our proposed method, the number of parts of the final feature maps of all local branches is set to 4.

Table 4. Influence of the number of parts on VeRi-776.

The Number of Parts	mAP	Top-1	Top-5
2	0.801	0.944	0.985
3	0.807	0.939	0.984
4	0.815	0.947	0.987
5	0.802	0.938	0.982

Vehicle sorting and attention map visualization. Figure 4 shows the qualitative results of our MRF-SAPL model on the vehicleID dataset, where each query image only has one target image in the gallery set. In Figure 4, the images on the left are the query images, and the images on the right are the Top-5 nearest neighborhoods from the gallery. Figure 5 shows the attention map visualization of the SAPL module in the LB_1 branch when the number of parts is 4. From Figure 5, we can observe that the four attention maps learned by the SAPL module focus on four different regions: the main area consisting of the lower part of the windshield and the hood, the roof area, the annual inspection mark area in the upper part of the windshield, and the fog lamp area. For the first row in Figure 4, the query image and the Top-3 image are two different vehicles belonging to the same manufacturer and model, with extremely similar appearances. The SAPL module accurately distinguishes them by focusing on the annual inspection mark area. For the third row in Figure 4, Top-4 and Top-5 have large color differences in the main area compared to the query, so they are ranked lower. Although Top-1 and the query are the same vehicle captured under different views, and Top-2 and Top-3 are extremely similar to the query, the SAPL module can distinguish them by focusing on the roof area and the main area, respectively. For the fourth row in Figure 4, the query and Top-1 are two different vehicles belonging to the same manufacturer and model, both of which were captured from a rear view. Top-2 and the query are the same vehicle captured from different perspectives. In this case, the SAPL module has difficulty distinguishing between Top-1 and Top-2 because there is no obvious difference in appearance information between Top-1 and the query. This demonstrates that MRF-SAPL is able to effectively distinguish vehicles with extremely similar appearances in most cases.

Figure 4. Visualization of the ranking list on the vehicleID dataset. The images in the first column are the vehicle images to query. The remaining images in each row are the Top-5 ranking results retrieved from the gallery that are most similar to the corresponding query image. The retrieved images with the same ID as the query image are shown with the green border, while the error samples are shown with the red border. Note: Some vehicle images in the VehicleID dataset contain Chinese characters for the shooting time and location, such as the characters in the top left corner of the Top-2 image in the second line, and their impact on vehicle recognition can be negligible.

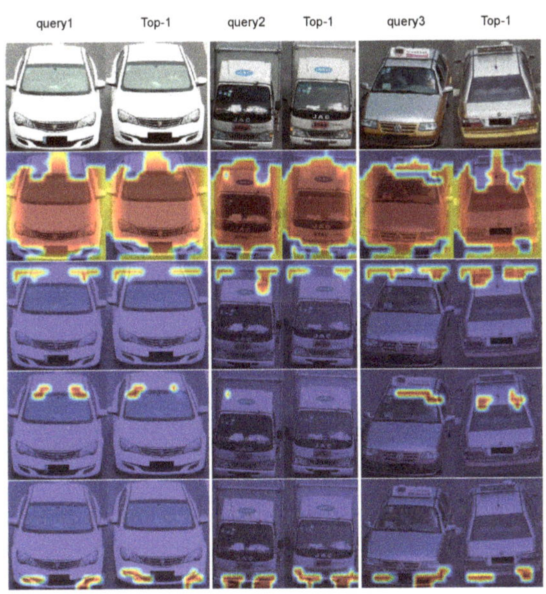

Figure 5. Visualization of attention maps. Each column displays four attention maps generated by the LB_1 branch of MRF-SAPL for a image. The first and second columns correspond to the attention maps of the query image and the Top-1 image in the first row of Figure 4, respectively. The third and fourth columns correspond to the attention maps of the query image and the Top-1 image in the second row of Figure 4, respectively. The fifth and sixth columns correspond to the attention maps of the query image and the Top-1 image in the third row of Figure 4.

5. Conclusions

In this paper, we propose a model for part-level feature learning, the Multi-Receptive Field Soft Attention Part Learning (MRF-SAPL) model. The model can learn fine-grained features at multiple semantic levels to effectively distinguish different vehicles with similar appearances. In particular, the soft-attention part learning module (SAPL) in this model does not require any part-related labels and can adaptively learn to localize the locations of the parts on the feature map to suppress severe spatial misalignments in vehicle Re-ID. Furthermore, we obtain parts with different semantic patterns by forcing the regions corresponding to the parts on the final feature map of a local branch to be as nonoverlapping as possible. Our Multi-Receptive Field Soft Attention Part Learning model achieves state-of-the-art performance on two public datasets.

Author Contributions: Conceptualization, X.P. and Y.Y.; methodology, X.P. and Y.Z.; validation, X.P., Y.Y. and Y.Z.; investigation, X.P. and Y.Z.; writing—original draft preparation, X.P.; writing—review and editing, Y.Y. and Y.Z.; visualization, X.P. All authors have read and agreed to the published version of the manuscript.

Funding: This research was funded by the Major Basic Research Project of the Natural Science Foundation of Shandong Province (Grant No. ZR2021ZD15).

Institutional Review Board Statement: Not applicable.

Data Availability Statement: All data included in this study are available upon request by contact with the corresponding author.

Conflicts of Interest: The authors declare no conflict of interest.

References

1. Shen, Y.; Xiao, T.; Li, H.; Yi, S.; Wang, X. Learning deep neural networks for vehicle re-id with visual-spatio-temporal path proposals. In Proceedings of the 2017 IEEE International Conference on Computer Vision (ICCV), Venice, Italy, 22–29 October 2017; pp. 1918–1927.
2. Liu, X.; Liu, W.; Mei, T.; Ma, H. PROVID: Progressive and multimodal vehicle reidentification for large-scale urban surveillance. *IEEE Trans. Multimed.* **2018**, *20*, 645–658. [CrossRef]
3. Zheng, A.; Lin, X.; Li, C.; He, R.; Tang, J. Attributes guided feature learning for vehicle re-identification. *arXiv* **2019**, arXiv:1905.08997.
4. He, L.; Sun, Z.; Zhu, Y.; Wang, Y. Recognizing partial biometric patterns. *arXiv* **2018**, arXiv:1810.07399.
5. Sun, Y.; Zheng, L.; Yang, Y.; Tian, Q.; Wang, S. Beyond part models: Person retrieval with refined part pooling (and a strong convolutional baseline). In Proceedings of the European Conference on Computer Vision (ECCV), Munich, Germany, 8–14 September 2018; pp. 480–496.
6. Fan, X.; Luo, H.; Zhang, X.; He, L.; Zhang, C.; Jiang, W. Scpnet: Spatial-channel parallelism network for joint holistic and partial person re-identification. In Proceedings of the Asian Conference on Computer Vision, Daejeon, Republic of Korea, 5–9 November 2018; Springer: Berlin/Heidelberg, Germany, 2018; pp. 19–34.
7. He, L.; Liang, J.; Li, H.; Sun, Z. Deep spatial feature reconstruction for partial person re-identification: Alignment-free approach. In Proceedings of the 2018 IEEE/CVF Conference on Computer Vision and Pattern Recognition, Salt Lake City, UT, USA, 18–23 June 2018; pp. 7073–7082.
8. Chen, H.; Lagadec, B.; Brémond, F. Partition and Reunion: A Two-Branch Neural Network for Vehicle Re-identification. In *CVPR Workshops*; IEEE: Piscataway, NJ, USA, 2019; pp. 184–192.
9. Wang, Z.; Tang, L.; Liu, X.; Yao, Z.; Yi, S.; Shao, J.; Yan, J.; Wang, S.; Li, H.; Wang, X. Orientation invariant feature embedding and spatial temporal regularization for vehicle re-identification . In Proceedings of the 2017 IEEE International Conference on Computer Vision (ICCV), Venice, Italy, 22–29 October 2017; pp. 379–387.
10. Gao, S.; Wang, J.; Lu, H.; Liu, Z. Pose-guided visible part matching for occluded person reid. In Proceedings of the 2020 IEEE/CVF Conference on Computer Vision and Pattern Recognition (CVPR), Seattle, WA, USA, 13–19 June 2020; pp. 11744–11752.
11. Zhou, Y.; Shao, L. Viewpoint-Aware Attentive Multi-View Inference for Vehicle Re-Identification. In Proceedings of the 2018 IEEE/CVF Conference on Computer Vision and Pattern Recognition, Salt Lake City, UT, USA, 18–23 June 2018; pp. 6489–6498.
12. Miao, J.; Wu, Y.; Liu, P.; Ding, Y.; Yang, Y. Pose-Guided Feature Alignment for Occluded Person Re-Identification. In Proceedings of the 2019 IEEE/CVF International Conference on Computer Vision (ICCV), Seoul, Republic of Korea, 27 October–2 November 2019; pp. 542–551.
13. Liu, K.; Xu, Z.; Hou, Z.; Zhao, Z.; Su, F. Further Non-local and Channel Attention Networks for Vehicle Re-Identification. In Proceedings of the 2020 IEEE/CVF Conference on Computer Vision and Pattern Recognition Workshops (CVPRW), Seattle, WA, USA, 14–19 June 2020; pp. 2494–2500.

14. Liao, S.; Hu, Y.; Zhu, X.; Li, S.Z. Person re-identification by local maximal occurrence representation and metric learning. In Proceedings of the 2015 IEEE Conference on Computer Vision and Pattern Recognition (CVPR), Boston, MA, USA, 7–12 June 2015; pp. 2197–2206.
15. Yang, Y.; Yang, J.; Yan, J.; Liao, S.; Yi, D.; Li, S.Z. Salient color names for person re-identification. In Proceedings of the Computer Vision—ECCV 2014: 13th European Conference, Zurich, Switzerland, 6–12 September 2014; Springer International Publishing: Berlin/Heidelberg, Germany, 2014; pp. 536–551.
16. Liao, S.; Li, S.Z. Efficient psd constrained asymmetric metric learning for person re-identification. In Proceedings of the 2015 IEEE International Conference on Computer Vision (ICCV), Santiago, Chile, 7–13 December 2015; pp. 3685–3693.
17. Khorramshahi, P.; Peri, N.; Chen, J.; Chellappa, R. The devil is in the details: Self-supervised attention for vehicle re-identification. In Proceedings of the Computer Vision–ECCV 2020: 16th European Conference, Glasgow, UK, 23–28 August 2020; pp. 369–386.
18. Zheng, W.; Gong, S.; Xiang, T. Reidentification by relative distance comparison. *IEEE Trans. Pattern Anal. Mach. Intell.* **2012**, *35*, 653–668. [CrossRef] [PubMed]
19. Kalayeh, M.M.; Basaran, E.; Gökmen, M.; Kamasak, M.E.; Shah, M. Human Semantic Parsing for Person Re-Identification. In Proceedings of the 2018 IEEE/CVF Conference on Computer Vision and Pattern Recognition, Salt Lake City, UT, USA, 18–23 June 2018; pp. 1062–1071.
20. Li, W.; Zhu, X.; Gong, S. Harmonious attention network for person re-identification. In Proceedings of the 2018 IEEE/CVF Conference on Computer Vision and Pattern Recognition, Salt Lake City, UT, USA, 18–23 June 2018; pp. 2285–2294.
21. Liu, X.; Zhao, H.; Tian, M.; Sheng, L.; Shao, J.; Yi, S.; Yan, J.; Wang, X. Hydraplus-net: Attentive deep features for pedestrian analysis. In Proceedings of the 2017 IEEE International Conference on Computer Vision (ICCV), Venice, Italy, 22–29 October 2017; pp. 350–359.
22. Sarfraz, M.S.; Schumann, A.; Eberle, A.; Stiefelhagen, R. A Pose-Sensitive Embedding for Person Re-Identification with Expanded Cross Neighborhood Re-Ranking. In Proceedings of the 2018 IEEE/CVF Conference on Computer Vision and Pattern Recognition, Salt Lake City, UT, USA, 18–23 June 2018; pp. 420–429.
23. Xu, J.; Zhao, R.; Zhu, F.; Wang, H.; Ouyang, W. Attention-aware compositional network for person re-identification. In Proceedings of the 2018 IEEE/CVF Conference on Computer Vision and Pattern Recognition, Salt Lake City, UT, USA, 18–23 June 2018; pp. 2119–2128.
24. Song, C.; Huang, Y.; Ouyang, W.; Wang, L. Mask-guided contrastive attention model for person re-identification. In Proceedings of the 2018 IEEE/CVF Conference on Computer Vision and Pattern Recognition, Salt Lake City, UT, USA, 18–23 June 2018; pp. 1179–1188.
25. Wang, H.; Peng, J.; Jiang, G.; Xu, F.; Fu, X. Discriminative feature and dictionary learning with part-aware model for vehicle re-identification. *Neurocomputing* **2021**, *438*, 55–62. [CrossRef]
26. Chen, T.; Liu, C.; Wu, C.; Chien, S. Orientation-aware vehicle re-identification with semantics-guided part attention network. In Proceedings of the Computer Vision–ECCV 2020: 16th European Conference, Glasgow, UK, 23–28 August 2020; pp. 330–346.
27. Liu, J.; Ni, B.; Yan, Y.; Zhou, P.; Cheng, S.; Hu, J. Pose transferrable person re-identification. In Proceedings of the 2018 IEEE/CVF Conference on Computer Vision and Pattern Recognition, Salt Lake City, UT, USA, 18–23 June 2018; pp. 4099–4108.
28. He, B.; Li, J.; Zhao, Y.; Tian, Y. Part-Regularized Near-Duplicate Vehicle Re-Identification. In Proceedings of the 2019 IEEE/CVF Conference on Computer Vision and Pattern Recognition (CVPR), Long Beach, CA, USA, 15–20 June 2019; pp. 3997–4005.
29. Li, Y.; He, J.; Zhang, T.; Liu, X.; Zhang, Y.; Wu, F. Diverse Part Discovery: Occluded Person Re-Identification with Part-Aware Transformer. In Proceedings of the 2021 IEEE/CVF Conference on Computer Vision and Pattern Recognition (CVPR), Nashville, TN, USA, 20–25 June 2021; pp. 2898–2907.
30. He, K.; Zhang, X.; Ren, S.; Sun, J. Spatial Pyramid Pooling in Deep Convolutional Networks for Visual Recognition. *IEEE Trans. Pattern Anal. Mach. Intell.* **2015**, *37*, 1904–1916. [CrossRef] [PubMed]
31. Zhao, H.; Shi, J.; Qi, X.; Wang, X.; Jia, J. Pyramid Scene Parsing Network. In Proceedings of the 2017 IEEE Conference on Computer Vision and Pattern Recognition (CVPR), Honolulu, HI, USA, 21–26 July 2017; pp. 6230–6239.
32. Szegedy, C.; Liu, W.; Jia, Y.; Sermanet, P.; Reed, S.E.; Anguelov, D.; Erhan, D.; Vanhoucke, V.; Rabinovich, A. Going deeper with convolutions. In Proceedings of the 2015 IEEE Conference on Computer Vision and Pattern Recognition (CVPR), Boston, MA, USA, 7–12 June 2015; pp. 1–9.
33. Tolstikhin, I.O.; Houlsby, N.; Kolesnikov, A.; Beyer, L.; Zhai, X.; Unterthiner, T.; Yung, J.; Steiner, A.; Keysers, D.; Uszkoreit, J.; et al. MLP-Mixer: An all-MLP Architecture for Vision. *Adv. Neural Inf. Process. Syst.* **2021**, *34*, 24261–24272.
34. Li, Y.; Yao, T.; Pan, Y.; Mei, T. Contextual Transformer Networks for Visual Recognition. *IEEE Trans. Pattern Anal. Mach. Intell.* **2023**, *45*, 1489–1500. [CrossRef] [PubMed]
35. He, K.; Zhang, X.; Ren, S.; Sun, J. Deep Residual Learning for Image Recognition. In Proceedings of the 2016 IEEE Conference on Computer Vision and Pattern Recognition (CVPR), Las Vegas, NV, USA, 27–30 June 2016; pp. 770–778.
36. Wang, G.; Yuan, Y.; Chen, X.; Li, J.; Zhou, X. Learning Discriminative Features with Multiple Granularities for Person Re-Identification. In Proceedings of the 26th ACM international conference on Multimedia, Seoul, Republic of Korea, 22–26 October 2018; pp. 274–282.
37. Lin, M.; Chen, Q.; Yan, S. Network In Network. *arXiv* **2014**, arXiv:1312.4400.

38. He, L.; Liu, W. Guided saliency feature learning for person re-identification in crowded scenes. In *Computer Vision—ECCV 2020*; Vedaldi, A., Bischof, H., Brox, T., Frahm, J.M., Eds.; Lecture Notes in Computer Science; Springer: Cham, Switzerland, 2020; pp. 357–373.
39. He, L.; Wang, Y.; Liu, W.; Zhao, H.; Sun, Z.; Feng, J. Foreground-aware pyramid reconstruction for alignment-free occluded person re-identification. In Proceedings of the 2019 IEEE/CVF International Conference on Computer Vision (ICCV), Seoul, Republic of Korea, 27 October–2 November 2019; pp. 8450–8459.
40. Chen, T.; Lee, M.; Liu, C.; Chien, S. Viewpoint-Aware Channel-Wise Attentive Network for Vehicle Re-Identification. In Proceedings of the 2020 IEEE/CVF Conference on Computer Vision and Pattern Recognition Workshops (CVPRW), Seattle, WA, USA, 14–19 June 2020; pp. 2448–2455.
41. Zhang, J.; Chen, J.; Cao, J.; Liu, R.; Bian, L.; Chen, S. Dual attention granularity network for vehicle re-identification. *Neural Comput. Appl.* **2022**, *34*, 2953–2964. [CrossRef]
42. Liu, X.; Zhang, S.; Huang, Q.; Gao, W. RAM: A Region-Aware Deep Model for Vehicle Re-Identification. In Proceedings of the 2018 IEEE International Conference on Multimedia and Expo (ICME), San Diego, CA, USA, 23–27 July 2018; pp. 1–6.
43. Lou, Y.; Bai, Y.; Liu, J.; Wang, S.; Duan, Li. Embedding Adversarial Learning for Vehicle Re-Identification. *IEEE Trans. Image Process.* **2019**, *28*, 3794–3807. [CrossRef] [PubMed]
44. Khorramshahi, P.; Kumar, A.; Peri, N.; Rambhatla, S.S.; Chen, J.; Chellappa, R. A Dual-Path Model with Adaptive Attention for Vehicle Re-Identification. In Proceedings of the 2019 IEEE/CVF International Conference on Computer Vision (ICCV), Seoul, Republic of Korea, 27 October–2 November 2019; pp. 6131–6140.
45. Teng, S.; Zhang, S.; Huang, Q.; Sebe, N. Viewpoint and scale consistency reinforcement for UAV vehicle re-identification. *Int. J. Comput. Vis.* **2021**, *129*, 719–735. [CrossRef]
46. Fu, X.; Peng, J.; Jiang, G.; Wang, H. Learning latent features with local channel drop network for vehicle re-identification. *Eng. Appl. Artif. Intell.* **2022**, *107*, 104540. [CrossRef]
47. Chen, Y.; Ke, W.; Lin, H.; Lam, C.; Lv, K.; Sheng, H.; Xiong, Z. Local perspective based synthesis for vehicle re-identification: A transformation state adversarial method. *J. Vis. Commun. Image Represent* **2022**, *83*, 103432. [CrossRef]
48. Liu, H.; Tian, Y.; Wang, Y.; Pang, L.; Huang, T. Deep relative distance learning: Tell the difference between similar vehicles. In Proceedings of the 2016 IEEE Conference on Computer Vision and Pattern Recognition (CVPR), Las Vegas, NV, USA, 27–30 June 2016; pp. 2167–2175.
49. Li, K.; Ding, Z.; Li, K.; Zhang, Y.; Fu, Y. Vehicle and Person Re-Identification with Support Neighbor Loss. *IEEE Trans. Neural Networks Learn. Syst.* **2022**, *33*, 826–838. [CrossRef] [PubMed]

entropy

MDPI

Article

Coupling Quantum Random Walks with Long- and Short-Term Memory for High Pixel Image Encryption Schemes

Junqing Liang [1], Zhaoyang Song [1], Zhongwei Sun [1], Mou Lv [2] and Hongyang Ma [3,*]

1 School of Information and Control Engineering, Qingdao University of Technology, Qingdao 266033, China
2 School of Environmental and Municipal Engineerin, Qingdao University of Technology, Qingdao 266033, China
3 School of Science, Qingdao University of Technology, Qingdao 266033, China
* Correspondence: hongyang_ma@aliyun.com

Abstract: This paper proposes an encryption scheme for high pixel density images. Based on the application of the quantum random walk algorithm, the long short-term memory (LSTM) can effectively solve the problem of low efficiency of the quantum random walk algorithm in generating large-scale pseudorandom matrices, and further improve the statistical properties of the pseudorandom matrices required for encryption. The LSTM is then divided into columns and fed into the LSTM in order for training. Due to the randomness of the input matrix, the LSTM cannot be trained effectively, so the output matrix is predicted to be highly random. The LSTM prediction matrix of the same size as the key matrix is generated based on the pixel density of the image to be encrypted, which can effectively complete the encryption of the image. In the statistical performance test, the proposed encryption scheme achieves an average information entropy of 7.9992, an average number of pixels changed rate (NPCR) of 99.6231%, an average uniform average change intensity (UACI) of 33.6029%, and an average correlation of 0.0032. Finally, various noise simulation tests are also conducted to verify its robustness in real-world applications where common noise and attack interference are encountered.

Keywords: image encryption; high pixel density; neural networks; quantum random walk

Citation: Liang, J.; Song, Z.; Sun, Z.; Lv, M.; Ma, H. Coupling Quantum Random Walks with Long- and Short-Term Memory for High Pixel Image Encryption Schemes. *Entropy* **2023**, *25*, 353. https://doi.org/10.3390/e25020353

Academic Editors: Oleg Sergiyenko, Wendy Flores-Fuentes, Julio Cesar Rodriguez-Quinonez and Jesús Elías Miranda-Vega

Received: 13 December 2022
Revised: 7 February 2023
Accepted: 9 February 2023
Published: 14 February 2023

1. Introduction

With the rapid development of Internet technology, more and more high-value data and information is being transmitted over the Internet, and therefore the security of data transmission is becoming more and more important. While ordinary data can be hidden and protected by classical encryption schemes such as DES [1] and AES [2], the information contained in an RGB image is represented by the pixel values. Because of the strong correlation between the neighbouring pixel values of RGB images and the amount of information stored in images, classical encryption schemes are often unable to achieve good encryption of image information, so the encryption of image information is separated from classical data encryption and becomes a separate research direction, focusing on image specific encryption schemes from the data information characteristics of images [3–8]. One very promising direction is the application of neural networks to image encryption. This is because cryptography places particular emphasis on the introduction of nonlinear transformations, which is a distinctive feature of neural networks, and, in addition to this, neural networks have characteristics such as ultra-fast parallel processing and operate in matrix form, all of which are extremely well suited to the field of image encryption, making neural networks increasingly interesting in the field of image encryption [9–11].

The LSTM [12] is a special type of recurrent neural network (RNN) [13] that uses the 'inner loop' of a neural network to preserve the contextual information of a time series, allowing the use of past signal data to infer an understanding of the current signal. Theoretically, RNN can retain information from any moment in time. However, in practice, the transfer of information tends to decay over long time intervals, and the effectiveness of the information is greatly reduced after a certain period of time. As a result, RNN is not well equipped to deal with the problem of long-term information dependence, resulting in a tendency to rely only on the most recent input information for inference. To overcome this problem, LSTM is proposed to solve the long-term dependency problem. In contrast to RNN, remembering the content of earlier moments is its default behaviour. Therefore, it does not require a significant cost specifically and works better.

Quantum computing is a new computing mode that follows the laws of quantum mechanics to regulate quantum information units for computing [14]. Quantum algorithm [15–18] is an algorithm based on quantum computation. By using the unique behavior of quantum mechanics, such as superposition, entanglement, and interference, some algorithms have achieved exponential acceleration compared with classical algorithms [17,19]. Quantum random walk (QW) is a quantum algorithm, which was first proposed by Aharonov et al. [20], including continuous time QW [21] and discrete time QW [22]. Compared with the classical random walk, the algorithm has a significant improvement in computational efficiency, and its time complexity is reduced from $O(n^2)$ to $O(n)$. On the basis of one-dimensional QW, Baryshnikov et al. studied the difference between two-dimensional and one-dimensional coordinate space, and expounded the advantages and unique properties of two-dimensional QW [23]. Although QW is a quantum algorithm, its probability matrix can be solved by classical computers, and the algorithm complexity is still O(n), which makes QW be able to be applied in classical computers in advance.

Both LSTM and QW have applications in image encryption. He et al. [24] proposed an OF-LSTMS that replaces the matrix operation in LSTM with an XOR operation to obtain an encrypted image after a single forward propagation. Yang et al. [25] studied the properties of one-dimensional QW and applied it to quantum image encryption for the first time. Abd et al. [26] analyzed the statistical properties of the probability distribution matrix of two-dimensional quantum walks and applied it to image encryption; Ma et al. [27] combined the discrete cosine transform (DCT) [28] and the probability matrix of alternating quantum walks (AQW) for image encryption, etc.

Although QW probability matrices have been widely used in the field of image encryption, they still have shortcomings and are too inefficient when dealing with high pixel images. The time complexity of the one-dimensional AQW probability matrix is $O(n)$, and the computational complexity of the AQW probability matrix is $O(n^2)$, which is still polynomial in time complexity, but the time consumed to generate the QW probability matrix is unacceptable in practical applications to encrypt high pixel value images. At the same time, we also found that the statistical properties required for the encryption of the QW probability matrix are not satisfactory, so when QW is used for encryption, other algorithms are often used to improve the encryption, e.g., Ma used a discrete cosine transform algorithm to perform further dislocation encryption in the DCT domain after applying QW to confuse the pixel values. This does not increase the encryption efficiency too much, but the use of separate algorithms for the scrambling and obfuscation phases nullifies the advantage of having an infinite key matrix for the QW, as it can only participate in one of the scrambling and obfuscation phases, and the two phases are independent of each other.

In order to optimize the statistical properties of the QW probability matrix and its performance on high pixel precision image encryption for better encryption, we propose an image encryption scheme that combines neural networks with quantum algorithms. By combining the QW with the LSTM, the initial matrix is generated using the QW probability matrix, and after training through the LSTM, a suitable prediction matrix is output as the key matrix for encryption according to the required pixel accuracy of the image to be encrypted. We show that this combination can improve the efficiency of the key matrix generation, and at the same time, because the QW probability matrix has strong randomness, the LSTM can not effectively find its pattern to predict, so the generated prediction matrix is also disordered, and has better statistical properties than the QW probability matrix for encryption, which can be better used as a key matrix for encryption. Section 2 of this paper presents the basics related to encryption schemes, including the study and analysis of LSTM and AQW. Section 3 presents specific encryption schemes. Section 4 presents the simulation and theoretical analysis of this paper for detecting the effectiveness of the encryption scheme and lists the comparison of similar schemes to the encryption scheme proposed in this paper. Section 5 concludes the work in this paper and also provides an outlook on the subsequent work. The most critical module of the LSTM is the cell state, which is represented by C_t, the current state at the current moment, and is generated by the state C_{t-1} at the previous moment together with the signal input x_t at the current moment, while C_t will continue to be passed to the next moment together with x_{t+1} to generate C_{t+1}.

2. Related Work and Background Knowledge

2.1. Long Short-Term Memory

LSTM is a type of Recurrent Neural Network (RNN) that has been widely used in various applications, such as speech recognition, natural language processing, and time series prediction. Unlike traditional RNNs, LSTMs have an internal memory cell that enables them to maintain information over a longer period of time, making them well-suited for tasks that require modeling sequential data with long-term dependencies.

The core component of an LSTM unit is its memory cell, which is responsible for maintaining information over a long period of time. The memory cell is controlled by three types of gates: the input gate, the forget gate, and the output gate. The input gate controls the flow of new information into the memory cell, the forget gate controls the amount of information retained from the previous time step, and the output gate controls the flow of information out of the memory cell and into the hidden state of the LSTM unit.

The LSTM architecture is derived from the equations that govern the behavior of the gates and the memory cell. At each time step, the input, forget, and output gates are computed using a sigmoid activation function, while the memory cell is updated using a tanh activation function. The equations governing the behavior of the LSTM unit are given by:

$$i_t = \sigma(W_{ix}x_t + W_{ih}h_{t-1} + b_i) \tag{1}$$

$$f_t = \sigma(W_{fx}x_t + W_{fh}h_{t-1} + b_f) \tag{2}$$

$$o_t = \sigma(W_{ox}x_t + W_{oh}h_{t-1} + b_o) \tag{3}$$

$$c_t = f_t \odot c_{t-1} + i_t \odot \tanh(W_{cx}x_t + W_{ch}h_{t-1} + b_c) \tag{4}$$

$$h_t = o_t \odot \tanh(c_t) \tag{5}$$

where x_t is the input at time step t, h_{t-1} is the hidden state at the previous time step, i_t, f_t, and o_t are the input, forget, and output gates at time step t, c_t is the memory cell at time step t, and σ and tanh are the sigmoid and hyperbolic tangent activation functions, respectively.

The LSTM architecture has proven to be highly effective in various applications, due to its ability to capture long-term dependencies and selectively forget or retain information. The equations presented here provide a foundation for understanding the behavior of LSTMs and for developing new models that incorporate LSTM units.

The chain structure diagram of the LSTM is shown in Figure 1, which illustrates the chain relationship between the three adjacent substructures and the composition of each LSTM substructure.

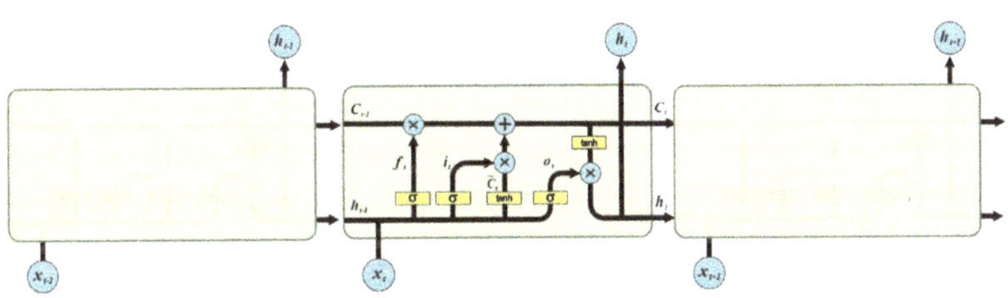

Figure 1. Chain model for LSTM.

2.2. *Quantum Random Walk*

This paper is based on the theory of discrete-time QW. The discrete-time QW consists of four main elements: the walker, the coins carried by the walker, the coin toss, and the walk rule.

The Hilbert space $\hat{H}$ of a one-dimensional discrete-time QW tensor consists of the walker position space H_w and the coin space H_Γ: $\hat{H} = H_w \otimes H_\Gamma$. In a QW, each step of the walk is determined by a unique coin flip operator Γ:

$$\Gamma = \begin{pmatrix} \cos\beta & \sin\beta \\ \sin\beta & -\cos\beta \end{pmatrix} \tag{6}$$

After the coin toss is completed, the movement of the walker is specified by the conditional displacement operator S_i: $S_i|\hat{x}\rangle = \left|\hat{x} + (-1)^\Gamma\right|, \Gamma \in 0,1$ The $|\hat{x}\rangle(\hat{x} \in Z)$ in the above equation forms the base vector of the walker's position space; the two base vectors $|0\rangle, |1\rangle$ form the coin space. We specify: when the coin state is $|0\rangle$, the walker is manipulated to move one unit in the forward direction; when the coin state is $|1\rangle$, the walker is manipulated to move one unit in the reverse direction.

In the AQW used in this paper, the walker controlled by the coin operator alternates between two arbitrarily chosen vertical directions $\tilde{x}$ and $\tilde{y}$, and the walking operator $\hat{U}$ for the whole QW process can be described as:

$$\hat{U} = \hat{S}_{\tilde{y}}(I \otimes H_\Gamma)\hat{S}_{\tilde{x}}(I \otimes H_\Gamma) \tag{7}$$

where $\hat{S}_{\tilde{y}}, \hat{S}_{\tilde{x}}$ are the displacement operators of the walker at each point on the $\tilde{x}$ and $\tilde{y}$ axes:

$$\begin{aligned} \hat{S}_{\tilde{y}} &= \sum_{\tilde{x},\tilde{y}}^{N} (|\tilde{x}, (\tilde{y}+1) \bmod \varpi, 0\rangle\langle\tilde{x}, \tilde{y}, 0|) \\ &+ \sum_{\tilde{x},\tilde{y}}^{N} (|\tilde{x}, (\tilde{y}-1) \bmod \varpi, 1\rangle\langle\tilde{x}, \tilde{y}, 1|) \\ \hat{S}_{\tilde{x}} &= \sum_{\tilde{x},y}^{N} (|(\tilde{x}+1) \bmod \varpi, \tilde{y}, 0\rangle\langle\tilde{x}, \tilde{y}, 0|) \\ &+ \sum_{\tilde{x},\tilde{y}}^{N} (|(\tilde{x}-1) \bmod \varpi, \tilde{y}, 1\rangle\langle\tilde{x}, \tilde{y}, 1|) \end{aligned} \tag{8}$$

where ϖ indicates the prescribed walking boundary.

Suppose the initial moment: The walker's location is $(0_{\tilde{x}}, 0_{\tilde{y}})$, and the coin is in the superposition state $H_\Gamma = \cos\alpha|0\rangle + \sin\alpha|1\rangle$; then, the initial moment system state is:

$$|\psi_0\rangle = |\varphi_0\rangle_w \otimes (\cos\alpha|0\rangle + \sin\alpha|1\rangle)_\Gamma \tag{9}$$

The system state after a T walk can be expressed as:

$$|\psi_T\rangle = \hat{U}^T|\psi_0\rangle \tag{10}$$

3. Algorithm Description

3.1. *The Encryption Process*

3.1.1. Preparation of Quantum Random Walk Probability Distribution Matrix

The data of the corresponding element in the matrix are the probability $P(\delta, \vartheta, T)$ of the walker appearing at the coordinates (δ_x, ϑ_y) of the location, as can be deduced from the above:

$$P(\delta, \vartheta, T) = \left|\left\langle \delta, \vartheta, 0 \middle| \hat{U}^T \middle| \psi_0 \right\rangle\right|^2 + \left|\left\langle \delta, \vartheta, 1 \middle| \hat{U}^T \middle| \psi_0 \right\rangle\right|^2 (\delta_x, \vartheta_y) \tag{11}$$

The resulting probability distribution matrix M and its four sub-matrices M_1, M_2, M_3, M_4 after equiproportional partitioning are as follows:

$$\boldsymbol{M} = \begin{pmatrix} P_{11} & \cdots & P_{1n} \\ \vdots & \ddots & \vdots \\ P_{11} & \cdots & P_{nn} \end{pmatrix}$$
$$\boldsymbol{M}_1 = \begin{pmatrix} P_{11} & \cdots & P_{1\frac{n}{2}} \\ \vdots & \ddots & \vdots \\ P_{\frac{n}{2}} & \cdots & P_{\frac{n}{2}\frac{1}{2}} \end{pmatrix} \boldsymbol{M}_2 = \begin{pmatrix} P_{1\frac{n}{2}} & \cdots & P_{1n} \\ \vdots & \ddots & \vdots \\ P_{\frac{n}{2}\frac{n}{2}} & \cdots & P_{\frac{n}{2}n} \end{pmatrix} \quad (12)$$
$$\boldsymbol{M}_3 = \begin{pmatrix} P_{\frac{n}{2}1} & \cdots & P_{\frac{n}{2}\frac{n}{2}} \\ \vdots & \ddots & \vdots \\ P_{n1} & \cdots & P_{n\frac{n}{2}} \end{pmatrix} \boldsymbol{M}_4 = \begin{pmatrix} P_{\frac{n}{2}} & \cdots & P_{\frac{n}{2}\frac{n}{2}} \\ \vdots & \ddots & \vdots \\ P_{\frac{n}{2}n} & \cdots & p_{nn} \end{pmatrix}$$

We set the walker to be at the center of the Hilbert space $\hat{H}$ tensed by H_w and H_c, so the four submatrices $\boldsymbol{M}_1, \boldsymbol{M}_2, \boldsymbol{M}_3, \boldsymbol{M}_4$ are centrosymmetric about the point $P_{\frac{n}{2}}$ in the final generation. To prevent the LSTM from learning the rule such that the statistical performance of the final generated key matrix is degraded, in this paper, only $\hat{\boldsymbol{M}} = \boldsymbol{M}_1$ is chosen as the required initial pseudo-random number matrix to participate in the encryption.

3.1.2. Preparing the Encryption Key Matrix

Step 1: Ensure the reproducibility of the LSTM across devices. (i) Fix the random seeds of each dependency library so that each function is called with the same initial value and random value each time it is trained by the LSTM. (ii) Presetting the dropout function in the LSTM to 0, i.e., not dropping any nodes of the neural network, to ensure that the network model is fixed each time. (iii) Fixed platforms as well as devices, taking the current mainstream pytroch framework as an example, which still cannot guarantee the accuracy of model reproduction under different CPU and GPU pairings, and also requires CUDA environment variable configuration, etc. in order to further reduce uncertainty.

Step 2: Generate the LSTM input vector. Divide $\hat{\boldsymbol{M}}$ by column:

$$\begin{pmatrix} P_{11} & \cdots & P_{1\frac{n}{2}} \\ \vdots & \ddots & \vdots \\ P_{\frac{n}{2}1} & \cdots & P_{\frac{n}{2}\frac{1}{2}} \end{pmatrix} \rightarrow \left(\varphi_1, \varphi_2, \cdots \varphi_{\frac{n}{2}-1}, \varphi_{\frac{n}{2}}\right) \quad (13)$$

$\hat{\boldsymbol{M}}'$ is obtained by Min-Max normalization of $\hat{\boldsymbol{M}}$:

$$\left(\varphi_1, \varphi_2, \cdots \varphi_{\frac{n}{2}-1}, \varphi_{\frac{n}{2}}\right) \longrightarrow \left(\xi_1, \xi_2, \cdots \xi_j \cdots \xi_{\frac{n}{2}}\right) \quad (14)$$

ξ_i is the vector to be input.

Step 3: Generate the key matrix required for encryption. Input the vectors ξ_i in matrix $\hat{\boldsymbol{M}}''$ into the LSTM in order for training, and set the LSTM prediction quantity as γ^2 to obtain the prediction matrix $\hat{\boldsymbol{M}}'''$:

$$\hat{\boldsymbol{M}}''' = \begin{pmatrix} \varpi_{11} & \cdots & \varpi_{1\gamma} \\ \vdots & \ddots & \vdots \\ \varpi_{\gamma 1} & \cdots & \varpi_{\gamma\gamma} \end{pmatrix} \quad (15)$$

Inverse normalization of $\hat{M}'''$ yields M_E:

$$\begin{pmatrix} \varpi_{11} & \cdots & \varpi_{1\gamma} \\ \vdots & \ddots & \vdots \\ \varpi_{\gamma 1} & \cdots & \varpi_{\gamma\gamma} \end{pmatrix} \longrightarrow \begin{pmatrix} \partial_{11} & \cdots & \partial_{1\gamma} \\ \vdots & \ddots & \vdots \\ \partial_{\gamma 1} & \cdots & \partial_{\gamma\gamma} \end{pmatrix} \quad (16)$$

In Figure 2, we show the comparison between the predicted data and the expected values formed from the accurate data after training the QW probability matrix as an LSTM training matrix. Subplot a shows the trend in randomness between predicted and expected values; subplot b shows the distribution between specific predicted and expected values.

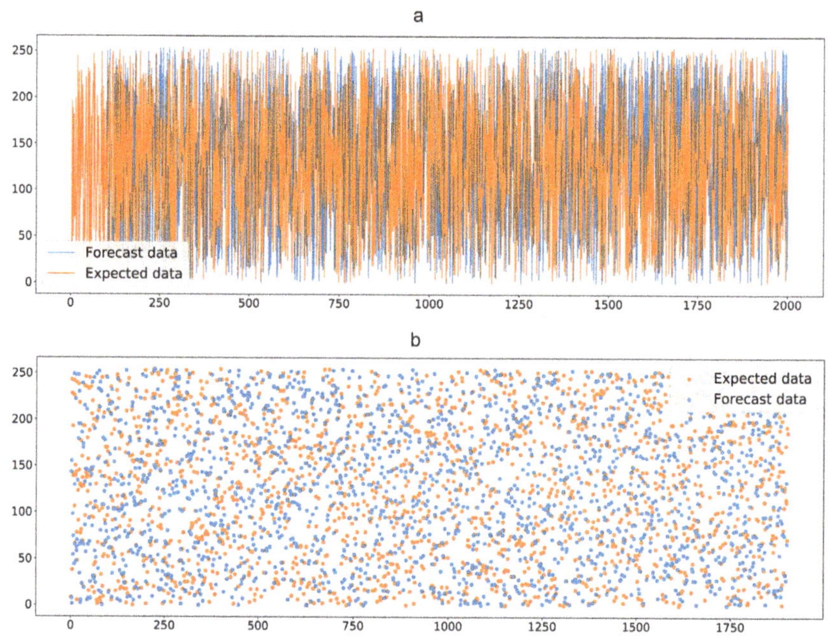

Figure 2. LSTM generation key matrix.

3.1.3. Image Encryption

The R, G and B channels in our proposed encryption scheme are performed separately, and our encryption algorithm is described in terms of $\gamma \times \gamma$ pixels of RGB image I corresponding to a grey-scale map in the form of matrix M_I.

Step 1: Hide the pixel information in M_I by obfuscating the pixel values. Here, we borrow the heteroskedastic algorithm to implement the obfuscation operation:

$$M_I' = M_I \oplus M_E \quad (17)$$

Step 2: Generate matrix $M_E' = M_E$, sort the index value matrix Ω of M_E' in order to obtain Ω', reorder the M_I' after the confusion operation according to the corresponding position in Ω', and achieve the dislocation of the image by destroying the relationship between adjacent pixel values to obtain M_I''. The schematic diagram of the dislocation algorithm is shown in Figure 3.

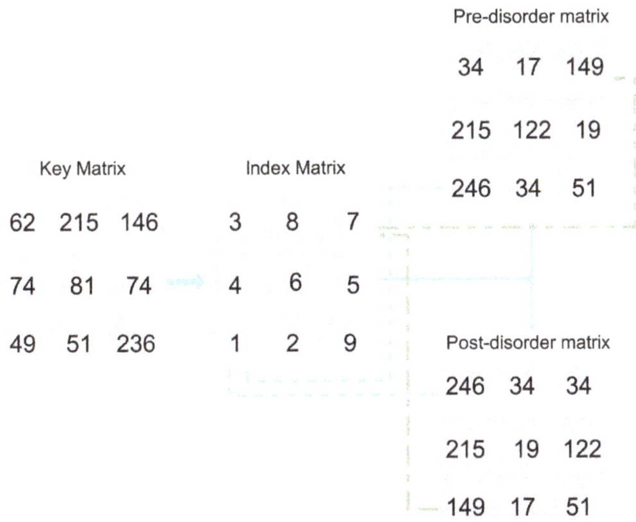

Figure 3. Encryption scheme—scrambling algorithm.

3.2. The Decryption Process

3.2.1. Preparing the Decryption Key Matrix

We use the probability distribution of the alternating quantum random walk algorithm at each grid point as the basis for generating the random number matrix required for encryption. The probability distribution matrix generated by the alternating quantum random walk has been shown to possess pseudo-randomness [22], i.e., the random number matrix $M' = M$ generated twice, provided that the initial parameters including α, β, ϖ are the same. Since we have removed the uncertainty and randomness from the LSTM, the M' is processed once according to the encryption process for M, and finally the prediction matrix generated by the LSTM is processed to obtain $M_D = M_E$.

3.2.2. Decryption of Encrypted Image

Step 1: The encrypted image M_I'' is obtained using the inverse permutation M_I'. This process is the inverse of the permutation operation, and the algorithm is shown in Figure 3:

Step 2: $M_I{}'$ for obfuscation reduction to obtain M_I.

3.3. Encryption and Decryption Algorithm Flow Chart

We show the key steps of our proposed image encryption scheme by means of a flowchart, including the generation of the QW probability density matrix, the process of generating the key matrix by LSTM, and the two key steps (scrambling, confusion) of the image encryption and decryption process using the key matrix, as shown in Figure 4.

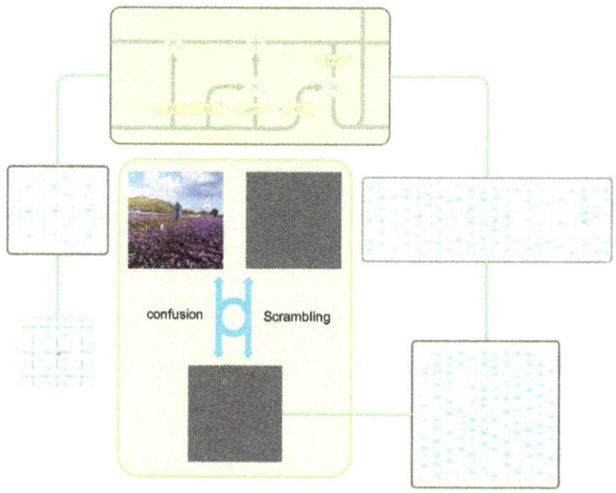

Figure 4. Encryption and decryption process.

4. Simulation and Analysis

To verify the resistance of the proposed scheme, three RGB images with a pixel size of 2000 × 2000 were encrypted and decrypted according to the proposed encryption scheme, and various statistical analyses were carried out on the encrypted images and the keys used, including histogram analysis, correlation analysis and information entropy analysis for the encrypted images; sensitivity analysis and key space analysis for the key matrix, etc.

4.1. Experimental Parameters and Encryption and Decryption Results

We use $\varpi = 240, \alpha = \frac{\pi}{23}, \beta = \frac{\pi}{41}$ as the start parameters of the QW to prepare a QW probability matrix of size 100 × 2000, and set the prediction length of the LSTM to 2000, i.e., to generate a key matrix of the same size as the RGB image to be encrypted. The encryption and decryption results are shown in Figure 5.

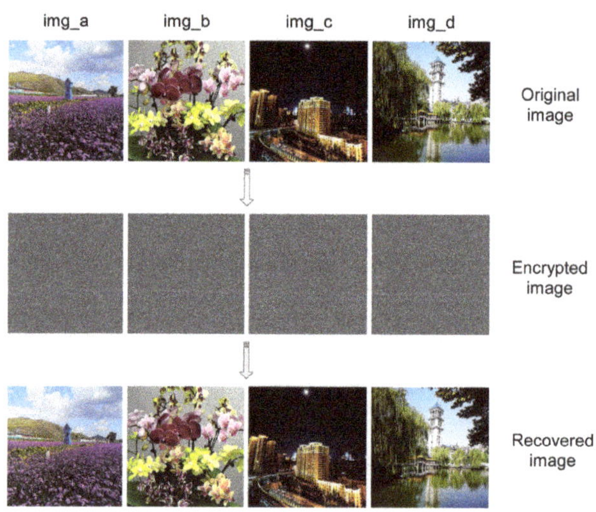

Figure 5. Image encryption before and after comparison.

4.2. The Statistical Analysis

4.2.1. Correlation Analysis

Adjacent pixel correlation R_{AB} is used to measure the degree of correlation of adjacent pixel values. Adjacent pixel values in RGB images often have strong correlations in horizontal, vertical and diagonal directions. Image encryption algorithms will destroy this correlation, and the degree of destruction can reflect the effect of encryption algorithms. The closer R_{AB} is to 0, the better the destruction effect is, and the more difficult it is to obtain image information through the relationship between adjacent pixels [27].

$$R_{AB} = \frac{\mathrm{cov}(A,B)}{\sqrt{D(A)}\sqrt{D(B)}} \tag{18}$$

where $\mathrm{cov}(A,B)$ is the covariance of A, B, and $\sqrt{D(A)}$ and $\sqrt{D(B)}$ are the standard deviations of A and B, respectively. In this paper, the horizontal, vertical, and diagonal correlations of the three RGB images of Lena, Lemon, and Sakur are compared before and after encryption. The correlation values for the three RGB images are shown in Table 1, and the specific pixel distribution information is shown in Figures 6 and 7.

Table 1. Pixel correlation analysis data.

Image	Channel	Horizontal	Vertical	Diagonal
Unencrypted (img_a)	Red	0.8846	0.8924	0.8297
	Green	0.9062	0.9146	0.8568
	Blue	0.9269	0.9272	0.8905
Encrypted (img_a)	Red	0.0006	0.0011	0.0032
	Green	0.0032	0.0027	0.0021
	Blue	0.0041	0.0016	0.0022
Unencrypted (img_b)	Red	0.9930	0.9944	0.9869
	Green	0.9940	0.9949	0.9897
	Blue	0.9927	0.9939	0.9876
Encrypted (img_b)	Red	0.0022	0.0011	0.0023
	Green	0.0021	0.0025	0.0014
	Blue	0.0009	0.0041	0.0013

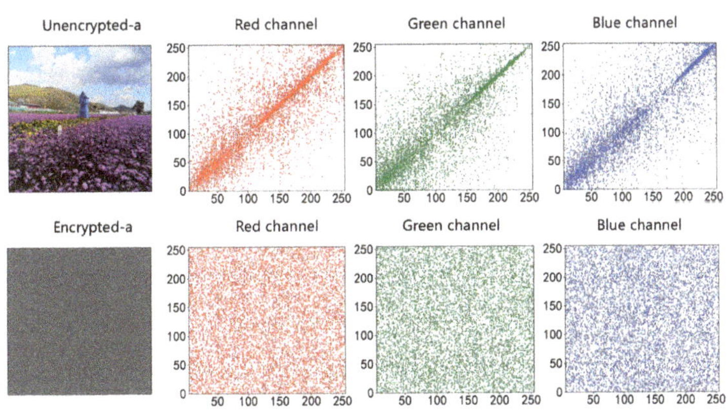

Figure 6. Comparison of correlation before and after img_a encryption.

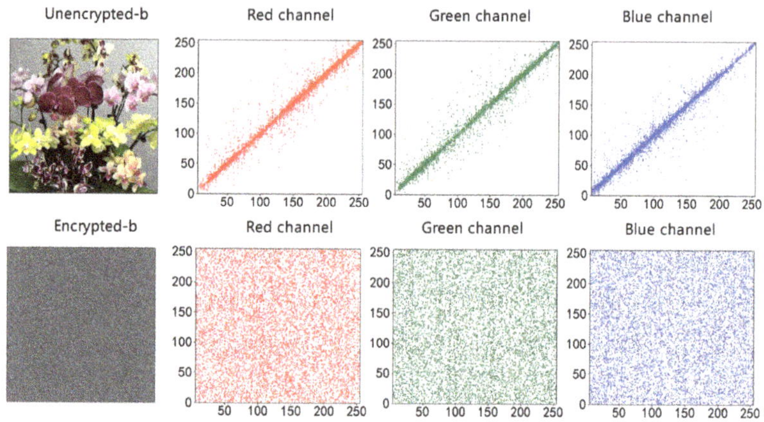

Figure 7. Comparison of correlation before and after img_b encryption.

4.2.2. Histogram Analysis

The histogram provides a visual representation of the statistical data of the pixel values in an RGB image. The histogram of a normal image usually has a distinct statistical pattern, and to resist statistical attacks [25], the histogram of an encrypted image must be as uniform and smooth as possible. The more such criteria are met, the more uniform the pixel distribution is, the less statistical information the image displays, the less information can be accurately predicted, and the more secure the image encryption scheme is [15]. In this paper, the histograms of the RGB three channels of Lena, Lemon, and Sakura images are analyzed separately, and the specific histograms are shown in Figures 8 and 9.

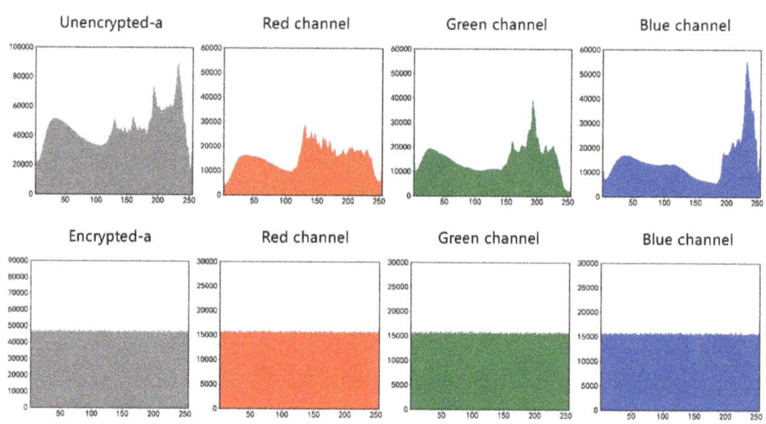

Figure 8. Comparison of histogram before and after img_a encryption.

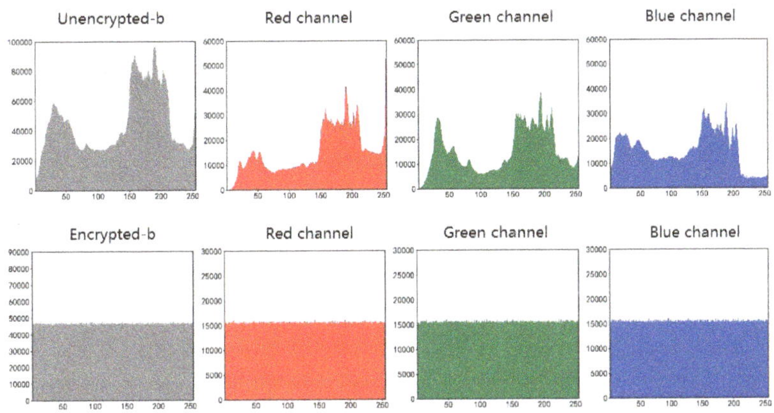

Figure 9. Comparison of histogram before and after img_b encryption.

4.2.3. Information Entropy Analysis

Information entropy H was proposed by Shannon, the father of information theory, to describe the uncertainty of the occurrence of each possible event of the information source. The pixel values of RGB images range from 0 to 255, so the information entropy $H \leq 8$. The closer the entropy value is to 8, the more information it carries and the more resistant it is to statistical attacks [11]. The formula for this is as follows:

$$H(m) = -\sum_{i=0}^{N-1} P(m_i) \log_2 P(m_i) \tag{19}$$

where m_i is the grey scale value and $P(x_i)$ is the probability of m_i occurrence. This paper analyzes the information entropy of the R, G, and B channels of the three different RGB images of Lena, Lemon, and Sakura. The relevant data are shown in Table 2.

Table 2. Entropy analysis.

Image	Channel	Image Entropy (bit)
Encrypted (img_a)	Red	7.9991
	Green	7.9996
	Blue	7.9989
Encrypted (img_b)	Red	7.9992
	Green	7.9992
	Blue	7.9994

4.2.4. Key Sensitivity Analysis

An effective key sensitivity means that a slight change in the key information will result in a significant change in the encrypted image. The ideal values of NPCR and UACI are 99.61% and 33.46%, respectively [29]. Higher calculated values of NPCI and UACI of an encryption scheme indicate that the encryption scheme is more resistant to differential attacks:

$$\Gamma(i,j) = f(x) = \begin{cases} 1, & \text{if } C_1(i,j) \neq C_2(i,j) \\ 0, & \text{otherwise} \end{cases} \tag{20}$$

$$NPCR = \frac{\sum_{i,j} \Gamma(i,j)}{\mathfrak{J} \times \mathfrak{R}} \times 100\% \tag{21}$$

$$UACI = \frac{1}{\Im \times \Re}\left[\sum_{i,j}\frac{C_1(i,j) - C_2(i,j)}{255}\right] \times 100\% \quad (22)$$

where , $\Im, \Re$ are the length and width of the encrypted image, $\Gamma(i,j)$ is the above equation, and C1, C2 are the images after encryption with different keys.

In this paper, the key sensitivity of the R, G and B channels of the RGB images of Lena, Lemon, and Sakura were analyzed separately, and the relevant data are shown in Table 3.

Table 3. Key sensitivity analysis.

Image	Channel	NPCR	UACI
ine	Red	99.6124%	33.4216%
img_a	Green	99.6088%	33.3657%
	Blue	99.6003%	34.2157%
ine	Red	99.6419%	33.6114%
img_b	Green	99.5986%	33.4268%
	Blue	99.6036%	33.5762%

4.2.5. The Key Space

The key space refers to the set of all possible keys used to generate the key and determines whether the encryption scheme can resist a brute-force attack. Cryptosystems with a key space size of 2^{128} are effective in resisting brute force attacks. The key space calculation for the scheme proposed in this paper is based on quantum effects. Since in quantum theory the position of a particle in a defined space is not deterministic, each position has its probability of existence, only with different probabilities, and this probability can be changed by specifying the size of the space for a QW and the initial walking direction and forward direction. As the walk direction takes values from 0 to 2π and the QW is extremely sensitive to accuracy, the change in probability is infinite as the accuracy of the computer increases, i.e., the key space established based on the QW is infinite.

4.2.6. Explicit Attack

- Known plaintext attack: The attacker can recover the key by obtaining the decrypted image and comparing it with the ciphertext image. Since the algorithm in this paper has a good diffusion effect, the difficulty of obtaining the key by this method is close to that of a direct brute force attack, so the encryption scheme in this paper can effectively resist known plaintext attacks.
- Selective plaintext attack: Assuming that the attacker has gained access to the encrypted machine, he can select an arbitrary number of plaintexts for the encryption algorithm under attack to encrypt and obtain the corresponding ciphertexts. The attacker's goal is to gain some information about the encryption algorithm through this process that will allow the attacker to more effectively crack messages encrypted by the same encryption algorithm (and associated key) in the future. In the worst case, the attacker can simply obtain the key used for decryption. This scheme is commonly used against public key encryption schemes. The keys in this scheme are not universal, i.e., they are changed periodically, even differently each time, making it impossible for an attacker to obtain valid information.

4.2.7. Time Complexity Analysis

The time complexity analysis of an encryption scheme is an important indicator to evaluate the excellence of an encryption scheme, which will directly affect the encryption efficiency. The time consumption of our proposed scheme consists of two parts, one is the time required to generate the key matrix, and the other is the completion of the image encryption by the key matrix. Although the efficiency of generating the pseudo random number matrix is important, it is not part of the time complexity of the encryption scheme as it is decoupled from the image encryption process. The encryption time complexity of our proposed scheme consists of a combination of pixel obfuscation and scrambling. The time complexity of this process is $O(n^2 + n\log n)$, as the time consumed by matrix permutation is $O(n^2)$. In summary, the encryption time complexity of our proposed scheme is $O(2n^2 + n\log n)$.

4.2.8. Noise Robustness Testing

During the transmission of image information over the network, information may be lost or misplaced due to packet loss, malicious attacks, and so on. We simulate the continuous loss of image information due to network fluctuations using Gaussian noise and pretzel noise. A malicious attack was simulated using partial block replacement of the encrypted image. Figure 10 shows the decrypted image of the Lena encrypted image with the addition of Gaussian noise, pretzel noise and a clipping attack.

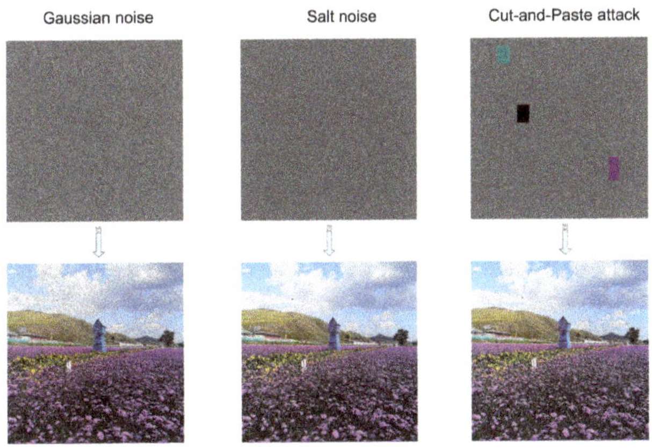

Figure 10. Comparison of histogram before and after img_b encryption.

4.3. Comparison of Encryption Schemes

In this section, we analyze and compare the use of QW alone, the encryption scheme proposed in this paper, and similar work in recent years in terms of the important measures of average relevance, information entropy, average NPCR, average UACI, and key space size to resist brute-force cracking, the data of which are presented in Table 4.

Table 4. The comparison in this article is for reference only as the images used in the different solutions are different and have different pixels. As the pixel sizes vary in each scenario, we have used the largest pixel images from their scenarios for comparison and selected their average values as a reference.

Scheme	NPCR (%)	UACI (%)	Correlation	Entropy (bit)	KeySpace
QW	93.14	32.36	0.0149	7.9947	$>2^{128}$
our	99.6109	33.6024	0.0032	7.9992	$>2^{128}$
[3]	99.6127	33.4471	0.0013		$>2^{128}$
[4]	99.6336	33.4636	0.0026	7.9937	$>2^{128}$
[5]	99.6326	33.4022	0041	7.9973	$>2^{128}$

5. Conclusions

We propose a more efficient encryption scheme for the current lack of encryption schemes for high pixel images in the field of image encryption. The probability density matrix generated by the quantum random walk is trained by exploiting the memory learning capability of the LSTM and the nonlinear nature of the quantum random walk. It can take advantage of the nearly infinite key space brought by the quantum random walk algorithm, and also solve the shortcomings of the low generation efficiency of the quantum random walk itself. At the same time, both the permutation and obfuscation processes of our scheme make use of the key space of the quantum random walk, avoiding the shortage of key space in a particular process.

Author Contributions: Conceptualization, J.L. and Z.S. (Zhaoyang Song); methodology, J.L. and Z.S. (Zhaoyang Song); software, Z.S. (Zhaoyang Song) and Z.S. (Zhongwei Sun); validation, M.L. and H.M.; formal analysis, Z.S. (Zhongwei Sun); investigation, Z.S. (Zhaoyang Song); resources, J.L.; data curation, J.L.; writing—original draft preparation, J.L.; writing—review and editing, Z.S. (Zhaoyang Song) and M.L.; visualization, Z.S. (Zhongwei Sun); supervision, M.L. and H.M.; project administration, M.L. and H.M.; funding acquisition, H.M.; image encryption, J.L. and Z.S. (Zhaoyang Song). All authors have read and agreed to the published version of the manuscript.

Funding: This research was funded by the Natural Science Foundation of Shandong Province, China (Grant No. ZR2021MF049), the Joint Fund of the Natural Science Foundation of Shandong Province (Grant No. ZR2022LLZ012), the Joint Fund of the Natural Science Foundation of Shandong Province (Grant No. ZR2021LLZ001), the project supported by the National Natural Science Foundation of China (Grant No. 11975132), the National Natural Science Foundation of China (Grant No. 12005110), and the Natural Science Foundation of Shandong Province, China (Grant No. ZR2022JQ04).

Institutional Review Board Statement: Not applicable.

Informed Consent Statement: Not applicable.

Data Availability Statement: The data and information supporting this study can be provided at the request of the corresponding author at reasonable request.

Conflicts of Interest: The authors declare that there is no conflict of interests regarding the publication of this paper.

References

1. Coppersmith, D. The Data Encryption Standard (DES) and its strength against attacks. *IBM J. Res. Dev.* **1994**, *38*, 243–250. [CrossRef]
2. Heron, S. Advanced encryption standard (AES). *Netw. Secur.* **2009**, *12*, 8–12. [CrossRef]
3. Wang, X.; Yang, J. A privacy image encryption algorithm based on piecewise coupled map lattice with multi dynamic coupling coefficient. *Inf. Sci.* **2021**, *569*, 217–240. [CrossRef]
4. Hua, Z.; Zhu, Z.; Chen, Y.; Li, Y. Color image encryption using orthogonal Latin squares and a new 2D chaotic system. *Nonlinear Dyn.* **2019**, *104*, 4505–4522. [CrossRef]
5. Chai, X.; Zhi, X.; Gan, Z.; Zhang, Y.; Chen, Y.; Fu, J. Combining improved genetic algorithm and matrix semi-tensor product (STP) in color image encryption. *Signal Process.* **2021**, *183*, 108041. [CrossRef]
6. Zhou, N.; Pan, S.; Cheng, S.; Zhou, Z. Image compression—Encryption scheme based on hyper-chaotic system and 2D compressive sensing. *Opt. Laser Technol.* **2016**, *82*, 121–133. [CrossRef]
7. Duan, C.-F.; Zhou, J.; Gong, L-H.; Wu, J.-Y.; Zhou, N.-R. New color image encryption scheme based on multi-parameter fractional discrete Tchebyshev moments and nonlinear fractal permutation method. *Opt. Lasers Eng.* **2022**, *150*, 106881. [CrossRef]
8. Chuman, T.; Sirichotedumrong, W.; Kiya, H. Encryption-then-compression systems using grayscale-based image encryption for JPEG images. *IEEE Trans. Inf. Forensics Secur.* **2018**, *14*, 1515–1525. [CrossRef]
9. Wang, X.-Y.; Li, Z.-M. A color image encryption algorithm based on Hopfield chaotic neural network. *Opt. Lasers Eng.* **2019**, *115*, 107–118. [CrossRef]
10. Chen, L.; Yin, H.; Huang, T.; Yuan, L.; Zheng, S.; Yin, L. Chaos in fractional-order discrete neural networks with application to image encryption. *Neural Netw.* **2020**, *125*, 174–184. [CrossRef]
11. Mani, P.; Rajan, R.; Shanmugam, L.; Joo, Y.H. Adaptive control for fractional order induced chaotic fuzzy cellular neural networks and its application to image encryption. *Inf. Sci.* **2019**, *491*, 74–89. [CrossRef]
12. Hochreiter, S.; Schmidhuber, J. Long Short-Term Memory. *Neural Comput.* **1997**, *9*, 1735–1780. [CrossRef]
13. Schmidhuber, J. Deep learning in neural networks: An overview. *Neural Netw.* **2015**, *61*, 85–117. [CrossRef]
14. Steane, A. Quantum computing. *Signal Process. Image Commun.* **2022**, *61*, 116891. [CrossRef]
15. Zhou, N.R.; Zhang, T.F.; Xie, X.W.; Wu, J.Y. Hybrid quantum–classical generative adversarial networks for image generation via learning discrete distribution. *IBM J. Res. Dev.* **2019**, *115*, 107–118. [CrossRef]
16. Wang, H.; Xue, Y.; Qu, Yi.; Mu, Xi.; Ma, H. Multidimensional Bose quantum error correction based on neural network decoder. *NPJ Quantum Inf.* **2022**, *8*, 134. [CrossRef]
17. Long, G.-L. Grover algorithm with zero theoretical failure rate. *Phys. Rev. A* **2001**, *64*, 107–118. [CrossRef]

18. Weinstein, Y.S.; Pravia, M.A.; Fortunato, E.M.; Lloyd, S.; Cory, D.G. Implementation of the Quantum Fourier Transform. *Phys. Rev. Lett.* **2001**, *86*, 1889–1891. [CrossRef]
19. Harrow, A.W.; Hassidim, A.; Lloyd, S. Quantum algorithm for linear systems of equations. *Phys. Rev. Lett.* **2009**, *103*, 150502. [CrossRef]
20. Aharonov, Y.; Davidovich, L.; Zagury, N. Quantum random walks. *Phys. Rev. A* **1993**, *48*, 107–118. [CrossRef]
21. Farhi, E.; Gutmann, S. Quantum computation and decision trees. *Phys. Rev. A* **1998**, *58*, 915–928. [CrossRef]
22. Watrous, J. Quantum simulations of classical random walks and undirected graph connectivity. *J. Comput. Syst. Sci.* **2001**, *62*, 376–391. [CrossRef]
23. Baryshnikov, Y.; Brady, W.; Bressler, A.; Pemantle, R. Two-dimensional quantum random walk. *J. Stat. Phys.* **2011**, *142*, 78–107. [CrossRef]
24. Zhao, Z.-P.; Zhou, S.; Wang, X.-Y. A new chaotic signal based on deep learning and its application in image encryption. *Acta Phys. Sin.* **2021**, *70*, 23. [CrossRef]
25. Yang, Y.-G.; Pan, Q.-X.; Sun, S.-J.; Xu, P. Novel image encryption based on quantum walks. *Sci. Rep.* **2015**, *5*, 107–118. [CrossRef]
26. Abd EL-Latif, A.A.; Abd-El-Atty, B.; Venegas-Andraca, S.E. Controlled alternate quantum walk-based pseudo-random number generator and its application to quantum color image encryption. *Phys. A Stat. Mech. Appl.* **2020**, *547*, 123869. [CrossRef]
27. Ma, Y.; Li, N.; Zhang, W.; Wang, S.; Ma, H. Image encryption scheme based on alternate quantum walks and discrete cosine transform. *Opt. Express* **2021**, *29*, 28338–28351. [CrossRef]
28. Lam, E.Y.; Goodman, J.W. A mathematical analysis of the DCT coefficient distributions for images. *IEEE Trans. Image Process.* **2000**, *9*, 1661–1666. [CrossRef]
29. Wu, Y.; Noonan, J.P.; Agaian, S. NPCR and UACI randomness tests for image encryption. *Cyber J. Multidiscip. J. Sci. Technol. J. Sel. Areas Telecommun.* **2011**, *1*, 31–38.

entropy

Article

An Infusion Containers Detection Method Based on YOLOv4 with Enhanced Image Feature Fusion

Lei Ju [1], Xueyu Zou [1,*], Xinjun Zhang [1], Xifa Xiong [1], Xuxun Liu [1,2,*] and Luoyu Zhou [1]

1 College of Electronic and Information Engineering, Yangtze University, Jingzhou 434023, China
2 College of Electronic and Information Engineering, South China University of Technology, Guangzhou 510641, China
* Correspondence: xyzou@sina.com (X.Z.); liuxuxun@scut.edu.cn (X.L.)

Abstract: The detection of infusion containers is highly conducive to reducing the workload of medical staff. However, when applied in complex environments, the current detection solutions cannot satisfy the high demands for clinical requirements. In this paper, we address this problem by proposing a novel method for the detection of infusion containers that is based on the conventional method, You Only Look Once version 4 (YOLOv4). First, the coordinate attention module is added after the backbone to improve the perception of direction and location information by the network. Then, we build the cross stage partial–spatial pyramid pooling (CSP-SPP) module to replace the spatial pyramid pooling (SPP) module, which allows the input information features to be reused. In addition, the adaptively spatial feature fusion (ASFF) module is added after the original feature fusion module, path aggregation network (PANet), to facilitate the fusion of feature maps at different scales for more complete feature information. Finally, EIoU is used as a loss function to solve the anchor frame aspect ratio problem, and this improvement allows for more stable and accurate information of the anchor aspect when calculating losses. The experimental results demonstrate the advantages of our method in terms of recall, timeliness, and mean average precision (mAP).

Keywords: object detection; YOLOv4; artificial intelligence; feature information

Citation: Ju, L.; Zou, X.; Zhang, X.; Xiong, X.; Liu, X.; Zhou, L. An Infusion Containers Detection Method Based on YOLOv4 with Enhanced Image Feature Fusion. *Entropy* **2023**, *25*, 275. https://doi.org/10.3390/e25020275

Academic Editors: Oleg Sergiyenko, Wendy Flores-Fuentes, Julio Cesar Rodriguez-Quinonez and Jesús Elías Miranda-Vega

Received: 2 December 2022
Revised: 24 January 2023
Accepted: 31 January 2023
Published: 2 February 2023

1. Introduction

Intravenous input is a very important treatment while infusion bottles and infusion bags are the most common containers for infusion. With the spread of the epidemic in recent years, many patients requiring infusions have gathered in hospitals, putting increasing pressure on the healthcare system. Detecting infusion containers can help healthcare professionals better understand the status of infusion recipients, which makes planning easier and reduces the stress of medical staff.

The detection of infusion containers plays an important role in reducing the pressure on medical personnel and is essentially part of the field of object detection [1]. There are traditional methods and deep learning in the field of object detection. With the development of deep learning, even in some specific areas traditional methods are as effective as deep learning, and the scalability and robustness of deep learning is making it increasingly mainstream. In hospital scenarios, individual infusion containers often overlap and are not at the same distance from the camera, which makes their detection difficult. In order to solve this problem here, we modify the neck part of the YOLO architecture by adding coordinate attention (CA) [2] after the backbone to effectively capture location and channel information, improve SPP to CSP-SPP to enhance the ability of feature fusion [3], and add the feature fusion module ASFF [4] after PANet to increase the depth of the network; this improved model is named NMYOLO. In addition, we adopt EIoU [5] instead of CIoU in the loss function to solve the problem of the ambiguous aspect ratio of anchor frame, resulting in more effective detection while maintaining the inference speed.

The main contributions of this paper are summarized as follows:

(1) The neck part of YOLOv4 [6] is improved by using replacing the original modules with several more effective modules. We have improved SPP to CSP-SPP, which enhanced the feature extraction capability of the model. We have also added CA and ASFF to obtain more image information. These improvements validate the scalability of the YOLO architecture and also lay the foundation for further research.
(2) The loss function of YOLOv4 is improved by replacing CIOU with EIOU in calculating the width-to-height ratio of the anchor box. This not only results in more stable and accurate prediction of boxes but also reduces the training time and calculation cost.

The content in this paper is structured as follows. Presented in Section 2 is the related work of our study. Section 3 describes the methods of YOLOv4 and Section 4 introduces the details of NMYOLO. A comparison between the specific parameters of the experiment and the final results is presented in Section 5. Finally, conclusions are drawn in Section 6.

2. Related Work

Target detection is the main component of computer vision. In this section, we review the solutions in deep learning for object detection.

Common deep learning object detection algorithms can be roughly divided into two categories: the first is the two-stage [7] detectors, which are also known as object detection models based on the candidate region proposal. The process of object detection involves, first, the generation of candidate regions on the image and extraction of the corresponding image features, which are then input into the classifier for judgment. Region-based convolutional neural network (RCNN), as reported by Grishick in 2014 [8], is the first of its kind in terms of two-stage detectors. On the basis of the original detector, Grishek proposed fast R-CNN [9] and faster R-CNN [10,11], both of which significantly reduced the time consumed by algorithmic reasoning with improved accuracy. The second categories is the one-stage [7] detectors, which began with You Only Look Once (YOLO) [12,13] proposed by Redmon et al. in 2016 and has gradually become the mainstream object detection algorithm after several years of development. Common one-stage detectors include YOLO series and the single shot multi-box detector (SSD) [14] proposed by Liu. One-stage detectors do not have the step of generating candidate box regions, so are much faster than two-stage detectors in inference, which allows the computational overhead to be reduced.

In recent years, the self-attention mechanism has also been widely used in target detection. In 2017, Vaswani et al. proposed transformer [15], a model that demonstrates that self-attention is very effective in deep learning. Wang et al. proposed the no-local network [16], a model that can capture long-range dependencies more easily. Hu et al. proposed SENet [17] in 2018, a module that can be easily added to other models and improve accuracy, which triggered the thinking about self-attention in the field of vision. Immediately after, Woo et al. proposed CBMA [18], a lightweight module that can also be easily integrated into other CNN architectures, which is divided into channel blocks and spatial blocks that can be used to regenerate feature maps.

3. The Methods of YOLOv4

3.1. CSP Structure

Cross stage partial network (CSPNet) [19] is characterized by the integration of feature mapping at the beginning and end of the network stages. Figure 1 shows the application of CSPNet on the ResNe(X)t [20] network structure.

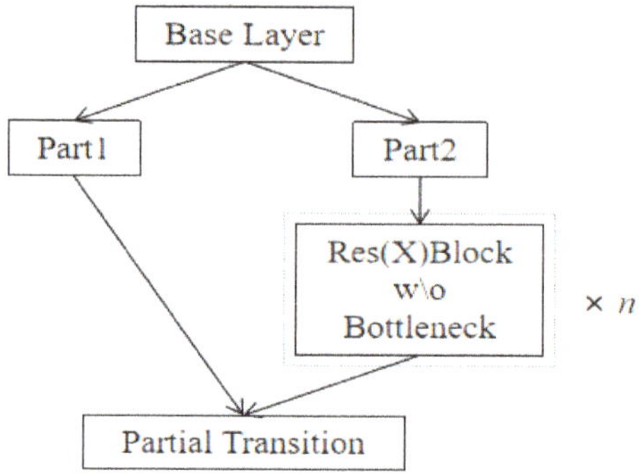

Figure 1. The structure of CSPResNe(X)t [19].

BaseLayer is the feature extraction layer in Figure 1, and the extracted feature information is divided into two parts, where the output of Part 2 is processed into a Res(X) module and the output feature map is spliced with the feature map of Part 1 to obtain the final output. This can reduce the amount of duplicated feature information in the network through cross-stage connection and, at the same time, improve the learning ability of the network to enhance the final result. This is the reason why both YOLOv4 and YOLOv5 choose the CSP-structured network as the backbone feature extraction network.

3.2. *YOLOv4*

YOLOv4 is an object detection model proposed by Bochkovskiy et al. in 2020 containing many improvements from YOLOv3. It includes the use of the mosaic data enhancement method, which solves the problem of difficult detection of small targets. The idea of CPSNet is absorbed to replace the backbone from Darknet53 to CSPDarknet53, and the cross-stage residual connection structure is used to obtain more effective feature extraction. The activation function of the backbone is replaced from Leakey_relu to Mish, while SPP [21] and PANet [22] are used in the feature pyramid module instead of feature pyramid networks (FPN) [23]. The overall structure for an input image size of 416 × 416 is shown in Figure 2.

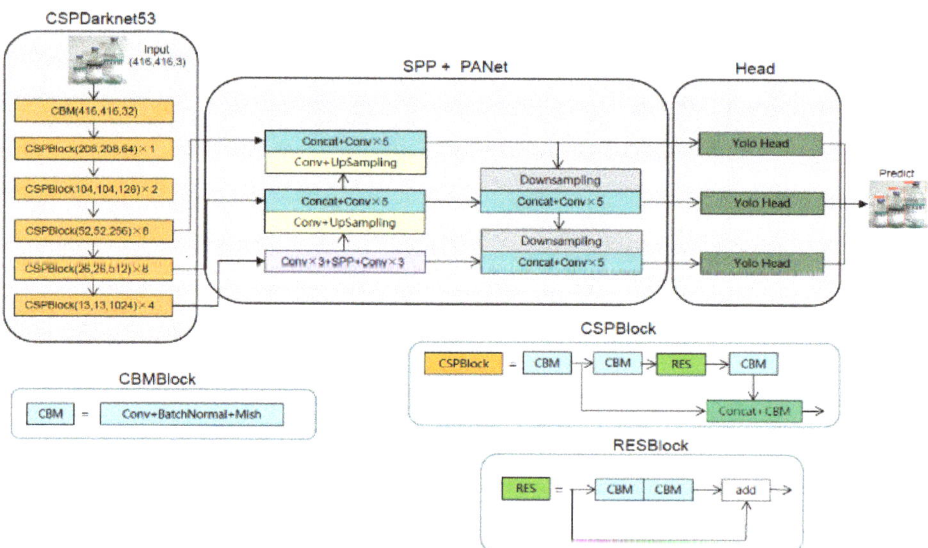

Figure 2. The structure of YOLOv4.

After the input image passes through the backbone feature extraction network, three different scales of 52 × 52, 26 × 26 and 13 × 13 features are output to the feature fusion layer, where the 13 × 13 scale features are enhanced by SPP module and then fused with the 52 × 52 and 26 × 26 scale features in PANet after upsampling and downsampling feature fusion concatenation. The features at different scales are extracted several times to obtain a better fusion effect. Finally, the fused features of different scales are input into three YOLO heads for prediction. In addition, YOLOv4 uses CIoU loss of bound box, which can be described as

$$\mathrm{CIoU} = \mathrm{IoU} - \frac{\rho^2\left(b, b^{gt}\right)}{d^2} - \alpha v \tag{1}$$

where $\rho(b,b^{\mathrm{gt}})$ is the distance between the prediction box and the center point of the ground truth box, d is the diagonal distance of the smallest rectangular box containing both the real and prediction boxes, and α and v are calculated as follows:

$$\alpha = \frac{v}{1 - \mathrm{IoU} + v} \tag{2}$$

$$v = \frac{4}{\pi^2}\left(\arctan\frac{w^{gt}}{h^{gt}} - \arctan\frac{w}{h}\right)^2 \tag{3}$$

where w^{gt} and h^{gt} represent the width and height of the ground truth box, respectively, while w and h represent the width and height of the prediction box, respectively.

4. The Detail of NMYOLO

Currently, CSPDarknet53, which is used in the overall structure of YOLOv4 to extract backbone network features, demonstrates good performance in object detection tasks, so it is retained in our proposed model. In this study, we have mainly improved the neck module, as most of the fusion processing for feature information is located in this module.

4.1. ASFF

Feature fusion is a very important component of the target detection task because the fusion of different-scale features is an important for improving the performance of model detection. Therefore, we choose to add a feature fusion module after PANet to improve the depth and detection ability of the model.

The structure of ASFF is shown in Figure 3. In our model, ASFF is added to PANet. The 52 × 52, 26 × 26 and 13 × 13 scales of feature maps are fused with other scales and then input to YOLO heads for prediction. In the process of feature fusion, ASFF uses the weight parameter to control the contribution of different feature maps, which also reflects the idea of attention, so it can help the network to better fuse the extracted high-level information and low-level information and thus improve the final detection capability.

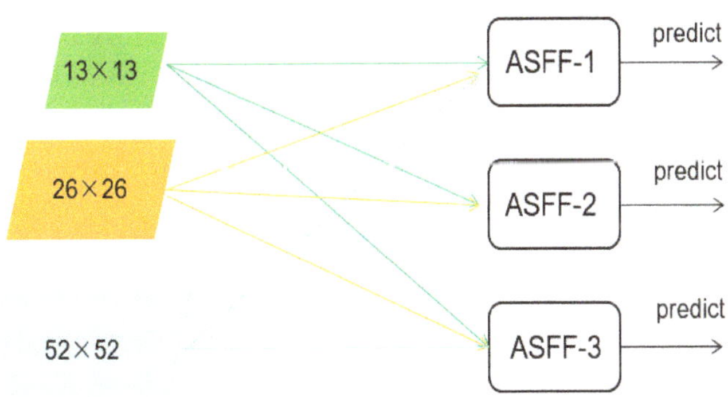

Figure 3. ASFF structure diagram.

4.2. Coordinate Attention

In the task of object detection, the effect of the improved attention mechanism on the final result is obvious. CA can capture the direction and position perception information while capturing the cross-channel information by embedding the position information in the picture into the channel attention. Therefore, CA can help the model to locate it more accurately and identify the target in the picture. At the same time, CA can improve the effect without occupying excessive computational overhead because it is a lightweight module. The structure of CA is shown in Figure 4.

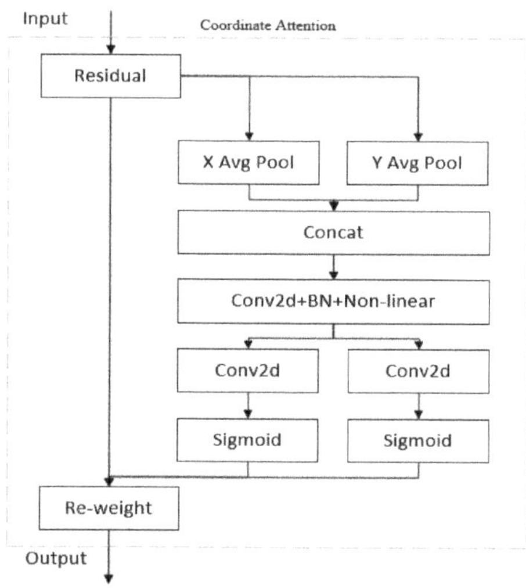

Figure 4. The structure of coordinate attention.

In CA, the input information is first passed through a residual structure, and the attentional feature information is subsequently and separately extracted according to the horizontal and vertical directions. Thus, the key *X*-axis and *Y*-axis position information of the input feature map is obtained. The formula is as follows:

$$Z_c^h(h) = \frac{1}{W} \sum_{0 \le i < W} x_c(h, i) \tag{4}$$

$$Z_c^w(w) = \frac{1}{H} \sum_{0 \le j < H} x_c(j, w) \tag{5}$$

where $x_c(h,c)$ represents the *c*-th input channel of height h, which is output $Z_c{}^h(h)$ after being encoded with a convolutional kernel of size (1,*W*). $x_c(j,w)$ represents the *c*-th input channel of width *w*, which is output as $Z_c{}^w(w)$ after being encoded by a convolutional kernel of size (1,*H*). Then, (4) and (5) are stitched together and fed into a convolutional module of 1×1, and the nonlinear data are then obtained through the activation function and then divided into two different sets of feature plots, which are defined as follows:

$$f = \delta(F_1([z^h, z^w])) \tag{6}$$

In (6), F_1 is a convolution of 1×1, while δ is a nonlinear activation step. After that, the output is separately fed into a convolutional module of 1×1, and the sigmoid is then used to gain the weight of attention.

$$g^h = \sigma(F_h(f^h)) \tag{7}$$

$$g^w = \sigma(F_w(f^w)) \tag{8}$$

In (7) and (8), f^h and f^w are the outputs of the previous step, F_h and F_w represent the corresponding 1×1 convolution, and σ is the sigmoid activation function. In addition, ${g_c}^h(i)$ and ${g_c}^w(j)$ are used as attention weights. Finally, the coordinate attention output features obtained after multiplying the initially input data with the horizontal and vertical weights are multiplied, and the final result can be written as

$$y_c(i,j) = x_c(i,j) \times g_c^h(i) \times g_c^w(j) \tag{9}$$

4.3. Improvements of SPP

Inspired by the CSP structure of the backbone network, the SPP structure after the backbone network is changed to a CSP-SPP structure in our model to better capture and fuse the featured information of the images. The SPP structure and the changed structure are shown in Figure 5.

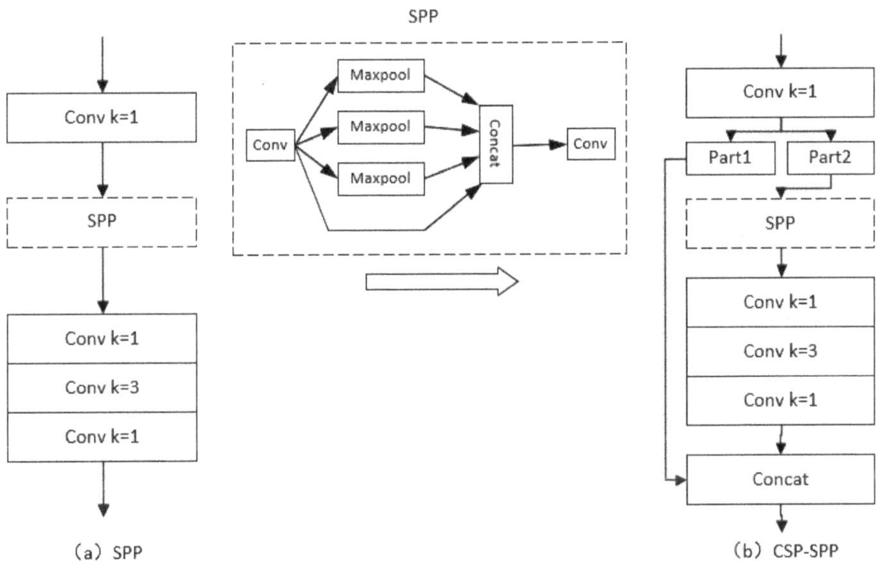

Figure 5. Improvement from SPP to CSP-SPP. (**a**) is the structure of SPP, (**b**) is the structure of CSP-SPP.

As shown in Figure 5, in the CSP-SPP structure, the output of the 1×1 convolutional module is divided into two parts before SPP, one of which enters the SPP structure normally, and then outputs after the convolutional module of 1×1, 3×3, and 1×1. The other part conducts concatenation with its output. This method can be used to map and connect the characteristics of different stages through cross-stage connections, effectively strengthening the learning ability of the network.

4.4. The Structure of NMYOLO

Based on the above improvements, we proposed NMYOLO. The final overall model also includes three parts: backbone, neck, and head. The overall network structure of the model is shown in Figure 6.

In NMYOLO, the size of input image is 416×416, and three feature maps of different sizes are generated after backbone, which first send to CA for processing to obtain positional attention information in different directions, thereby further improving the feature extraction ability of the main target of the network. Then, the feature map of 13×13 scale is fed into CSP-SPP for processing, and the 13×13 scale feature map is captured and first fused with feature information.

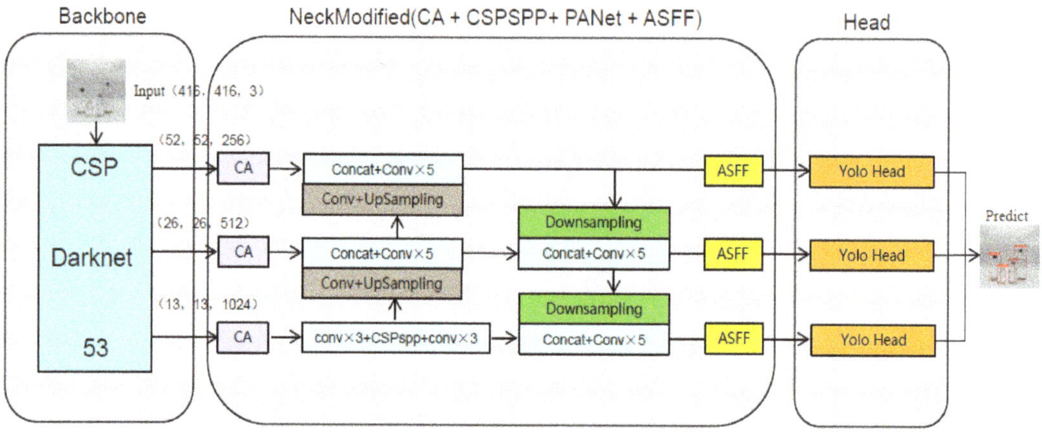

Figure 6. The structure of NMYOLO.

Following this, the feature map of three scales is sent to PANet for feature fusion, and information feature extraction of different scales is realized, and ASFF processing is then carried out such that a group of features can contain information after the fusion of other scale feature maps, and the feature maps of different scales finally obtained by the fusion are input to YOLO head for output. Because the final output scale sizes are 52 × 52, 26 × 26, and 13 × 13, NMYOLO demonstrates good performance in detecting targets of different scale sizes.

In addition, unlike the prediction box loss function CIoU in YOLOv4, the loss function used by NMYOLO for prediction box classification is EIoU. With the same consideration of the overlapping area of the bounding box and the distance of the center point, treatment of the aspect ratio by EIoU involves calculating the true difference between the individual width and height and its confidence, while CIoU only calculates the difference between its overall aspect ratio. EIoU takes a more comprehensive view and is able to obtain more stable and accurate anchor frame information to speed up training and improve detection results. EIoU is shown in (10):

$$\begin{aligned} L_{\text{EIoU}} &= L_{\text{IoU}} + L_{dis} + L_{asp} \\ &= 1 - \text{IoU} + \frac{p^2(b,b^{gt})}{d^2} + \frac{p^2(w,w^{gt})}{w_{\min}^2} + \frac{p^2(h,h^{gt})}{h_{\min}^2} \end{aligned} \quad (10)$$

where IoU is the ratio of intersection and union between prediction box and true box. In addition, $\rho(w,w^{gt})$ and $\rho(h,h^{gt})$ represent the distance between the predicted and true width–height center point, while $w_{\min}$ and $h_{\min}$ are the width and height of the minimum add-in box that covers both the prediction box and the true box.

4.5. Evaluation Metrics

We divided the model evaluation into subjective and objective evaluation metrics. For subjective evaluation, the detection effect graph of each model is output, with observation of its detection effect and whether there is any wrong or missing detection. For objective evaluation, recall, mean average precision (mAP), giga floating point of operations (GFLOPs), and frame per second (FPS) of each model are used as evaluation indicators.

Among them, the GFLOPs indicator is used to measure the model complexity, and recall and mAP calculation are as follows:

$$\text{Recall} = \frac{\text{TP}}{P} \quad (11)$$

$$\text{mAP} = \frac{\sum_{q=1}^{n} \text{AP}_q}{n} \tag{12}$$

Equation (11) is the recall rate calculation formula, where TP indicates the number of positive samples that are correctly predicted as positive. P represents the number of samples in which all predictions are positive. Equation (12) is the formula for mAP calculation, with n representing the total number of categories, and APq representing the average precision of class q. The mAP values at IoU thresholds of 0.5 and 0.75 are usually denoted by mAP50 and mAP75, respectively, and mAP0.5:0.95 is used to characterize the statistical average of mAP IoU thresholds starting from 0.5 and increasing sequentially by 0.05 up to 0.95.

5. Experiments and Results

5.1. Dataset Preparation

There are no datasets on infusion bottles and infusion bags available online, so we took initiative in establishing such a dataset [24]. In this study, a total of 9959 pictures of infusion bottles and infusion bags were taken, collected, and sorted. These include images of single infusion containers and images of multiple infusion containers overlapping each other, while some of these images were taken by adjusting the camera aperture light and dark to simulate environmental changes, and distracting factors such as transparent glasses and common water glasses for drinks were added to some other images. After completing the basic data acquisition and labeling, we apply random masking to a portion of the images. This increases the complexity of our dataset and will allow us to more thoroughly evaluate the effect of the tested models. There are 7661 images for training, 852 images for validation, and 946 images for testing. After obtaining the dataset images, we used Labelimg for annotation. There are five classes in our dataset, inf_bot and inf_bag are infusion bottles and infusion bags, respectively, while bot, sprite, and cola are interference classes used to enhance robustness. The details are shown in Figures 7 and 8, while some examples of the dataset images are shown in Figure 9.

For YOLOv4, the anchor box needs to be set in advance, so we adopted the k-means [25] clustering method for the dataset, producing 9 groups of anchor boxes, namely [(12,16), (19,36), (40,28)], [(36,75), (76,55), (72,146)], and [(142,110), (192,243), (459,401)]. Among these, the first three sets of anchor boxes correspond to the output of the 13 $\times$ 13 scale, and the middle three and the last three correspond to the output of the 26 $\times$ 26 and 52 $\times$ 52 scales, respectively. In addition, we conducted some preprocessing of the dataset before training [26].

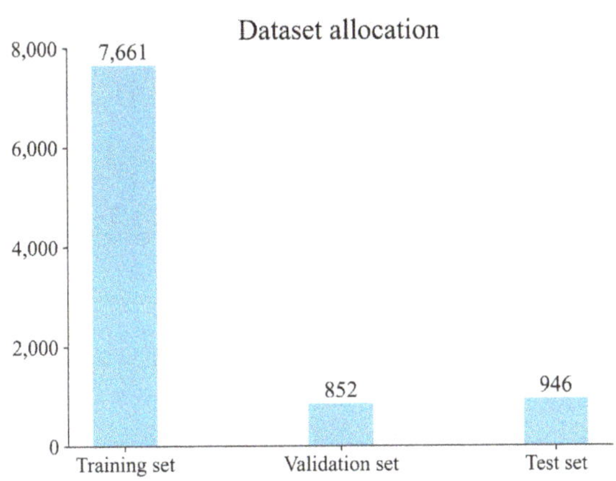

Figure 7. Dataset allocation.

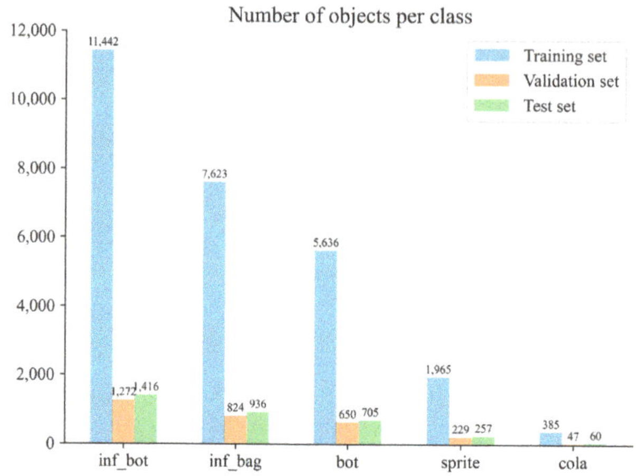

Figure 8. The number of objects per class.

Figure 9. Some pictures from the dataset.

5.2. Environment

The training and inference process of all the models in our study is completed on RTX3060, the deep learning framework used in our models is pytorch1.7, and the CUDA version is 11.2. The specific hyperparameter settings during model training in this study are shown in Table 1. In order to ensure the fairness of the experiment, the model involved in the ablation experiment uses the hyperparameters shown in Table 1.

Table 1. List of hyperparameters.

Hyperparameters	Value
Optimizer	SGD
Learning rate	0.02
Momentum	0.937
Input shape	416 × 416
Weight decay	0.0005
Training Epochs	100
Cos_lr	True
Mosaic	True

5.3. Results and Analysis

In order to verify the effectiveness of the improvements made in this study, the improvements are gradually added on the basis of YOLOv4, and the final results are compared one by one for evaluation.

From Table 2, we see that mAP50 is 3.78% higher for our model than for YOLOv4, precision is improved by 3.55% and recall is improved by 9.91%. Each assessed indicator shows a corresponding improvement. This fully reflects the ability of NMYOLO to detect objects in the face of relatively complex situations, such as occlusion and overlap between infusion sets. Due to the addition of some detection and feature fusion modules, GFLOPs is increased by 1.88 frames while FPS is decreased by 2.79 frames when compared with YOLOv4. However, this still meets the frame rate requirements for video transmission.

Table 2. Comparison of the effects of ablation experiments. And ✗ means we will not add such a module to the baseline, ✓ means we will add it.

CA	ASFF	CSP-SPP	Loss	Precision	Recall	mAP50	GFLOPs	FPS
✗	✗	✗	CIoU	92.65	78.59	91.43	29.89	39.08
✓	✗	✗	CIoU	94.76	81.03	92.91	29.90	38.26
✓	✓	✗	CIoU	93.12	88.18	93.57	31.68	36.92
✓	✓	✓	CIoU	94.84	87.18	94.52	31.68	36.89
✓	✓	✓	EIoU	96.20	88.50	95.21	31.77	36.29

In Table 3, the three indicators of recall, precision, and mAP50 of our model are 96.20%, 66.80%, and 95.21%, respectively, which are the best among the commonly used one-stage models that are compared. Meanwhile, the best value of mAP0.5:0.95 is 73.40%, which gets by YOLOv8m.The best value of mAP75 is 68.23% of YOLOv5m.

Table 3. The comparison between our model and other related one-stage models.

Methods	Input size	Precision	Recall	mAP0.5:0.95	mAP50	mAP75	FPS
SSD	416 × 416	88.76	74.38	42.70	83.58	39.36	/
YOLOv3	416 × 416	91.52	74.87	50.30	85.73	50.54	36.50
YOLOv3-spp	416 × 416	/	/	60.10	88.41	65.03	36.07
YOLOv4	416 × 416	92.65	78.59	62.00	91.43	61.03	39.08
YOLOv5m	416 × 416	93.28	81.77	66.10	91.97	68.23	72.03
YOLOv8m	416 × 416	91.60	87.60	73.40	94.40	/	69.04
YOLOX	416 × 416	93.86	85.12	66.50	93.67	68.16	56.03
YOLOv7	416 × 416	95.76	84.80	65.20	94.14	67.10	38.54
NMYOLO	416 × 416	96.20	88.50	66.80	95.21	67.90	36.29

In Figure 10, when the recall is below 0.6, there are four categories with a precision close to 1, but after the recall is greater than 0.8, the precision decreases rapidly. In addition, the maximum value of each class recall is hardly close to 1. This shows that it is more difficult to detect objects in this dataset than it is to detect them correctly. It also shows that if we want to continue to improve mAP, we need to improve the recall of the model.

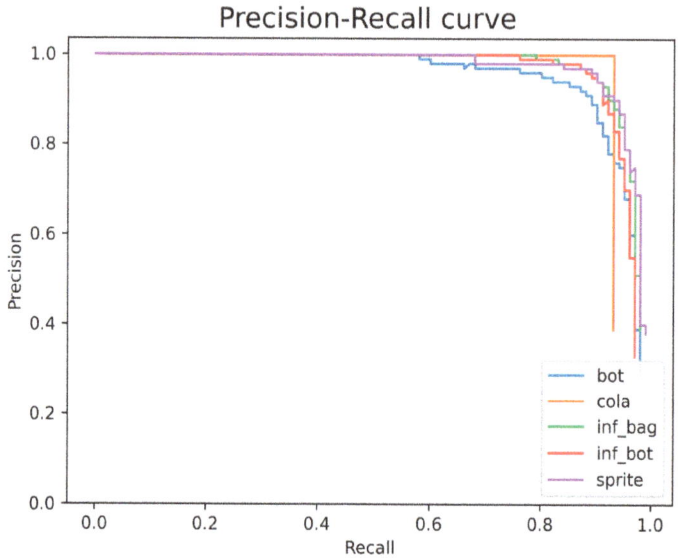

Figure 10. Precision–Recall curve of each class.

In order to facilitate the reader to directly and effectively see the improvements associated with our model, some of the predicted images of YOLOv4 are selected for comparison with some of the predicted images of NMYOLO in Figure 11.

The first column in Figure 11 is the original image, the second column is the result from YOLOv4 detection, and the third column is the result from NMYOLO detection. Among them, regarding the first line near the overlapping target scene, the original picture has three infusion bags and one infusion bottle; YOLOv4 detected two of these infusion bags and the one infusion bottle, thus missing one infusion bag, and NMYOLO detected all four targets, which shows that the NMYOLO has better detection efficacy in the face of overlapping occlusion targets. Regarding the second line of the distant target scene, NMYOLO and YOLOv4 both identify all three targets, but YOLOv4 wrongly identifies one corner of the box as a bottle, whereas there are no detection errors in the case of NMYOLO, indicating it has stronger stability in detecting distant targets. The third behavior is a complex scenario of overlapping multiple infusion bottles, and it is clear that NMYOLO has a higher detection rate for different placements and overlaps of infusion devices when faced with particularly complex infusion container scenarios.

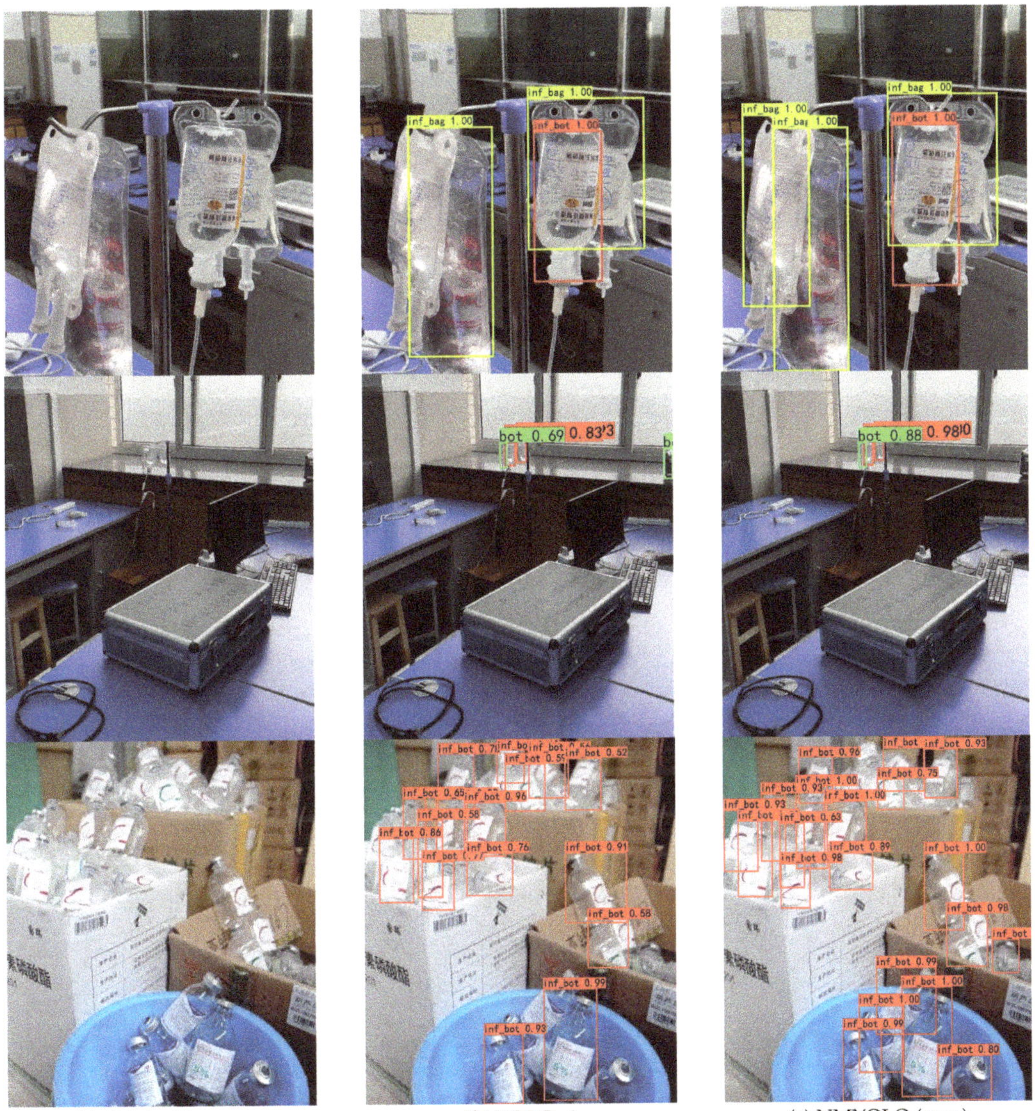

(**a**) Original images (**b**) YOLOv4 (**c**) NMYOLO (ours)

Figure 11. Original images and examples of detection results.

6. Conclusions

To solve the difficult problem of detecting medical infusion containers under dense occlusion and in complex environment scenarios, we proposed the novel method NMYOLO. In this study, we enhance the depth of the neural network in the model by adding ASFF and CA, and improve the information fusion capability of the model. We also modify SPP to CSP-SPP to make the model obtain more information features, and use EIOU to make the model more stable. These improvements make NMYOLO have better detection. In addition, NMYOLO is shown to have better performance compared with other mainstream one-stage detection models.

Although NMYOLO has served our purpose, we still need to discuss what its shortcomings are. The disadvantage of the proposed model is the reduction in the detection frame rate compared with YOLOv4, so one idea for subsequent improvement is to change the method of lightweight backbone or reduce some convolution modules that are not very important to reduce the number of parameters in the model. Moreover, we can replace some modules or change the architecture of the model to reduce the size of the model and ensure detection accuracy.

Author Contributions: Conceptualization, L.J., X.Z. (Xinjun Zhang) and X.X.; Formal analysis, L.J.; Funding acquisition, L.Z.; Investigation, L.J., X.Z. (Xinjun Zhang) and X.X.; Methodology, L.J.; Project administration, X.Z. (Xueyu Zou) and X.L.; Software, L.J.; Supervision, X.Z. (Xueyu Zou) and L.Z.; Validation, L.J. and X.Z. (Xueyu Zou); Writing—original draft, L.J.; Writing—review and editing, L.J., X.Z. (Xueyu Zou) and X.L. All authors have read and agreed to the published version of the manuscript.

Funding: This research was funded by National Outstanding Youth Science Fund Project of Nation-al Natural Science Foundation of China, grant number 61901059.

Institutional Review Board Statement: Not applicable.

Data Availability Statement: Not applicable.

Conflicts of Interest: The authors declare no conflict of interest.

Abbreviations

YOLO	You only look once
CSP	Cross stage partial
NMYOLO	Neck-modfied YOLO
SENet	Squeeze-and-excitation network
CBMA	Convolutional block attention module
SPP	Spatial pyramid pooling
ASFF	Adaptively spatial feature fusion
PANet	Path aggregation network
RCNN	Region-based convolutional neural network
SSD	Shot multi-box detector
CA	Coordinate attention
FPN	Feature pyramid networks
IOU	Intersection of union
EIOU	Efficient IOU
CIOU	Complete IOU
mAPx	mean average precision while confidence at 0.x
SGD	Stochastic gradient descent
CNN	Convolutional neural networks

References

1. Cheng, M.M.; Hou, Q.B.; Zhang, S.H.; Rosin, P.L. Intelligent visual media processing: When graphics meets vision. *J. Comput. Sci. Technol.* **2017**, *32*, 110–121. [CrossRef]
2. Hou, Q.; Zhou, D.; Feng, J. Coordinate attention for efficient mobile network design. In Proceedings of the IEEE/CVF Conference on Computer Vision and Pattern Recognition, Nashville, TN, USA, 20–25 June 2021; pp. 13713–13722.
3. Aloysius, N.; Geetha, M. A review on deep convolutional neural networks. In Proceedings of the International Conference on Communication and Signal Processing (ICCSP), Melmaruvathur, India, 6–8 April 2017; pp. 0588–0592.
4. Liu, S.; Huang, D.; Wang, Y. Learning spatial fusion for single-shot object detection. *arXiv* **2019**, arXiv:1911.09516.
5. Zhang, Y.F.; Ren, W.; Zhang, Z.; Jia, Z.; Wang, L.; Tan, T. Focal and efficient IOU loss for accurate bounding box regression. *Neurocomputing* **2022**, *506*, 146–157. [CrossRef]
6. Bochkovskiy, A.; Wang, C.Y.; Liao, H.Y.M. Yolov4: Optimal speed and accuracy of object detection. *arXiv* **2020**, arXiv:2004.10934.
7. Wu, X.; Sahoo, D.; Hoi, S.C. Recent advances in deep learning for object detection. *Neurocomputing* **2020**, *396*, 39–64. [CrossRef]
8. Girshick, R.; Donahue, J.; Darrell, T.; Malik, J. Rich feature hierarchies for accurate object detection and semantic segmentation. In Proceedings of the IEEE Conference on Computer Vision and Pattern Recognition, Columbus, OH, USA, 23–28 June 2014; pp. 580–587.

9. Girshick, R. Fast r-cnn. In Proceedings of the IEEE International Conference on Computer Vision, Boston, MA, USA, 7–12 June 2015; pp. 1440–1448.
10. Ren, S.; He, K.; Girshick, R.; Sun, J. Faster R-CNN: Towards Real-Time Object Detection with Region Proposal Networks. *IEEE Trans. Pattern Anal. Mach. Intell.* **2017**, *39*, 1137–1149. [CrossRef] [PubMed]
11. Khasawneh, N.; Fraiwan, M.; Fraiwan, L. Detection of K-complexes in EEG waveform images using faster R-CNN and deep transfer learning. *BMC Med. Inform. Decis. Mak.* **2022**, *22*, 297. [CrossRef] [PubMed]
12. Redmon, J.; Divvala, S.; Girshick, R.; Farhadi, A. You only look once: Unified, real-time object detection. In Proceedings of the IEEE Conference on Computer Vision and Pattern Recognition, Las Vegas, NV, USA, 27–30 June 2016; pp. 779–788.
13. Khasawneh, N.; Fraiwan, M.; Fraiwan, L. Detection of K-complexes in EEG signals using deep transfer learning and YOLOv3. *Clust. Comput.* **2022**, 1–11. [CrossRef]
14. Liu, W.; Anguelov, D.; Erhan, D.; Szegedy, C.; Reed, S.; Fu, C.Y.; Berg, A.C. Ssd: Single shot multibox detector. In Proceedings of the European Conference on Computer Vision, Amsterdam, The Netherlands, 8–16 October 2016; Springer: Cham, Switzerland; pp. 21–37.
15. Vaswani, A.; Shazeer, N.; Parmar, N.; Uszkoreit, J.; Jones, L.; Gomez, A.N.; Polosukhin, I. Attention is all you need. *Adv. Neural Inf. Process. Syst.* **2017**, *30*, 6000–6010.
16. Wang, X.; Girshick, R.; Gupta, A.; He, K. Non-local neural networks. In Proceedings of the IEEE Conference on Computer Vision and Pattern Recognition, Salt Lake City, UT, USA, 18–23 June 2018; pp. 7794–7803.
17. Hu, J.; Shen, L.; Sun, G. Squeeze-and-excitation networks. In Proceedings of the IEEE Conference on Computer Vision and Pattern Recognition, Salt Lake City, UT, USA, 18–23 June 2018; pp. 7132–7141.
18. Woo, S.; Park, J.; Lee, J.Y.; Kweon, I.S. Cbam: Convolutional block attention module. In Proceedings of the European Conference on Computer Vision, Munich, Germany, 8–14 September 2018; pp. 3–19.
19. Wang, C.Y.; Liao, H.Y.M.; Wu, Y.H.; Chen, P.Y.; Hsieh, J.W.; Yeh, I.H. CSPNet: A new backbone that can enhance learning capability of CNN. In Proceedings of the IEEE/CVF Conference on Computer Vision and Pattern Recognition Workshops, Seattle, WA, USA, 13–19 June 2020; pp. 390–391.
20. Xie, S.; Girshick, R.; Dollár, P.; Tu, Z.; He, K. Aggregated residual transformations for deep neural networks. In Proceedings of the IEEE Conference on Computer Vision and Pattern Recognition, Honolulu, HI, USA, 21–26 July 2017; pp. 1492–1500.
21. He, K.; Zhang, X.; Ren, S.; Sun, J. Spatial pyramid pooling in deep convolutional networks for visual recognition. *IEEE Trans. Pattern Anal. Mach. Intell.* **2015**, *37*, 1904–1916. [CrossRef] [PubMed]
22. Liu, S.; Qi, L.; Qin, H.; Shi, J.; Jia, J. Path aggregation network for instance segmentation. In Proceedings of the IEEE Conference on Computer Vision and Pattern Recognition, Salt Lake City, UT, USA, 18–23 June 2018; pp. 8759–8768.
23. Lin, T.Y.; Dollár, P.; Girshick, R.; He, K.; Hariharan, B.; Belongie, S. Feature pyramid networks for object detection. In Proceedings of the IEEE Conference on Computer Vision and Pattern Recognition, Honolulu, HI, USA, 21–26 July 2017; pp. 2117–2125.
24. Rivera-Castillo, J.; Flores-Fuentes, W.; Rivas-López, M.; Sergiyenko, O.; Gonzalez-Navarro, F.F.; Rodríguez-Quiñonez, J.C.; Básaca-Preciado, L.C. Experimental image and range scanner datasets fusion in SHM for displacement detection. *Struct. Control Health Monit.* **2017**, *24*, e1967. [CrossRef]
25. Krishna, K.; Murty, M.N. Genetic K-means algorithm. *IEEE Trans. Syst. Man Cybern. Part B (Cybern.)* **1999**, *29*, 433–439. [CrossRef] [PubMed]
26. Flores-Fuentes, W.; Alba-Corpus, I.Y.; Sergiyenko, O.; Rodríguez-Quiñonez, J.C. A structural health monitoring method pro posal based on optical scanning and computational models. *Int. J. Distrib. Sens. Netw.* **2022**, *18*, 15501329221112606. [CrossRef]

entropy MDPI

Article

Image Registration for Visualizing Magnetic Flux Leakage Testing under Different Orientations of Magnetization

Shengping Li [1], Jie Zhang [1,*], Gaofei Liu [2], Nanhui Chen [1], Lulu Tian [1], Libing Bai [1] and Cong Chen [1]

[1] School of Automation Engineering, University of Electronic Science and Technology of China, Chengdu 611731, China

[2] China Petroleum Pipeline Insptection Technologies Co., Ltd., Langfang 065000, China

* Correspondence: zhj06_19@uestc.edu.cn

Abstract: The Magnetic Flux Leakage (MFL) visualization technique is widely used in the surface defect inspection of ferromagnetic materials. However, the information of the images detected through the MFL method is incomplete when the defect (especially for the cracks) is complex, and some information would be lost when magnetized unidirectionally. Then, the multidirectional magnetization method is proposed to fuse the images detected under different magnetization orientations. It causes a critical problem: the existing image registration methods cannot be applied to align the images because the images are different when detected under different magnetization orientations. This study presents a novel image registration method for MFL visualization to solve this problem. In order to evaluate the registration, and to fuse the information detected in different directions, the mutual information between the reference image and the MFL image calculated by the forward model is designed as a measure. Furthermore, Particle Swarm Optimization (PSO) is used to optimize the registration process. The comparative experimental results demonstrate that this method has a higher registration accuracy for the MFL images of complex cracks than the existing methods.

Keywords: nondestructive testing; Magnetic Flux Leakage; solenoid modal; image registration; mutual imformation; PSO

Citation: Li, S.; Zhang, J.; Liu, G.; Chen, N.; Tian, L.; Bai, L.; Chen, C. Image Registration for Visualizing Magnetic Flux Leakage Testing under Different Orientations of Magnetization. *Entropy* **2023**, *25*, 167. https://doi.org/10.3390/e25010167

Academic Editors: Oleg Sergiyenko, Wendy Flores-Fuentes, Julio Cesar Rodriguez-Quinonez and Jesús Elías Miranda-Vega

Received: 14 November 2022
Revised: 9 January 2023
Accepted: 12 January 2023
Published: 13 January 2023

1. Introduction

Magnetic Flux Leakage (MFL) detection is widely used in the nondestructive testing of defects in ferromagnetic components [1–4]. The inversion for the profile of a defect is the most interesting part of the related research. In existing studies of 3D profiles, the effect of the surface profile of the defect on the defect depth is often not considered [5,6]. However, the surface profile of the defect severely affects the distribution of the MFL field in space. Existing profile reconstruction methods mainly use data collected using magnetization in a single direction [2,7], which is effective when there are no edges that are parallel to the magnetization direction, or complex corners in the defects. However, cracks are prone to more complex signal coupling, or they are missing due to their small width and complex surface profile, which cannot be collected completely from a single direction, and they must be magnetized from multiple directions to obtain complete information about the surface profile of the defect [8,9]. (Figure 1 shows the different MFL field distributions of a V-shaped defect under different orientations of magnetization.) Traditional MFL testing needs to scan the same area several times in different directions. In order to analyze the MFL signal characteristics of the acquired defects, a method is required first to align the acquired MFL images under different orientations of magnetization (DOM).

Image registration is defined as aligning images acquired at different times, distinct viewpoints, and where valuable information is conveyed in more than one image [10]. The MFL images captured under DOM are difficult to spatially align in actual application because of the displacement and rotation of sensors. However, no published papers discuss the MFL image registration of defects captured under DOM, to the authors' knowledge.

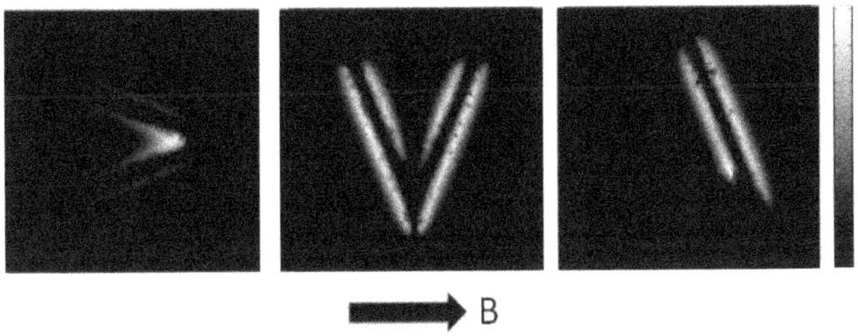

Figure 1. The distribution of MFL field with DOM (Simulated by COMSOL).

The general image registration methods (e.g., medical image registration and remote sensing image registration) are mainly classified as being intensity-based and feature-based [11]. Intensity-based methods use optimization methods to find the optimal value of the alignment. Different pixel intensity-based measures are applied to evaluate the registration of each iteration of the optimization process [12,13]. However, such methods cannot consider the inconsistency of the magnetic field distribution under DOM. They can only match the parts of the image with similar intensity distributions, because such methods are based on the assumption that the corresponding structures in the registered images would have similar intensities [10]. Additionally, the MFL field distribution detected under DOM varies greatly, making the shape alignment methods inapplicable [14].

The feature-based approaches use distance-based measures to match the features extracted from the input images. The methods require the presence of features: such as centerlines, outlines, corners, etc. Additionally, the images must possess a relatively clear corresponding point mapping [13,15,16], which would also be affected by the distortion of MFL because the distribution of MFL images varies (such as the first and the second image in Figure 1).

The above analysis shows that the currently used image registration methods do not apply to the MFL images captured under DOM. This paper proposes an adaptive registration method to register the multidirectional magnetized MFL images of surface defects (as in Figure 2). The proposed method combines the MFL forward model and the multi-model image alignment method. Firstly, the PSO optimization method updates the image transform parameters and generates a new defect shape according to the aligned images. Then, the MFL field distribution of the new reconstructed shape is generated via the use of the solenoid model [8,17]. The similarity between the generated MFL field distribution and the acquired reference image is calculated as the judgment of optimization. Using the above process, the optimal alignment parameters are achieved, and the registration of the MFL images under DOM is completed.

There are three contributions to this paper:

1. A new registration method for MFL images detected under DOM is designed.
2. The solenoid model is first used in MFL image registration.
3. The comparative experiment is carried out, and the proposed method shows a higher accuracy than the traditional methods.

This paper is organized as follows. Section 2: The proposed method is introduced. Section 3: The experimental setup. Section 4: Presents the results and discussion. Section 5: The conclusions.

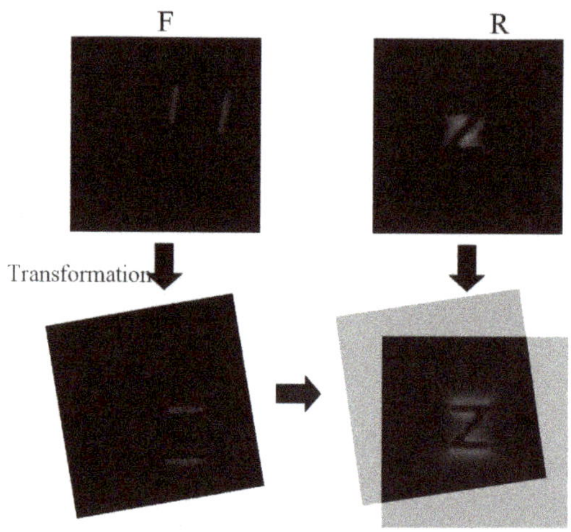

Figure 2. Registration Schematic of MFL images under DOM.

2. Methodology

Most conventional registration methods can be described as an optimization process. During each step of the optimization algorithm, a new parameter of the transformation function is updated, which is applied to the floating image (F) by spatially aligning it with the reference image (R). Additionally, the registration performance is accessed using a measure of similarity between the reference and the transformed floating image. The procession is shown in Figure 2.

Herein, the reference image (R) and floating image (F) are captured by the magneto-optical image (MOI) system, which is shown in Figure 3. The polarized light passes through the magneto-optical film (MOF), which rotates due to the Faraday rotation effect, and the rotation angle θ can be calculated by Equation (1). The rotated light reflected by a mirror under the MO film would be filtered by the polarizer and captured by the CMOS. The acquired gray image depicts the normal components of the MFL field.

$$\theta = VBL \tag{1}$$

where V is the Verdet constant of the MOF, and B is the intensity of the introduced magnetic field parallel to the direction of the light. L denotes the distance that the light travels through the MOF.

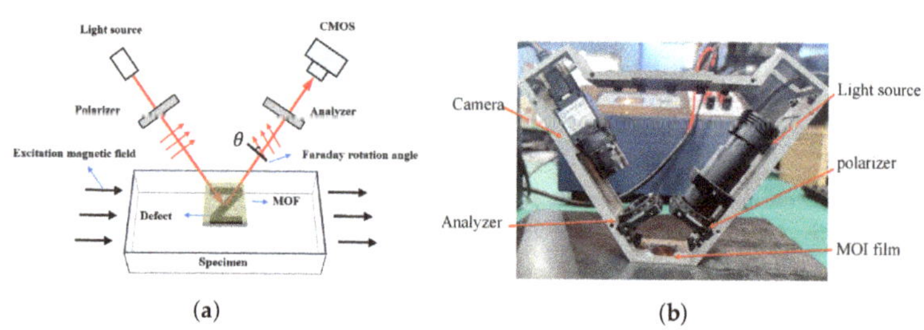

Figure 3. Magneto-optical imaging: (**a**) Schematic of MOI, (**b**) The MOI system.

The MFL image detection under DOM can be expressed as in Figure 4. The angle between the MOI system and the magnetic yoke is fixed to keep the direction of the external magnetic field in the captured images. Additionally, the performance of registration would be better when the rotation angle comes to the position where the images are complementary. (Complete information about defects can be obtained at an angle of about 80°∼90°).

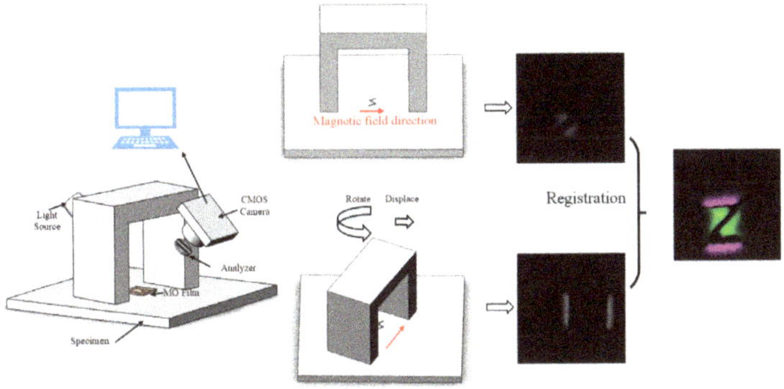

Figure 4. Schematic of the procession.

Like the prevailing algorithms, the proposed registration algorithm consists of an iterative trial-and-error process that attempts to optimize a given transformation function after a limited number of iterations [16,18]. The process of the proposed registration method can be described, as shown in Figure 5. The first step is in segmenting the original images R and F to R_s and F_s. Then, Particle Swarm Optimization (PSO) is used to maximize the similarity by optimizing the transformation of in-plane parameters. In each iteration of PSO, five steps are needed:

1. Producing a new registration parameter;
2. Transforming the F_s according to the registration parameter and aligning it to R_s;
3. Fusing the R_s and the transformed F_s, then reconstructing a shape of crack (I_t);
4. Generating a new distribution (I_g) of the crack (I_t);
5. Calculating the similarity between R and I_g.

After the iterations, the parameter of the optimal similarity is output as the final result of registration.

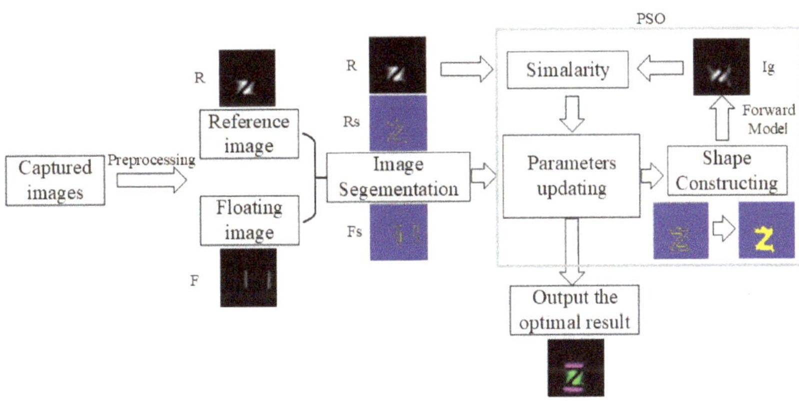

Figure 5. The proposed method.

2.1. Preprocessing

A rough registration to limit the translation is needed to reduce the computation time and to avoid the registration parameters escaping from the feasible solution space (a high similarity may occur, even if the images are not correctly registered).

Because the background of the MO image is clear, it is easy to find the location of the defect. Firstly, obtain the projection along the row and column direction from the R and the F. Then, the cosine distance of the projections is minimized by adjusting the displacement along the row and column. As seen in Figure 6, the projection of the defect in two directions would partially overlap, and the cosine distance is used to assess the degree of overlap, which is calculated as

$$cos(\Theta) = \frac{Rp \cdot Fp}{||Rp|| \cdot ||Fp||} = \frac{\sum_{i=1}^{n} (Rp_i \cdot Fp_i)}{\sqrt{\sum_{i-1}^{n} Rp_i^2}\sqrt{\sum_{i-1}^{n} Fp_i^2}} \quad (2)$$

where Rp and Fp are the vectors resulting from the projections of R and F along the row and column directions, respectively, n is the number of elements in Rp and Fp, and i is denoted as the traversal of each element.

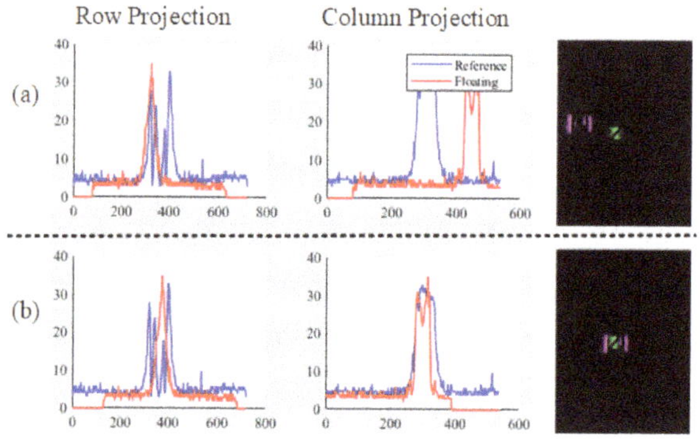

Figure 6. The comparison of graphs along the row and column directions; (**a**) Before transformation, (**b**) after transformation.

The displacement range in the iteration of the following optimization algorithm depends on the width of the magnetic field distribution in the image. (In the subsequent process, only the displacement part of the image will be shown; take Figure 7 as an example).

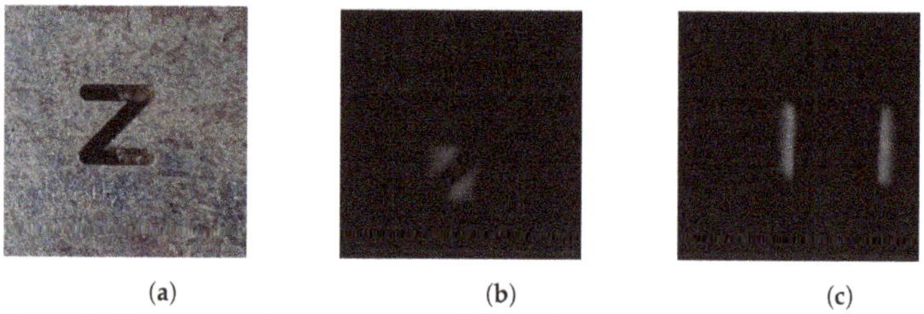

Figure 7. The example image images. (**a**) The original image of the defect. (**b**) The first direction of magnetization. (**c**) The second direction of magnetization (the angle between c and b is about 85°∼90°).

2.2. Segmentation

The first step is in segmenting the original images to extract the partial edge of the defects. The MFL field distribution among the edge of defects is usually sharp because the magnetic line leaks out from the edge, and the intensity decreases as the distance grows. Such a distribution could be presented by the gray intensities of the captured images. The Laplacian of the Gaussian (LoG) filter, which could be approximated by using the

Difference of Gaussian (DoG), shows a good performance in identifying the edge of the defect [19,20]. The output image (O) can be calculated by:

$$O(x,y) = \nabla^2(I(x,y) * G(x,y)) \quad (3)$$

where the I is the input image, G is the Gaussian filter, and ∇^2 is the Laplacian.

However, it is easy to identify the inner and outer edges of the MFL field while detecting the border. Considering that the intensity of the MFL field is inversely proportional to the square of the distance between the defects and other parts, there has always been a peak beside the edge of defects, which can be used as a further judgment. Here, we set a limit distance that the detected point is to the peaks along the direction of magnetization. This is retained when the extracted points of LOG are within the set distance from the peak, and it is discarded if it is outside the set distance. Additionally, the result is presented in Figure 8.

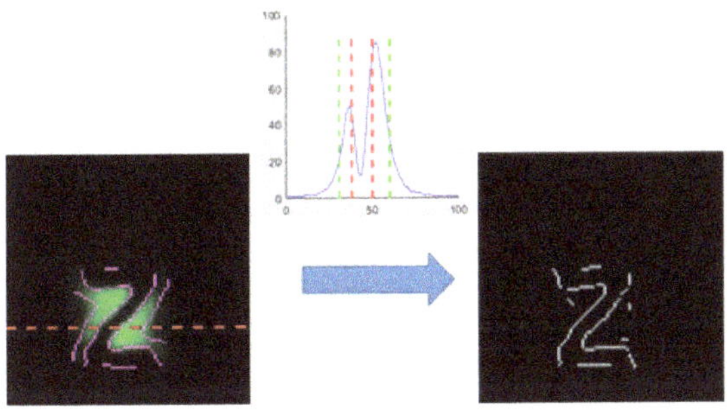

Figure 8. The segmentation of captured image.

2.3. *Particle Swarm Optimization Algorithm*

This study uses the PSO algorithm to update the transformation function. In PSO, each particle represents a possible solution to the optimization task, and the particles tend to cluster where the optimum solution is.

Here, the rigid geometric transformation is applied. The floating image is registered to the reference image with a global transformation. The registration parameter is a three-dimensional vector, which consists of one rotation angle θ (unit: degrees) and two translation distances t_x,t_y (unit: millimeters). The transformation matrix of image coordinates P_F to P_R from the image F to image R can be shown as:

$$P_R = M \times P_F \quad (4)$$

where

$$M = \begin{pmatrix} \cos\theta & -\sin\theta & (1-\cos\theta)t_x + t_y \times \sin\theta \\ \sin\theta & \cos\theta & (1-\cos\theta)t_x - t_y \times \sin\theta \\ 0 & 0 & 1 \end{pmatrix} \quad (5)$$

Therefore, the search space and the position of each particle of PSO can be represented by a three-dimensional vector $x_i = (t_x, t_y, \theta)$ [21,22]. The task performed in each particle includes:

- Transforming the image F_s.
- Constructing the shapes of defects.
- Generating the MFL field distribution (I_g) of this defect.
- Calculating the similarity between the image R and I_g.

The new position can be updated according to each iteration's local and globally optimal solutions. The own personal best solution so far by the particle $p_{id}(t)$ and the global best position of any particle in the swarm $P_{gd}(t)$ so far are found. They are called the "individual best position" and the "global best position". The new velocity and position of each particle can be updated according to $p_{id}(t)$ and $P_{gd}(t)$:

$$v_i(t+1) = w(t) \times v_i(t) + c_1 \times \text{Rand} \times (P_{id}(t) - x_{id}(t)) + c_2 \times \text{Rand} \times \left(P_{gd}(t) - x_i(t)\right) \tag{6}$$

$$x_i(t+1) = x_i(t) + v_i(t+1) \tag{7}$$

where c_1 and c_2 are two constants parameters. c_1 can change the step size of a particle moving toward its individual optimal position. c_2 can change the step size for moving toward the global optimal position. (Here, for a better global optimum search, c_2 is set to be slightly larger than c_1. In this study, c_1 used for an image of size 100×100 is 0.5, and c_2 is 0.7, for reference only). *Rand* is a random value limited by $(0, 1)$. $w(t)$ is called the inertia weight, which influences the current velocity according to the particle's previous velocity, and is updated in the iterations by:

$$w(t) = (w_{\max} - w_{\min}) \times \frac{(t_{\max} - t)}{t_{\max}} + w_{\min} \tag{8}$$

where $w_{\max}$ and $w_{\min}$ denote the maximum and minimum of inertial weight, respectively. $t_{\max}$ is the maximum iteration number.

The positions of the particles should be restricted in $[X_{\min}, X_{\max}]$, $[Y_{\min}, Y_{\max}]$ to avoid particles escaping from the feasible solution space. Additionally, the domains are calculated in Section 2.1.

2.4. Opening Shape Reconstruction

After transformation, the F_s would be partially overlaid with R_s, and then they are fused to construct the shapes of defects, which would be used to generate the distribution of the MFL field in the later process. First, two images are overlaid (performing *OR* operations on the corresponding pixel points) to generate F_{temp}, and *Closing* (morphology) is used to connect the adjacent connected domains, so that all of the regions in the image are connected. The *Closing* can be presented as:

$$F_{\text{temp}} \bullet K = (F_{\text{temp}} \oplus K) \ominus K \tag{9}$$

where the K is the structural element, and

$$F_{\text{temp}} \ominus K = \{x, y \mid (K)_{xy} \subseteq F_{\text{temp}}\} \tag{10}$$

$$F_{\text{temp}} \oplus K = \{x, y \mid (K)_{xy} \cap F_{\text{temp}} \neq \varnothing\} \tag{11}$$

The closed area would then be detected and filled, to generate the shape of the defect.

2.5. Solenoid Modal for the Visualization of MFL

The solenoid modal is adjusted to generate the distribution of the magnetic field of the constructed shape. It is based on the theory of ampere molecular currents, which are arranged neatly and closely along the magnetization direction in the specimen. Then, the molecular currents form a series of solenoids in the specimen, as shown in Figure 9.

When a solenoid is truncated, a semi-infinite solenoid model can be established to simulate the MFL on the surface of the defect, according to Biot–Savart's law. The intensity can be calculated by:

$$d\mathbf{H}_{ls} = \frac{M_{de}}{4\pi} \frac{\mathbf{r}_s}{r_s^3} ds \tag{12}$$

where M_{de} is the effective component of M_d in the normal direction at the defective surface, and M_d is the magnetization. ds is an element size at the surface of the defect, and $\mathbf{r}_s$ is a vector from the pole of the solenoid to the point of the field.

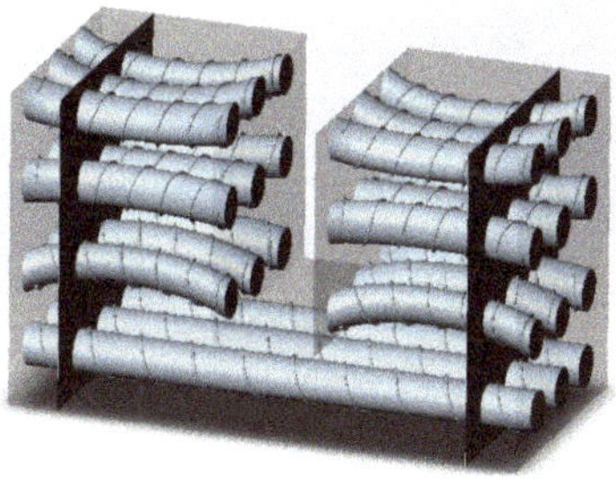

Figure 9. The solenoid in specimen.

The above model is constructed on the assumption that the magnetization on the defect surface is uniform. However, the directions of each solenoid change due to the interaction force for actual complex defects, which makes the MFL field distribution more complicated. The magnetization of the specimen is assumed to be quasi-saturated, and the interaction force among the solenoids should be introduced to improve the model. Since the force is proportional to the magnetic field intensity, the intensity of the interaction field can be calculated by:

$$H_{\text{inter}} = \frac{M_{de}}{4\pi} a\tan\left(\frac{r_s ds}{r_s^3}\right) \tag{13}$$

The interaction between the solenoids causes the solenoids to deviate from the direction of the excitation field, and the deviation angle can be calculated by:

$$\theta = a\tan\left(\frac{H_{\text{inter}}}{H_{ex}}\right) \tag{14}$$

where H_{ex} is the magnetic field intensity in the specimen, and here, it is set according to the material of the specimen.

The interaction of the solenoids on the defect boundary is shown in Figure 10. When the interaction of the solenoids on the defect surface is not considered, the solenoids will be uniformly arranged, and their generated leakage magnetism will be uniformly distributed (Figure 10a). When the interaction between the solenoids at the defect surface is considered, the leakage magnetization generated by the solenoids at the end surface will interact with each other, resulting in the deflection of the solenoids at the end surface. In Figure 10b, we can see three solenoids, A, B, and C, at the corner of the defect. Since A and B have magnetic leakage at the end face, and C is not broken, B will be repelled by A and deflected toward C, creating a bend in B here. The influence of each solenoid at the end face by the surrounding solenoids is calculated, and this is how the solenoid model is set up. This approach has obvious advantages in the calculation of the inhomogeneous distribution of the leakage of the magnetic field at the corners of complex defects.

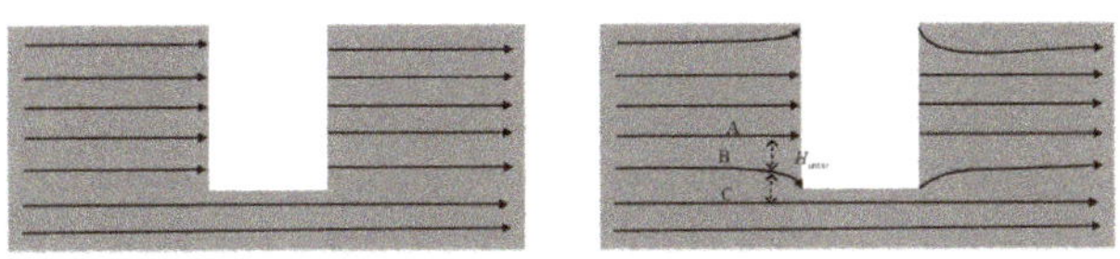

(a) (b)

Figure 10. Schematic of solenoid interaction. (**a**) No interactions; (**b**) With interactions.

The applicability to the calculation accuracy of the model for complex defects is improved by modifying the direction of the solenoids. Then, the angle between the direction of the excitation field and the image should be fixed to achieve the right distribution of MFL. The depth of defects was set according to the thickness of the tested specimen and the maximal intensity of the originally captured images. Therefore, the solenoid distribution of the defective edges can be calculated. The intensity of image I_g can be derived by the calculated magnetic field strength, and the formula is determined by:

$$I = I_0 \sin^2(VBL) \tag{15}$$

where I_0 is the maximal intensity of captured images.

The comparison between different methods is shown in Figure 11. Compared with the widely used magnetic dipole model (MDM), the present model has a more obvious advantage in considering the signal differences in defect corner coupling and edges (several places circled in Figure 11c have uneven signal distributions when the defect shape is complex, which cannot be fitted by MDM, while the solenoid model performs this better). Therefore, the solenoid model can better approximate the actual MFL image.

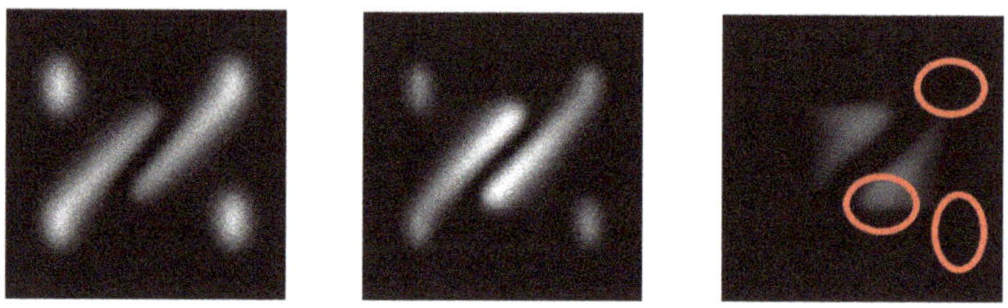

(**a**) (**b**) (**c**)

Figure 11. Comparison of magnetic dipole model and solenoid model. (**a**) Magnetic dipole model, (**b**) Solenoid model, (**c**) The actual MOI image.

2.6. Similarity Measure

Now that the images (R and I_g) for assessing registration are ready, this section should apply a coefficient to measure the similarity between the images to evaluate the registration.

Mutual-Information (MI) is the most popular and widely studied similarity metric in intensity-based registration [21,23]. It is developed from information entropy to describe the relationship between two images. Take two random variables as an example, A and B, with marginal probability distributions, $P_A(a)$ and $P_B(b)$; joint probability distribution, $P_{AB}(a,b)$, are statistically independent if

$$P_{AB}(a,b) = P_A(a) \bullet P_B(b) \tag{16}$$

they are maximally dependent if they are related by a one-to-one mapping T:

$$P_A(a) = P_B(T(a)) = P_{AB}(a, T(a)) \tag{17}$$

MI, $I(A,B)$ measures the degree of dependence of A and B by measuring the distance between the joint distribution $P_{AB}(a,b)$ and the distribution associated with the case of complete independence $P_A(a) \bullet P_B(b)$.

$$MI(A,B) = \sum_{a,b} p(a,b) \log \frac{p(a,b)}{p(a)p(b)} \tag{18}$$

If the images are geometrically aligned, the MI of the image intensity values for the corresponding pixel pairs should be maximum. This criterion is very general and robust. It can be applied automatically without prior segmentation, because no assumptions about the relationship between the two image intensities are made. These properties are appropriate for calculating the similarity between the generated and original graphics, so here, we use MI as a registration criterion.

The whole algorithm of the registration of visualizing MFL testing under DOM is presented in Algorithm 1.

Algorithm 1 Registration

Input: Captured MFL images R and F
Output: The optimal registration parameters (θ, t_x, t_y);
Segmeting the images to R_s and F_s;
Setting the particles and iterations number of PSO;
repeat
    Updating a new registration parameter;
    Transforming the F_s;
    Reconstructing a shape of crack (I_t);
    Generating a new distribution (I_g) of (I_t);
    Calculating the similarity between R and I_g;
    Recording the global and local optimal positions.
until The iteration of PSO finishes.

Based on the method designed above, we take the first two simulated images in Figure 1 with the first two magnetization directions differing by 90° as an example; the image size is 100 × 100, and the relative displacement of the two images in the row and column directions is 0 pixels. The number of particles in the optimization algorithm is set to 5, and the number of iterations is set to 100. We can obtain the registration results, as shown in Figure 12a. The convergence process of the iterations is shown in Figure 12b, which shows that the similarity reaches its best after 33 iterations, with a similarity of 1.1802. The registration result is a rotation angle of 92.0510°, with a displacement of 0.46805 pixels along the row direction, and −0.64315 pixels along the column direction. This registration result has some errors with the actual results, but it shows that the method is iterative convergent.

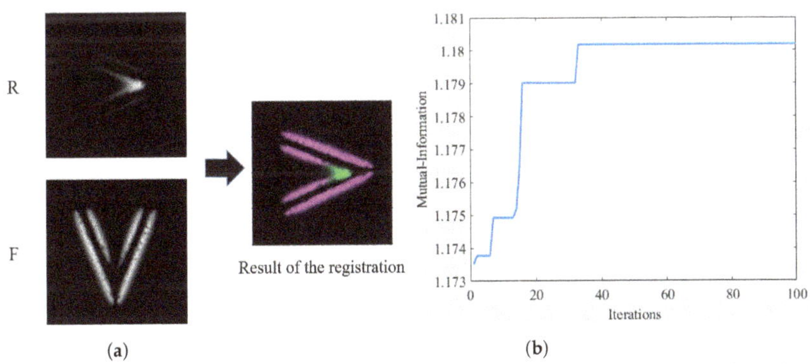

Figure 12. Example of operational feasibility. (**a**) Results of registration, (**b**) Convergence of iterations

3. Experiment Setup

3.1. The Settings of the Experimental Equipment

In order to verify the proposed methods, the experimental platform is set up (shown in Figure 13). The electric magnetic yoke is the magnetizing excitation source to generate a uniform excitation magnetic field. The MOI system is fixed in the middle of the magnetic yoke, making the angle between the excitation direction and the captured image constant.

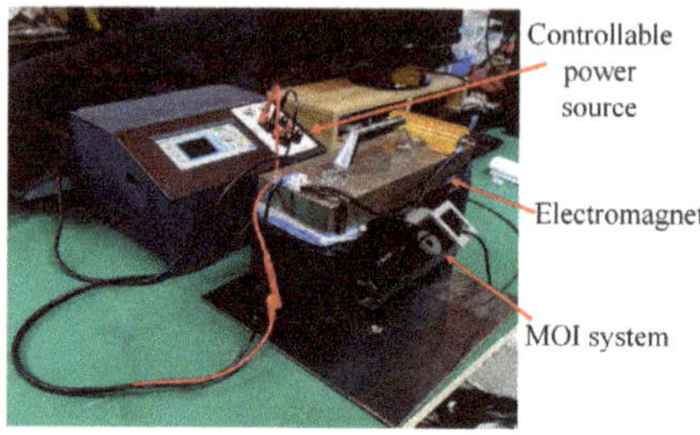

Figure 13. The platform of MFL image detection.

Additionally, the MO film is a $(BiTm)_3(GaFe)_5O_{12}$ single-crystal thin film [24,25] provided by the State Key Laboratory of Electronic Thin Films and Integrated Devices, University of Electronic Science and Technology of China, and the maximum Verdet constant $2.595 \times 10^{-4}/(Oe\mu m) \cdot 10\ \mu m$.

3.2. Experiment Samples

The test samples include two artificial and two natural cracks. The artificial samples include a z-shaped (Figure 7a) and t-shaped cracks (Figure 14a). The natural sample includes two fatigue cracks, the first is a crack with one-line (Figure 14b), and the other one is a crack with three-line coupling (Figure 14c).

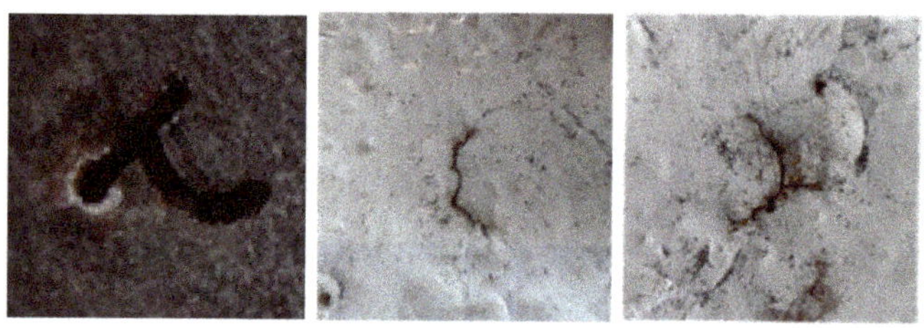

(a) (b) (c)

Figure 14. The samples: (**a**) Manual t-shaped crack, (**b**) Natural one-line crack, (**c**) natural three-line crack.

4. Results and Discussion

4.1. Experiment Results

The images in Figure 7 are used to verify the method, and the result is shown in Table 1. The variation of the parameter of the transformation matrix leads to different shapes of defects, therefore generating different distributions of the MFL field. As can be seen from the last column of Table 1, when the transformation matrix comes to the correct position, the MI similarity of the generated image (I_g) and the reference image (R) is optimal. We can also find that the constructed shape is the most similar to the shape of the actual defect when the similarity is optimal, which can be used to evaluate the defect. The MI similarity here shows a significant difference between the incorrect and correct registrations, and the better it is registered, the bigger the similarity coefficient is. Because of the difference in the shape of the fused defects, the magnetic field distribution deduced from the defects will also have significant differences. Meanwhile, we can see that the reconstructed shape is closest to the actual defect morphology when in the correct alignment state, and the MFL distribution calculated by the solenoid model has the highest similarity to the original image, which also proves the correctness of the solenoid model for the defect MFL calculation at the same time. Therefore, the present method can also be regarded as a reconstruction of the defect surface profile by fusing multiple sources of the MFL image data.

However, some significant factors also affect the result of the calculation of similarity between images.

1. The segmenting method for the original images R and F is significant. The better the edge of defects is detected, the better the shape of defects is constructed.
2. The fusion of segmented images and the method of shape construction directly affects the MFL field distribution of I_g.
3. The stability and accuracy of the MFL distribution forward model.

Table 1. The similarity between the I_g and R changes according to the deviation of F_s.

-	Angle + 30°	X + 10	Y + 10	Correct
R_s				
F_s	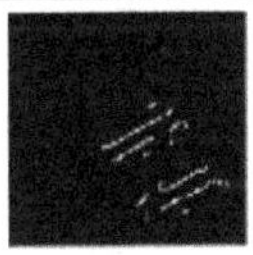	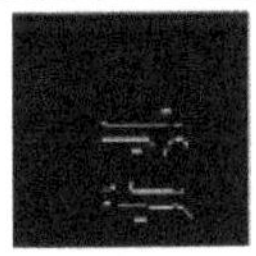	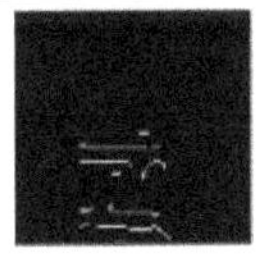	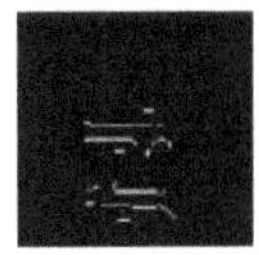
F_{temp}	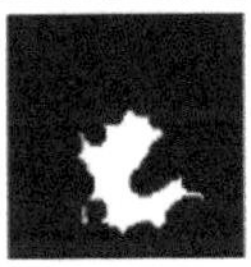			
I_g	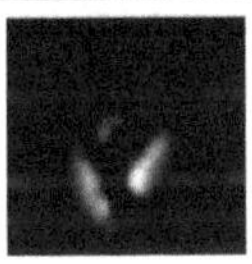		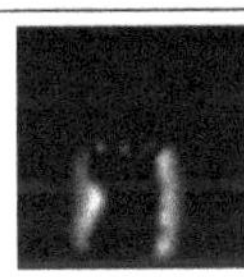	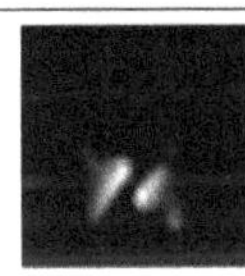
Similarity	0.6446	0.7503	0.6589	0.9200

4.2. Robustness of the Proposed Method

The robustness of the method can be evaluated through comparisons with the feature-based, intensity-based, and manual registration methods (shown in Table 2).

The feature-based approach requires the presence of features: such as corners, outlines, and some particular points. Because of the smooth edge and the distortion, it is challenging to extract valid corresponding feature points. At the same time, the obtained features cannot be matched with each other because the magnetic field distribution generated by different magnetization directions cannot correspond to each other. All these problems make it impossible to register the images correctly.

The deviations in the magnetic field distribution also affect the intensity-based registration. The images are only matched according to intensity because the shapes of defects and magnetic field distortions cannot be considered. In the registration, calculating the intensity similarity optimum only leads to the match of the regions in two images that are close in intensity, which does not contribute to the correct registration of the images.

Manual registration shows a good performance because it is based on the existing knowledge of the shapes of the defects and the relative orientations of the images. However, manual registration cannot be employed in inspection applications with large data and for defects of unknown shape and location. This is why a method is needed for the scenarios described in this paper. In this set of comparisons, manual registration is only used as a standard group to evaluate the results of the registration of images.

The last row of Table 2 shows the experimental results of the proposed method, which can achieve a better result compared to feature-based and intensity-based registration methods, and it is more consistent with the manual one. For cracks, it can achieve a high registration accuracy due to the prominent structural characteristics of the magnetic field distribution, and the error of the crack registration mainly occurs in the translation. The error in defect reconstruction and the difference between the actual data and the model derivation make a certain deviation in the registration results, which needs to be enhanced and solved in the subsequent study.

Table 2. The result of image registration from different methods. (The feature-based and intensity-based results come from MATLAB 2019b image registration app: the surf and multimodal intensity module).

–	T	Z	*OneLine*	*ThreeLine*
R			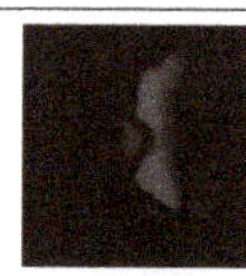	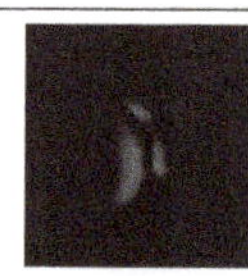
F	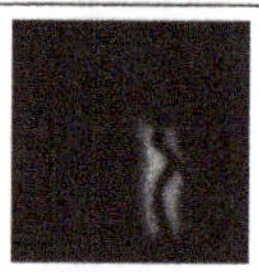			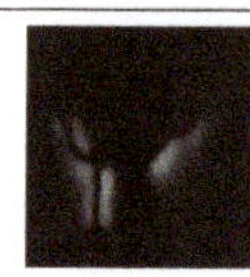
Feature-Based		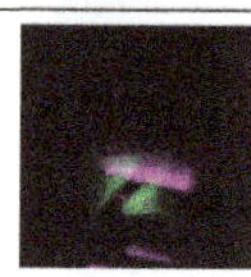	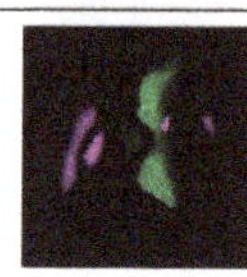	
Intensity-Based	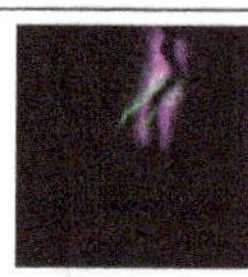	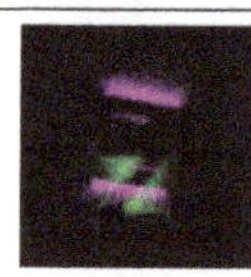	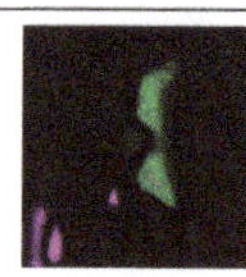	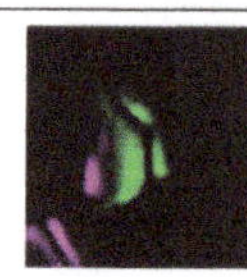
Manual	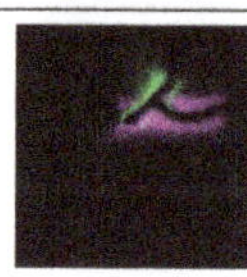	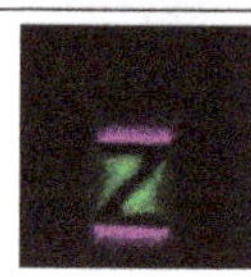	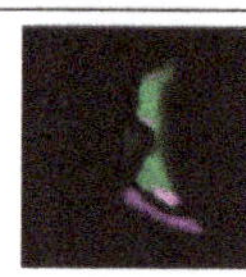	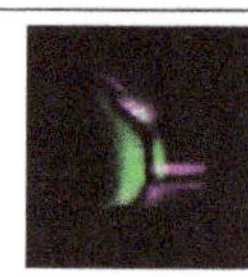
Proposed	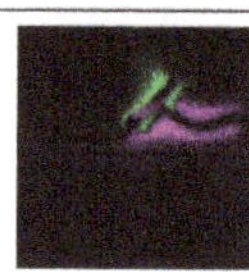	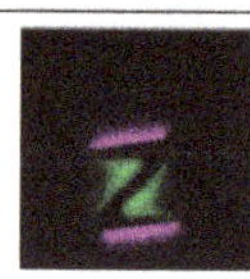	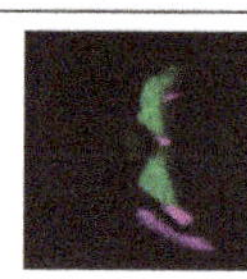	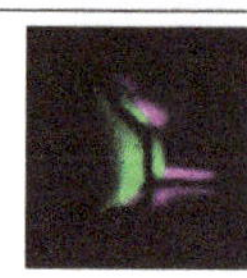

Here, we use the correlation coefficient to calculate the registered F for comparison with the manually registered F to evaluate the registration results of different methods. Equation (19) is the calculation of the correlation coefficient:

$$Cor = \frac{\sum_{i=1}^{N}\left(Fm_i - \overline{Fm}\right)\cdot\left(Fa_i - \overline{Fa}\right)}{\sqrt{\sum_{i=1}^{N}\left(Fm_i - \overline{Fm}\right)^2 \cdot \sum_{i=1}^{N}\left(Fa_i - \overline{Fa}\right)^2}} \cdot 100\% \quad (19)$$

where Fm is the manually registered F and Fa is the F registered using other methods. N denotes all the pixels in the graph, and i denotes the transversal of each element of it. The closer the Cor is to 100%, the closer is the result of the registration method used to the manual registration result.

As shown in Figure 15, by comparing the registration results of different methods with the manual registration results, it can be seen that the feature-based and intensity-based methods are less similar to the results of the manual registration methods, and they can even be considered to be completely unusable. Additionally, the proposed method has a clear advantage in that it can have better registration results for all of the cracks tested here.

It can be observed from the above results that the proposed method can be used for the MFL information fusion of cracks under DOM, which helps in the subsequent analysis of defects.

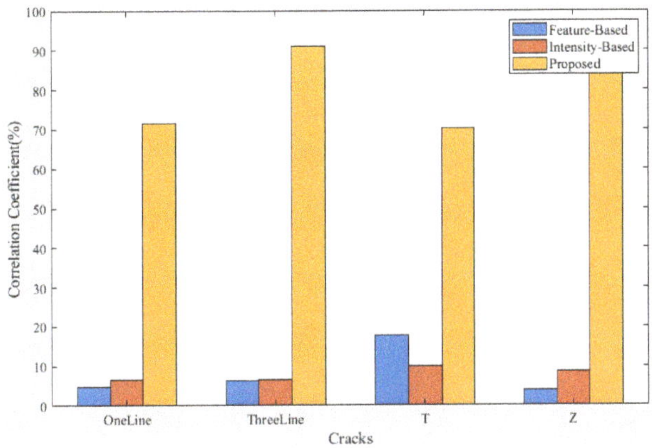

Figure 15. Comparison of correlation coefficients with manual registration results.

5. Conclusions

This study presented a registration method for MFL field visualization under DOM. It solved the problem of the mismatch between the distorted images captured under DOM by considering the 2D shape reconstruction of defects, and the application of the forward model. The solenoid model for MFL visualization is first applied to analyze and to generate the MFL field distribution. The experimental results showed that the proposed method performs better in MFL image registration than do the currently used methods. The proposed method can be applied in MFL detection engineering applications for crack detection. Future research could focus on image segmentation, shape reconstruction, and improving the calculation accuracy of the forward model. Additionally, it could be noticed that such a registration structure could also be used in multimodel situations, such as registering images captured using different NDT methods.

Author Contributions: Conceptualization, S.L. and J.Z.; Methodology, S.L. and N.C.; Software, S.L. and N.C.; Validation, S.L., G.L., L.T. and L.B.; Formal analysis, S.L. and J.Z.; Investigation, S.L., G.L., N.C. and L.B.; Resources, J.Z. and L.B.; Data curation, S.L. and C.C.; Writing—original draft, S.L. and N.C.; Writing—review & editing, S.L. and L.T.; Visualization, L.T.; Supervision, J.Z.; Project administration, J.Z.; Funding acquisition, J.Z. and L.B. All authors have read and agreed to the published version of the manuscript.

Funding: This work was supported in part by the National Natural Science Foundation of China under Grant U2030205, Grant 62003075, Grant 61903065, 62003074 and Sichuan Science and Technology Planning Project 2022JDJQ0040.

Institutional Review Board Statement: Not applicable.

Conflicts of Interest: The authors declare no conflict of interest.

References

1. Hosseingholizadeh, S.; Filleter, T.; Sinclair, A.N. Evaluation of a Magnetic Dipole Model in a DC Magnetic Flux Leakage System. *IEEE Trans. Magn.* **2019**, *55*, 6200407. [CrossRef]
2. Huang, S.; Peng, L.; Wang, Q.; Wang, S.; Zhao, W. An opening profile recognition method for magnetic flux leakage signals of defect. *IEEE Trans. Instrum. Meas.* **2019**, *68*, 2229–2236. [CrossRef]
3. Wu, Z.; Deng, Y.; Liu, J.; Wang, L. A Reinforcement Learning-Based Reconstruction Method for Complex Defect Profiles in MFL Inspection. *IEEE Trans. Instrum. Meas.* **2021**, *70*, 2506010. [CrossRef]
4. Peng, X.; Liu, H.; Siggers, K.; Liu, Z. Automated Box Data Matching for Multi-Modal Magnetic Flux Leakage Inspection of Pipelines. *IEEE Trans. Magn.* **2021**, *57*, 6200410. [CrossRef]
5. Han, W.; Xu, J.; Wang, P.; Tian, G. Defect profile estimation from magnetic flux leakage signal via efficient managing particle swarm optimization. *Sensors* **2014**, *14*, 10361–10380. [CrossRef] [PubMed]
6. Lu, S.; Liu, J.; Wu, J.; Fu, X. A Fast Globally Convergent Particle Swarm Optimization for Defect Profile Inversion Using MFL Detector. *Machines* **2022**, *10*, 1091. [CrossRef]
7. Ravan, M.; Amineh, R.K.; Koziel, S.; Nikolova, N.K.; Reilly, J.P. Sizing of 3-D arbitrary defects using magnetic flux leakage measurements. *IEEE Trans. Magn.* **2010**, *46*, 1024–1033. [CrossRef]
8. Cheng, Y.; Wang, Y.; Yu, H.; Zhang, Y.; Zhang, J.; Yang, Q.; Sheng, H.; Bai, L. Solenoid model for visualizing magnetic flux leakage testing of complex defects. *NDT E Int.* **2018**, *100*, 166–174. [CrossRef]
9. Zhang, J.; Liu, X.; Xiao, J.; Yang, Z.; Wu, B.; He, C. A comparative study between magnetic field distortion and magnetic flux leakage techniques for surface defect shape reconstruction in steel plates. *Sens. Actuators A Phys.* **2019**, *288*, 10–20. [CrossRef]
10. Oliveira, F.P.M.; Tavares, J.M.R.S. Medical image registration: A review. *Comput. Methods Biomech. Biomed. Eng.* **2017**, **17**, 73–93. [CrossRef] [PubMed]
11. Zhang, J.; Ma, W.; Wu, Y.; Jiao, L. Multimodal Remote Sensing Image Registration Based on Image Transfer and Local Features. *IEEE Geosci. Remote Sens. Lett.* **2019**, *16*, 1210–1214. [CrossRef]
12. Inglada, J.; Giros, A. On the possibility of automatic multisensor image registration. *IEEE Trans. Geosci. Remote Sens.* **2004**, *42*, 2104–2120. [CrossRef]
13. Guan, S.Y.; Wang, T.M.; Meng, C.; Wang, J.C. A review of point feature based medical image registration. *Chin. J. Mech. Eng. Engl. Ed.* **2018**, *31*, 76. [CrossRef]
14. Yi, J.; Zhang, S.; Cao, Y.; Zhang, E.; Sun, H. Rigid shape registration based on extended hamiltonian learning. *Entropy* **2020**, *22*, 539. [CrossRef]
15. Lee, I.; Seo, D.C.; Choi, T.S. Entropy-based block processing for satellite image registration. *Entropy* **2012**, *14*, 2397–2407. [CrossRef]
16. Chakraborty, S.; Pradhan, R.; Ashour, A.S.; Moraru, L.; Dey, N. Grey-wolf-based Wang's demons for retinal image registration. *Entropy* **2020**, *22*, 659. [CrossRef]
17. Wang, Y.; Cheng, Y.; Bai, L.; Zhang, J.; Yu, H.; Alimey, F.J. Solenoid Model for the Magnetic Flux Leakage Testing Based on the Molecular Current. *IEEE Trans. Magn.* **2018**, *54*, 6203014. [CrossRef]
18. Savva, A.D.; Economopoulos, T.L.; Matsopoulos, G.K. Geometry-based vs. intensity-based medical image registration: A comparative study on 3D CT data. *Comput. Biol. Med.* **2016**, *69*, 120–133. [CrossRef]
19. Ulupinar, F.; Medioni, G. Refining edges detected by a LoG operator. *Comput. Vis. Graph. Image Process.* **1990**, *51*, 275–298. [CrossRef]
20. Wu, W. Paralleled Laplacian of Gaussian (LoG) edge detection algorithm by using GPU. In Proceedings of the Eighth International Conference on Digital Image Processing (ICDIP 2016), Chengu, China, 20–22 May 2016; p. 1003309.
21. Chen, Y.W.; Lin, C.L.; Mimori, A. Multimodal medical image registration using particle swarm optimization. In Proceedings of the 2008 Eighth International Conference on Intelligent Systems Design and Applications ISDA 2008, Kaohsiung, Taiwan, 26–28 November 2008; Volume 3, pp. 127–131.
22. Jin, J.; Wang, Q.; Shen, Y. High-performance medical image registration using improved particle swarm optimization. In Proceedings of the 2008 IEEE Instrumentation and Measurement Technology Conference,Victoria, BC, Canada, 12–15 May 2008; pp. 736–740.
23. Pluim, J.P.W.; Maintz, J.B.A.A.; Viergever, M.A. Mutual-information-based registration of medical images: A survey. *IEEE Trans. Med. Imaging* **2003**, *22*, 986–1004. [CrossRef]
24. Zhang, D.; Yang, Q.; Hao, J.; Jiang, Y.; Wang, M.; Du, S.; Syvorotka, I.I.; Zhang, H. Effect of lattice mismatch on the laser-induced damage thresholds of $(BiTm)_3(GaFe)_5O_{12}$ thin films. *Appl. Surf. Sci.* **2018**, *473*, 235–241. [CrossRef]
25. Zhang, D.; Yang, Q.; Wang, M.; Du, S.; Jiang, Y.; Syvorotka, I.I.; Zhang, H. Effect of substrate defects on LIDT of $(BiTm)_3(GaFe)_5O_{12}$ films grown by LPE. *Appl. Surf. Sci.* **2018**, *484*, 169–174. [CrossRef]

entropy

MDPI

Article

Scale Enhancement Pyramid Network for Small Object Detection from UAV Images

Jian Sun [1], Hongwei Gao [2,3,*], Xuna Wang [2] and Jiahui Yu [4,5,*]

1 School of Graduate, Shenyang Ligong University, Shenyang 110159, China
2 School of Automation and Electrical Engineering, Shenyang Ligong University, Shenyang 110159, China
3 China State Key Laboratory of Robotics, Shenyang Institute of Automation, Chinese Academy of Sciences, Shenyang 110016, China
4 Department of Biomedical Engineering, Zhejiang University, Hangzhou 310058, China
5 Innovation Center for Smart Medical Technologies & Devices, Binjiang Institute of Zhejiang University, Hangzhou 310053, China
* Correspondence: ghw1978@sohu.com (H.G.); jiahui.yu@port.ac.uk (J.Y.)

Abstract: Object detection is challenging in large-scale images captured by unmanned aerial vehicles (UAVs), especially when detecting small objects with significant scale variation. Most solutions employ the fusion of different scale features by building multi-scale feature pyramids to ensure that the detail and semantic information are abundant. Although feature fusion benefits object detection, it still requires the long-range dependencies information necessary for small objects with significant scale variation detection. We propose a simple yet effective scale enhancement pyramid network (SEPNet) to address these problems. A SEPNet consists of a context enhancement module (CEM) and feature alignment module (FAM). Technically, the CEM combines multi-scale atrous convolution and multi-branch grouped convolution to model global relationships. Additionally, it enhances object feature representation, preventing features with lost spatial information from flowing into the feature pyramid network (FPN). The FAM adaptively learns offsets of pixels to preserve feature consistency. The FAM aims to adjust the location of sampling points in the convolutional kernel, effectively alleviating information conflict caused by the fusion of adjacent features. Results indicate that the SEPNet achieves an AP score of 18.9% on VisDrone, which is 7.1% higher than the AP score of state-of-the-art detectors RetinaNet achieves an AP score of 81.5% on PASCAL VOC.

Keywords: object detection; unmanned aerial vehicles; small objects; feature fusion

Citation: Sun, J.; Gao, H.; Wang, X.; Yu, J. Scale Enhancement Pyramid Network for Small Object Detection from UAV Images. *Entropy* **2022**, *24*, 1699. https://doi.org/10.3390/e24111699

Academic Editors: Wendy Flores-Fuentes, Oleg Sergiyenko, Julio Cesar Rodriguez-Quinonez and Jesús Elías Miranda-Vega

Received: 12 October 2022
Accepted: 17 November 2022
Published: 21 November 2022

1. Introduction

UAVs have the advantages of low operational cost, high mobility, and multiple viewpoints, thus promoting the application of drone object detection [1,2] in many fields, such as power line detection [3], precision agriculture [4], and environmental monitoring [5,6]. Under the positive influence of maturity of hardware devices and the availability of training datasets, deep learning has achieved unprecedented success because of its impressive ability to learn representation from data. At present, UAV image detection algorithms are generally based on convolutional neural networks (CNNs), such as ResNet [7], DenseNet [8], and ConvNet [9]. Due to CNNs' strong local perception and inductive biases, a series of breakthroughs have been made in computer vision tasks, such as object detection [10], semantic segmentation [11,12], human–robot interaction [13], etc. Although deep learning has made significant progress in natural image detection, aerial image detection of state-of-the-art object detectors, such as YOLO [14] and RetinaNet [15], still needs to be more satisfactory in terms of both accuracy and efficiency.

There exist some significant differences between nature images (e.g., PASCAL VOC [16]) and UAV images (e.g., VisDrone [17]), leading to two major challenges of object detection. The first challenge is that high-resolution UAV images tend to contain small objects (area < 32^2 pixels) and are generally sparsely distributed, as Figure 1a depicts. The features of small objects are difficult to describe because the small scale of the target is featured by fewer pixels, which is likely to cause information to gradually disperse or even vanish when they pass through a deep network. Sparse objects in images with a wide field of view are easier to be confused with complex backgrounds. Second, extreme object scale variation and special UAV perspectives can be present, as Figure 1b depicts. The UAV images of large-scale scenes are affected by the variety of altitudes and perspectives of UAVs. When UAVs shoot at lower altitudes, objects become more negligible. Objects become smaller when UAVs shoot at higher altitudes. Lengthening the perspectives also causes distant objects to become smaller. Even objects of the same class may differ several times in scale.

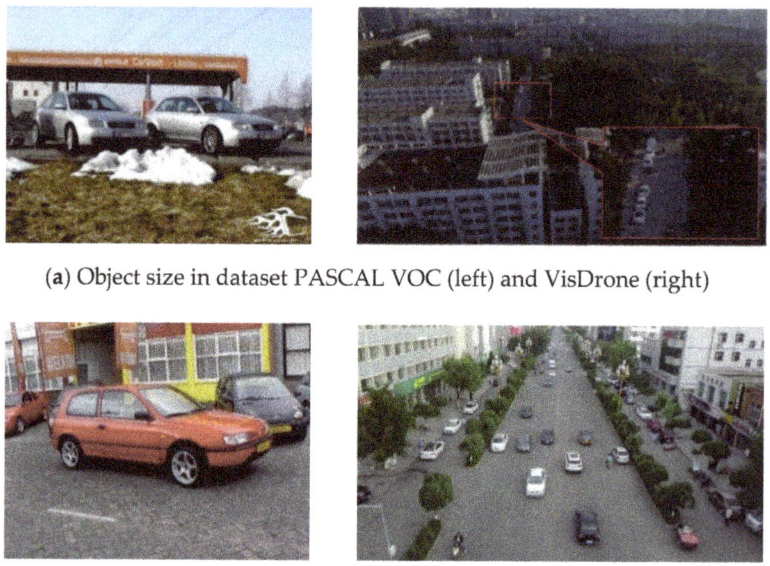

(**a**) Object size in dataset PASCAL VOC (left) and VisDrone (right)

(**b**) Various scales in dataset PASCAL VOC (left) and VisDrone (right)

Figure 1. Compared with natural scene images, UAV images from VisDrone show great challenges. (**a**) Object with a small size and sparse distribution in a UAV image. (**b**) The particular perspective of the UAV makes the aerial image come in extremely varying scales.

One way to address the challenges above is to use the cutting strategy [18,19]. The high-resolution image is dealt with as small patches and then fed separately into the network for prediction. However, such methods may require repeated computation of features, resulting in higher computation and memory requirements. In addition, multi-scale feature fusion [20,21] enriches difficulty discerning object feature representations by integrating deep and shallow features while adding less computational cost. The other line of effort aims to expand the receptive field using stacking atrous convolutions with different atrous rates or convolutional filters with different sizes [22,23], which is also an effective way to improve object detection performance. Some methods use an attention mechanism [24,25] to highlight helpful information from small targets while suppressing useless information. The attention mechanism can improve the detection performance of most existing CNN-based methods while introducing very little computation.

This paper proposes a scale enhancement pyramid network, namely SEPNet, to improve UAV image detection performance by mitigating the inconsistency in gradient computation of the adjacent layers. Our algorithm mainly consists of two core modules. We notice that the deep network is effective in detecting complex scenes. However, the deep network loses essential details in forward propagation. Although the number of network layers deepens, the receptive field becomes more significant. The single receptive field makes the detector suffer contextual limits. Based on this observation, we designed a lightweight context enhancement module (CEM) core consisting of a multi-scale dilated convolution branch and a pyramidal convolution branch. Unlike most existing methods, we combined multi-scale dilated and pyramidal convolution to model the global relationships for objects of various scales instead of artificially designed complicated decoder networks. In addition, to enhance network performance, multi-scale features are generally used to fuse information at different levels to obtain more powerful representations, and direct fusion between different levels destroys feature consistency in gradient computation, which makes features obtained after the CEM module weaken the expressive representation. We used the feature alignment module (FAM) to automatically learn the correlation between two feature layers and keep them aligned. Our SEPNet is based on one-stage detectors.

The main contributions of this paper are summarized as follows:

1. We propose a SEPNet to solve small object and multi-scale object detection difficulties in UAV images.
2. We propose the CEM to produce more salient context information by combining special groups of atrous convolutions and group convolutions and redistribution to the top of FPN, thereby improving the feature representation of objects at different scales.
3. We add the FAM that learns transformation offsets of pixels to preserve the aggregate feature space translation invariance and address the feature inconsistency issue for FPN, avoiding small objects being drowned in feature conflicts. To continue improvement, we introduce channel attention to refine pre-aggregated features while making the network focus on the target area rather than the broad background.
4. We validate the proposed two components and SEPNet on two datasets. Compared to the baseline model, RetinaNet, our component can significantly improve performance, from 21.3% to 23.5% on the VisDrone dataset. Furthermore, our SEPNet outperforms the popular detector CornerNet [26] by 1.5%.

2. Related Work

In this section, we briefly review the recent representative work on object detection, feature fusion architecture design, and the attention mechanism of convolutional networks.

2.1. Object Detection

With the development of deep learning, remarkable progress has been achieved in object detection. The mainstream object detectors based on deep learning can be divided into one-stage detectors and two-stage detectors. The significant difference between the two network architectures is that two-stage detectors first generate region proposals and then apply a convolutional network to classify and regression each region proposal. In contrast, one-stage detectors skip the proposal stage and manually set priority boxes. Two-stage methods, such as Faster RCNN [27], maintain an advantage in precision, but the speed is not satisfactory due to the need to obtain region proposals before detection. One-stage methods, such as Single Shot MultiBox Detector (SSD) [28], improve detection speed at the cost of accuracy drop. Recently, anchor-free methods were proposed. Compared to anchor-based methods, anchor-free methods replace complex anchor designs by capturing features of object centers or key points. CenterNet [29] generates heatmaps (distribution of important information in the feature map) to obtain the target center coordinates and adjust the center offset. Fully convolutional one-stage object detection (FCOS) [30], feature selective anchor-free module (FSAF) [31], and FoveaBox [32] drop prior anchor settings and directly encode and decode the bounding boxes as anchor points. This detects all positive sample points, and the positive samples point to boundary distances of the bounding box. Anchor-free methods are not constrained by predefined anchors and reduce hyperparameters and forward inference time. However, these intensive prediction tasks are prone to noise interference, resulting in many false positives.

2.2. Feature Fusion

Object detection in UAV images is a challenging problem due to small objects [33,34] and extreme scale variation. FPN [35] is an efficient way to alleviate the problem arising from small objects and object scale variation. In the deep network, low-level features generally lack semantic information and are rich in geometric details. In contrast, high-level features are the opposite of low-level features. FPN builds a feature pyramid to extract and fuse multi-scale features through the top-down pathway and lateral connections. The path aggregation network (PANet) [36] adds an extra bottom-up path on the top of FPN. EfficientDet [37] proposes a bidirectional feature pyramid network (BiFPN), which is a weighted bidirectional FPN used to perform fast feature fusion. Giraffedet [38] enriches multi-level contextual information through bottom-up skip-layer connection and sufficient cross-scale connection between different levels. Apart from network structure improvement, some other works [39,40] are devoted to enhancing contextual information. They generally combine multiple branches with different kernel sizes and dilated convolutions to effectively capture long-range information without reducing spatial resolution. To solve the problem of feature misalignment during high-level and low-level fusion, feature-aligned pyramid networks (FaPN) [41] achieve implicit compensation with deformable revolution to enhance feature consistency. The above methods effectively fuse different levels of semantic and location information and achieve great success but ignore the problem of feature inconsistency when dealing with different input features.

2.3. Attention Mechanism

The attention mechanism is recognized as a potential means to enhance deep CNNs since it allows the network to selectively focus on the most important regions of an image while ignoring the ones with irrelevant parts. Currently, attention mechanisms are prevalent in various tasks, such as machine translation [42], object detection [43], and semantic segmentation [44]. More recently, multiple attention mechanisms have provided benefits in visual studies to improve convolutional network expression ability. Squeeze-and-excitation networks (SENet) [45] are typical channel attention mechanisms. They can adaptively recalibrate channel-wise response with global contextual information by signals aggregated from feature maps. Efficient channel attention networks (ECANet) [46] employ the one-dimensional convolution layer to determine channel interaction and reduce the attention module parameters. Still, the information captured by the one-dimensional convolutional is inefficient. Selective kernel networks (SKNet) [47] apply multiple branches with different kernel sizes to adaptively adjust the receptive field, effectively increasing the flexibility of the network. Beyond channel attention, non-local neural networks (non-local) [48] deploy self-attention as a generalized global operator to capture the long-range dependencies. Non-local can effectively capture global features of spatial sequences and are more friendly for video detection. Convolutional block attention modules (CBAM) [49] and bottleneck attention modules (BAM) [50] introduce channel and spatial attention to allow the network to generate weights of different channels and spatial automatically, highlighting the location and category information of the network. Furthermore, SANet [51] propose efficient shuffle attention, which can effectively combine spatial and channel attention through shuffle units to enrich the network with deep information. In contrast, our work focuses on the correlation of channels between different levels of features to further integrate information at different scales of the feature map.

3. Method

The overall architecture of SEPNet is shown in Figure 2. We first use ResNet to build our backbone network as the feature extractor. Each pyramidal feature map (denoted C2, C3, C4, C5) extracted by ResNet is followed by an additional 1×1 convolution to compress channels. Then, these feature maps are used to build a feature pyramid for multi-scale detection. We input C5 into a CEM module and concatenate it with P5 to obtain rich semantic information. We also use the FAM module to learn the correlation of pre-fused features, preventing important information from being destroyed when features are aggregated. It is worth noting that we use the concatenate operation instead of the sum operation for bottom-to-top feature fusion. After addressing top-to-bottom (denoted P2, P3, P4, P5) and bottom-to-top (denoted N2, N3, N4, N5) feature fusion, we will describe the implementation details of the main modules in the following sections.

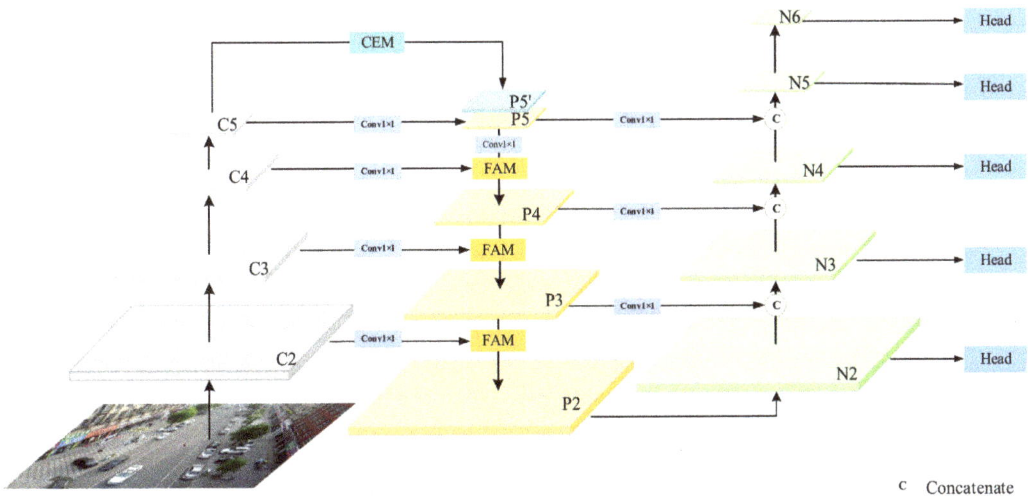

Figure 2. The overall architecture of the proposed SEPNet.

3.1. Context Enhancement Module

As we all know, with the deepening of the network layer, the features lose spatial information, and the ability to express features is weakened. In addition, due to the fixed convolution operation, the features lack the contextual information necessary for object detection at different scales. To extract high-level information, atrous spatial pyramid pooling (ASPP) [52] uses atrous convolutions of different dilation rates to capture the context at multiple scales. Although ASPP can encode multi-scale information and proves effective in semantic segmentation, we believe that the uniform resolution obtained by atrous convolution alone is not enough for UAV detection. For this reason, we are inspired by PyConv [53] and propose a context enhancement module (CEM), which aims at optimizing the deeper layer features to avoid the propagation of lost information features in FPN. CEM injects rich context information into the top of the feature pyramid network to enhance object feature representation, as shown in Figure 3.

The critical components in CEM include atrous spatial pyramid convolutions and grouped pyramidal convolutions. To better explain our CEM, we use a graph to show standard convolution, atrous convolution, and grouped convolution, as shown in Figure 4.

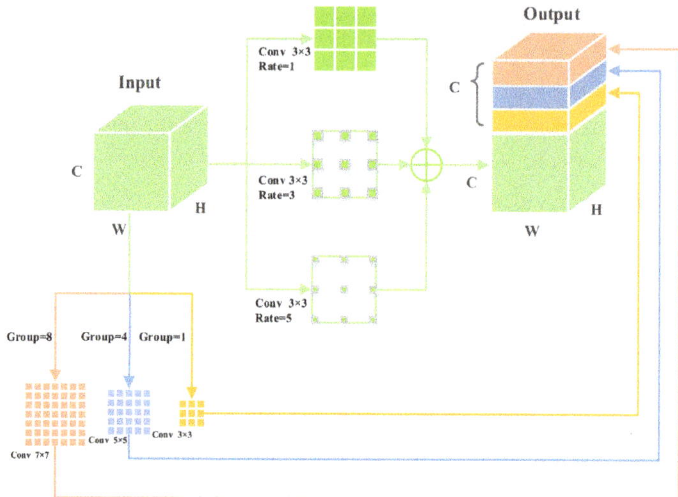

Figure 3. The CEM structure consists of two branches. One branch is processed by dilated convolutions with rates of 1, 3, and 5. The other is processed by grouped convolutions divided into groups 1, 4, and 8, respectively. Finally, two branches are processed by concatenating.

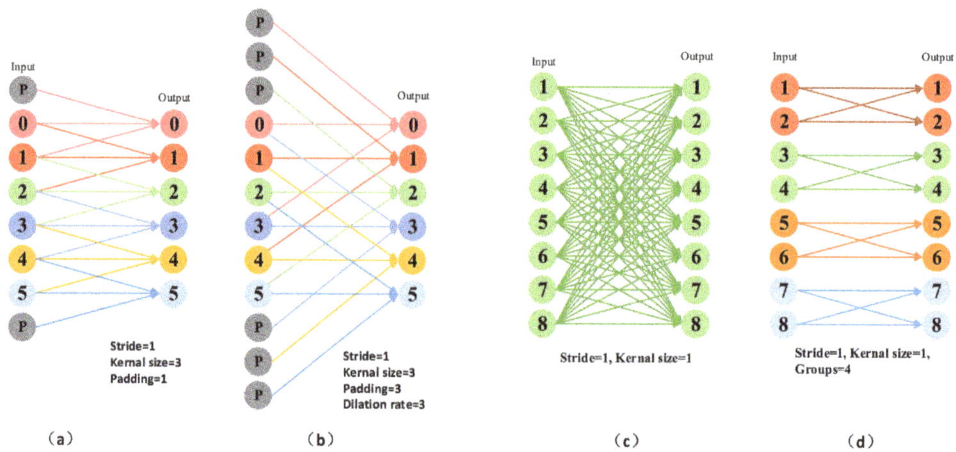

Figure 4. Different convolution visualization results. (**a**) is standard convolution, kernel size is 3 × 3 with padding 1, and the stride is 1. (**b**) represents atrous convolution, kernel size is 3 × 3 with dilation rates 3, padding is 3, and stride is 1. (**c**) is the standard convolution, and the kernel size is 1 × 1. (**d**) shows the grouped convolution is split into four groups.

We use a one-dimensional expansion to demonstrate the different convolutions used in our CEM components. The 3 × 3 convolution allows for the efficient extraction of local features, and the underlying architecture is optimized for it. The 1 × 1 convolution mainly serves to integrate information between feature channels. The advantage of atrous convolution is that it can increase the receptive field without reducing the feature resolution. The characteristic of grouped convolution is that the computational complexity decreases with the number of groups increasing.

Having understood the purpose and core components of CEM, we describe it in a more rigorous mathematical formulation and explain why it is beneficial for the network.

Specifically, let us first consider an input feature $X \in \mathbb{R}^{C\times H\times W}$, where C, H, and W indicate the channel number, spatial height, and width. CEM performs three parallel convolutions with different atrous rates to enlarge the receptive field without adding extra kernel parameters. The formula for the three parallel atrous convolutions with different atrous rates is as follows:

$$O_d = \sum_{k=1}^{N} \sum_{a=1}^{N} D_{k,2a-1}(X), \tag{1}$$

In Equation (1), where $X \in \mathbb{R}^{C\times H\times W}$ is the input feature, $O_d \in \mathbb{R}^{C\times H\times W}$ is the output feature, where $D_{k,2a-1}(\cdot)$ means the atrous convolution, k, a denotes the filter size and the dilation rates, respectively, and N represents the number of atrous convolutions. We add three different sets of $D_{k,2a-1}(\cdot)$ to obtain the intermediate output O_d.

Considering that the atrous convolution loses detailed information, we add different groups of convolutions to supplement the different levels of detailed information. In addition, we also apply different sizes of convolution kernels to obtain different spatial resolutions, effectively alleviating object scale variation in UAV images. Grouped convolution is lightweight and efficient, adding a small amount of extra computation to improve performance. We use three levels of different kernel sizes: 3×3, 5×5, and 7×7, and the corresponding grouping depths are 1, 4, and 8, respectively. It can be formulated as follows:

$$O_g = Concate\left(\left[G_{k,g}(X), G_{k,g}(X), G_{k,g}(X)\right]\right), \tag{2}$$

$G_{k,g}(\cdot) \in \mathbb{R}^{C/3\times H\times W}$ is grouped convolution, k and g correspondingly denote the filter size and the split into different groups, and $Concate(\cdot)$ means the concatenation operation. O_g is the concatenation of grouped convolution operations of different groups.

Finally, we concatenate O_d and O_g to obtain semantically rich output features. The CEM formula is defined as:

$$O = conv(Concate([O_g, O_d])), \tag{3}$$

$conv(\cdot)$ is 1×1 convolution. We apply a 1×1 convolution to reduce the feature maps to the same as the X. Note that in this architecture, when we connect the input and output, there are multiple branching paths to obtain different levels of receptive fields. Our CEM uses a sizeable receptive field to capture semantic information and a small receptive field to capture location information. Therefore, the CEM module can effectively deal with object scale changes.

3.2. Feature Alignment Module

We noticed that the main reason for the poor detection of small objects in aerial image detectors is that the location information obtained by the fusion of adjacent feature layers is inaccurate, and small objects are susceptible to location deviation. To this end, we introduce the FAM to add modulated deformable convolution and channel attention based on FPN.

First, let us review the FPN structure, as shown in Figure 5. In FPN, high-level features use up-sampling operations and fuse with the feature maps at low-level features, enabling the low-level feature to obtain high-level semantic information. The resulting features are naturally endowed with different levels of contextual information. However, the significant problem is that merging adjacent layer features without special processing destroys feature consistency at scale and semantic levels. We introduce the FAM module to solve this problem. The structure of FAM is shown in Figure 6.

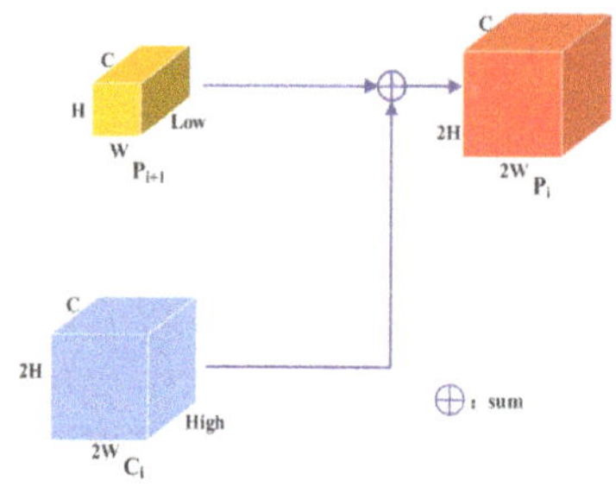

Figure 5. The structure of FPN.

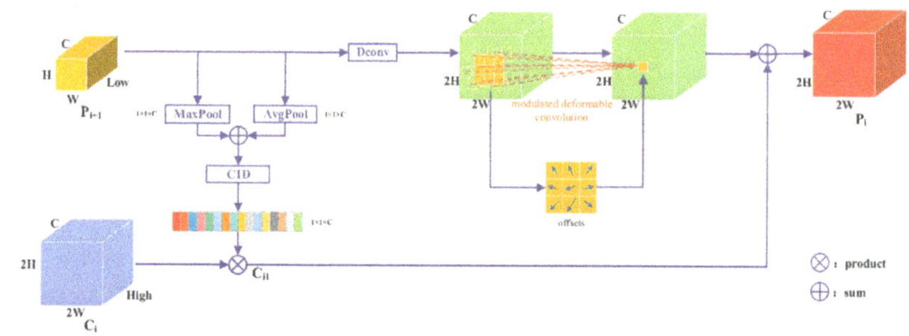

Figure 6. The structure of FAM.

Next, we introduce the core parts of FAM in detail. Our survey found that traditional convolution cannot make adaptive changes when adjacent features are fused due to fixed operation rules. Deformable convolutions [54] learn offsets for the convolution sampling points with freeform sampling grids, and the aim is to make the receptive field adaptively zoomed. Due to this characteristic, it is widely used for feature alignment or dealing with dense spatial transformations and can learn according to the actual scene of the data. Formally, the deformable convolution operation is defined as follows:

$$Y(P) = \sum_{n=1}^{K} W_n \times X(P + P_n + \Delta P_n), \tag{4}$$

where $X \in \mathbb{R}^{C \times H \times W}$ is the input feature map, $Y(P) \in \mathbb{R}^{C \times H \times W}$ is the out feature map, and K and n refer to the size of the kernel and the index, respectively. W_n, P, and P_n are the nth weight, indices of the center, and the nth prespecified offset, respectively. ΔP_n is the additional learnable offset. Since the learnable offset ΔP_n is typically fractional, we use the bilinear interpolation difference to obtain the position of the ΔP_n in the feature map.

To further enhance the feature alignment ability, modulated deformable convolution [55] adds an adjustment mechanism based on deformable convolution, which can effectively adjust the offset of the perceptual input features. The modulated deformable convolution is defined in Equation (5):

$$Y(P) = \sum_{n=1}^{K} W_n \times X(P + P_n + \Delta P_n) \cdot \Delta m_n, \tag{5}$$

where Δm_n is the modulation scalar for the nth location. FAM uses modulated deformable convolution to learn offsets after the up-sampling of high-level features.

Furthermore, we pass the channel information of high-level features to low-level features through channel attention to inject the low-level features with semantic information. SENet pioneered channel attention, with consists of two parts: squeeze and excitation.

SENet uses global average pooling to recalibrate the channel-wise relationship adaptively. This operation can then be expressed as:

$$Y_i = (sigmoid(W_2 \times ReLU(W_1 F_{avg}(X)))) \times X, \quad (6)$$

$$k = \left| \frac{\log_2(C)}{\gamma} + \frac{b}{\gamma} \right|, \quad (7)$$

where W_1 and W_2 represent the fully connected layers, $Y_i \in \mathbb{R}^{C \times H \times W}$ is the result of the channel attention output, $F_{avg}(\cdot)$ is global average pooling, and sigmoid represents activation function and aim to normalize the data. SENet uses two fully connected layers to learn channel weights. In order to reduce the complexity of the model, dimensionality reduction operations are performed, which bring some negative effects. We use one-dimensional convolution of size k instead of full connection, and k represents the range of channel learning. The size of k can be obtained by Formula (7), where C is the channel number, and γ and b are the two adjustable variables in the non-linear mapping. We set γ and b to 2 and 1, respectively.

$$C_{i1} = (sigmoid(C1D_k\ (F_{avg}\ (P_{i+1}) + F_{max}\ (P_{i+1})))) \times C_i, \quad (8)$$

where $C1D_k(\cdot)$ is the one-dimensional convolution of size k, $C_i \in \mathbb{R}^{C \times 2H \times 2W}$ is a high-level feature, $P_{i+1} \in \mathbb{R}^{C \times H \times W}$ is a low-level feature, and F_{max} is global max pooling. $C_{i1} \in \mathbb{R}^{C \times 2H \times 2W}$ is the result of the attention output. Different from FPN, our FAM uses learnable deconvolution to enlarge feature map resolution instead of up-sampling and then uses modulated deformable convolution adaptively learned feature offset to align spatial features. FAM method can be written as:

$$P_i = Y(Deconv(P_{i+1})) + conv(C_{i1}), \quad (9)$$

where C_i and P_{i+1} are the inputs of two adjacent feature layers, $Y(\cdot)$ represents the modulated deformable convolution, $P_i \in \mathbb{R}^{C \times H \times W}$ is the output of FEM, and $Deconv(\cdot)$ means deconvolution. We perform the $Deconv(\cdot)$ operation on the low resolution P_{i+1} to obtain higher-resolution features. FAM suppresses inconsistencies in gradient computation by modulating deformable convolution before feature aggregation. In addition, we obtain the channel attention of high-level semantic features to low-level features.

4. Experiments

In this section, we first introduce the dataset and implementation details. Then, we conduct ablation studies to prove the effectiveness of each model. In addition, we compare the proposed SEPNet with other methods and provide detailed and abundant analyses of the experiments provided to understand our framework better. Finally, we present a visual analysis of the detection results, which shows that the problems of small objects and significant scale changes in SEPNet are indeed alleviated.

4.1. The Dataset and Evaluation Metrics

To evaluate the proposed method, we conduct quantitative experiments on aerial image datasets VisDrone 2019 and PASCAL VOC 2007/12, respectively.

VisDrone2019: The drone platform acquires the dataset and contains different weather and light conditions representing common scenarios in our daily lives. The image scale of the dataset is approximately 2000 × 1500 pixels. The VisDrone 2019 has 10 object classes and consists of 6471 training images, 548 validation images, and 1610 testing images.

PASCAL VOC2007/12: The PASCAL VOC 2007/12 is the standard object detection dataset with 20 object classes and includes 22,136 training images and 5000 validation images. We train models on PASCAL VOC2007/12 train-val sets and report results on the VOC2007 test set with a total of 4952 images.

For VisDrone, we follow the standard MS COCO [56] protocol where average precision (AP) is measured by averaging multiple intersection over union (IoU) [57] thresholds to evaluate the performance. We use AP, AP50, AP75, APs (area < 32^2 pixels), APm (32^2 < area < 96^2 pixels), and APl (area > 96^2 pixels) as the metrics to measure precision; AP50 and AP75 are computed at the single complete intersection over union (CIoU) [58] threshold 0.5 and 0.75 overall categories. For PASCAL VOC, we use mean of average precision (mAP) to evaluate our model, and the CIoU threshold is set to 0.5.

4.2. Data Augmentation

Data augmentation only processes the input image without changing the network structure or adding extra parameters. Therefore, it hardly adds extra computation and can be applied to various computer vision tasks. In SEPNet, we use a combination of geometric augmentations (such as horizontal flipping, random cropping of the images, resizing, etc.) and photometric augmentations (such as brightness adjustment, contrast adjustment, saturation adjustment, and adding noise to images) in data augmentation. In addition, we follow the training practices below: Most images are large in VisDrone, resulting in the disappearance of small target features after down-sampling by the deep network. Therefore, input images are uniformly divided into four patches without overlapping during training and inference. Each patch is fed into the network for further precise detection. Meanwhile, the original images are also forwarded to the network to detect large objects and prevent the clipped target from being undetectable. Finally, the detection results of each patch and the original image are combined to obtain the final result. The image is divided into a four patches strategy, as shown in Figure 7.

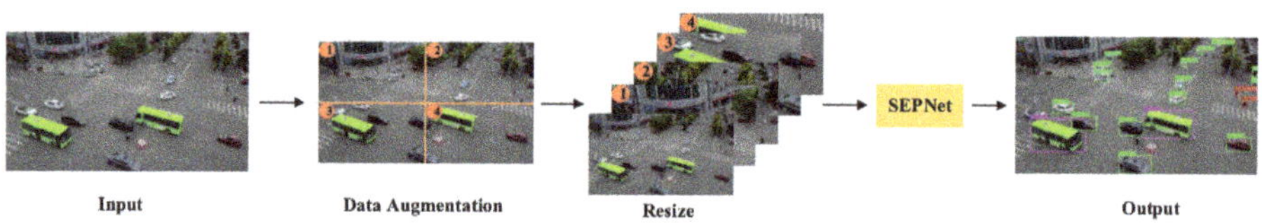

Figure 7. In the data augmentation method, input images are uniformly divided into 4 patches without overlapping.

4.3. Implementation Details

For most experiments, we trained and evaluated the models on a machine with 1 NVIDIA RTX 3090 GPU, CUDA 11.1, and implemented the proposed SEPNet on Pytorch 1.70. Our experiments were conducted on VisDrone and PASCAL VOC datasets, respectively. We selected object detectors RetinaNet as our baseline model, and ResNet pretrained in ImageNet was used as the backbone.

In the training phase, we applied the stochastic gradient descent (SGD) optimizer with a batch size of 32 images per GPU. Weight decay and momentum were set to 0.0005 and 0.9. We trained our models for 150 epochs, with the initial learning rate set to 0.001, decaying by 10 separately at epochs 90 and 120, and the resolution size of the input image was set to 800 × 800. On PASCAL VOC, the epochs were set to 200, and the learning rate was set to 0.005 and decreased 0.1 times after the 90th and 150th rounds.

The loss function for classification was the focal loss [15], and the smooth L1 [59] was used for regression. The overall training objective was:

$$Loss = \frac{1}{N_{POS}} \sum_i L^i_{cls} + \frac{1}{N_{POS}} \sum_j L^j_{reg}, \tag{10}$$

where N is the number of matched positive samples, L^i_{cls} and L^j_{reg} stand for the classification loss and regression loss, respectively, N_{POS} is the number of positive samples, i are all positive and negative samples, and j are all positive samples. For data augmentation, we adopted the same method as that in Section 4.2. During the inference process, bounding box regression was the crucial step. IoU measures the positional relationship between the predicted box and the ground-truth box. However, IoU has the problems of slow convergence and inaccurate regression when detecting small objects. Therefore, IoU was replaced by CIoU loss. Unlike IoU, CIoU considers bounding box overlap size, center point distance, and aspect ratio. IoU is defined as shown in equation:

$$IoU = \frac{|A \cap B|}{|A \cup B|}, \tag{11}$$

where A and B are the ground-truth box and predicted box. Penalty term can be represented as:

$$R_{CIoU} = \frac{\rho^2(b, b^{gt})}{c^2} + \alpha v, \tag{12}$$

where b and b^{gt} are the central points of the predicted box and ground-truth box, $\rho(\cdot)$ denotes the Euclidean distance, and c is the diagonal length of the smallest enclosing box covering the two boxes. v measures the consistency of the aspect ratio as follows:

$$v = \frac{4}{\pi^2}\left(\arctan\frac{w^{gt}}{h^{gt}} - \arctan\frac{w}{h}\right)^2, \tag{13}$$

where w^{gt} and h^{gt} are the width and height of the ground-truth box, and w and h denote the width and height of the predicted box. α is a positive trade-off parameter, as seen in Equation (14):

$$\alpha = \frac{v}{(1 - IoU) + v}. \tag{14}$$

The loss function can be defined as:

$$CIoU = 1 - IoU + R_{CIoU}. \tag{15}$$

4.4. Ablation Study

In this section, we conducted ablation experiments to analyze the effectiveness of each component and compared them with the baseline model RetinaNet on the VisDrone dataset.

We gradually applied data augmentation, CEM, and FAM to the baseline model to verify its effectiveness and compare it with the baseline model. At the same time, we analyzed why each component can improve network performance.

Ablation study results on the VisDrone test set are shown in Table 1, and the IoU threshold for non-maximum suppression was set to 0.5. We can observe that our method significantly improved object detection performance, especially for small objects. Specifically, data enhancement saw a 1.1% AP increase without introducing additional parameters; CEM and FAM improved the baseline method by 0.6% AP and 0.5% AP and introduced 2.3M and 2.1M parameters, respectively. Combining three strategies improved baseline model detection performance from 21.3% to 23.5% AP when using ResNet-50 as the backbone. In addition, our strategy significantly improved small object detection by 2.2% AP, only adding 4.4M parameters. The above experimental results demonstrate that the CEM component can effectively supplement contextual information of deep networks to improve scale variation detection performance. It was also verified that the FAM embedded in the baseline model is helpful for the fusion of adjacent features and effectively improves the detection results of small objects. At the same time, our data augmentation strategy can effectively avoid the problem of losing small object information during down-sampling, so it can improve the detection accuracy of small objects.

Table 1. Ablation study results on VisDrone. RetinaNet was selected as the baseline, and we gradually added our components to the baseline to verify the effectiveness of each component. "DA" represents the data augmentation.

Backbone	DA	CEM	FAM	AP	AP_s	AP_m	AP_l	Params
		Baseline		21.3	11.2	32.2	47.5	37.8 M
	√			22.4	12.3	32.9	48.1	37.8 M
ResNet-50		√		21.9	11.8	32.7	48.3	40.1 M
			√	21.8	11.9	32.5	47.7	39.9 M
	√	√	√	23.5	13.5	33.8	48.9	42.2 M

To verify the generalization ability of proposed method, two components were trained and tested on the PASCAL VOC dataset. We gradually added each component to the baseline model and analyzed the accuracy and number of parameter relationships using ResNet-19, ResNet-50, ResNet-101, and ResNet-152 as the backbone network, respectively. The experimental results are shown in Figure 8.

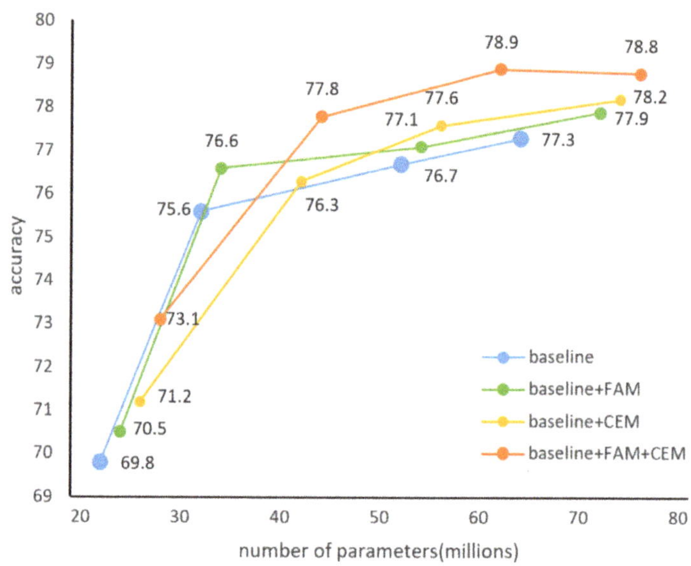

Figure 8. Analysis of the relationship between accuracy and number of parameters in the PASCAL VOC test set.

In the PASCAL VOC test set, for ResNet-19 as the backbone network, the detection accuracy was increased by 1.4% and 0.7% after adding CEM and FAM components, respectively. Combining the use of CEM and FAM components, accuracy was increased by 3.3%, and the number of parameters was increased by 4.4M. For ResNet-50 as the backbone network, combining two components improved baseline model detection performance from 75.6% to 77.8%. For ResNet-101 as the backbone network, each component also improved the model's accuracy. It is worth noting that when the backbone network was switched from ResNet-101 to ResNet-152, combining the two components into the baseline model, the accuracy no longer increased.

These experiments prove that our two components achieve significant improvements by introducing fewer additional parameters and can adapt to different datasets, indicating their effectiveness and generality.

4.5. Comparisons with Other Methods

Regarding VisDrone and PASCAL VOC, we compared the performance of our SEPNet with other popular one-stage detectors and two-stage detectors. The experimental results are shown in Table 2.

Table 2. Comparison of our method with other state-of-the-art methods for object detection on the VisDrone test set.

Method	Backbone	AP	AP_{50}	AP_{75}
One-stage:				
RetinaNet [15]	Res101	11.8	21.4	11.6
CenterNet [29]	ResNext-101-64x4d	14.2	19.3	15.5
RefineDet512 [60]	VGG-16	14.9	28.8	14.1
FPN [35]	VGG-16	16.5	32.2	14.9
CornerNet [26]	Hourglass-104	17.4	34.1	15.8
Two-stage:				
Cascade R-CNN [61]	ResNet101	16.1	31.9	15.0
Light-RCNN [62]	ResNet101	16.5	32.8	15.1
Ours:				
SEPNet	ResNext-101	18.9	34.8	16.7

In this experiment, we used the training set of VisDrone for training and the test set for validation. Table 2 shows the comparison of our proposed method with some current popular methods. Our SEPNet outperformed Cascade R-CNN and Light-RCNN by 2.8% and 2.4%, respectively. Compared with existing one-stage methods, our SEPNet outperformed CornerNet by 1.5%, 0.7%, and 0.9% on AP, AP50, and AP75, respectively.

In addition to the contrast experiments on VisDrone2019, we also conducted experiments on PASCAL VOC to verify the generalization of SEPNet. We reported results on the PASCAL VOC test set. The experimental results are shown in Table 3.

Table 3. Results on the PASCAL VOC test set. Comparison with the other state-of-the-art methods, ours is better.

Method	Backbone	Train	Test	mAP/%
One-stage:				
RFBNet [63]	VGG16	VOC2007 + 2012	VOC2007	76.8
SSD300 [28]	VGG16	VOC2007 + 2012	VOC2007	77.1
SSD512 [28]	VGG16	VOC2007 + 2012	VOC2007	78.5
DSSD [64]	ResNet-101	VOC2007 + 2012	VOC2007	78.6
CenterNet [29]	ResNet-101	VOC2007 + 2012	VOC2007	78.7
YOLO v3 [65]	Darknet-53	VOC2007 + 2012	VOC2007	79.4
FCOS [30]	ResNet-101	VOC2007 + 2012	VOC2007	80.1
CenterNet [29]	DLA	VOC2007 + 2012	VOC2007	80.7
Two-stage:				
Fast R-CNN [59]	VGG16	VOC2007 + 2012	VOC2007	70.0
Faster R-CNN [27]	ResNet-101	VOC2007 + 2012	VOC2007	76.4
R-FCN [66]	ResNet-101	VOC2007 + 2012	VOC2007	80.5
Ours:				
SEPNet	ResNet-101	VOC2007 + 2012	VOC2007	81.5

We compared our SEPNet with popular detectors in the PASCAL VOC test set. The experimental results show that our SEPNet outperforms the advanced one-stage detection algorithms DSSD and CenterNet by 2.9% and 0.8%, respectively. Compared to the two-stage algorithms Faster R-CNN and R-FCN, our SEPNet also increased by 5.1% and 1%, respectively. The experimental observations on the PASCAL VOC test dataset maintained a consistent improvement with the experimental results on the VisDrone dataset, which demonstrates that our method has similar generalization ability to other datasets and can be applied to different scenes.

To further demonstrate the effectiveness of the proposed SEPNet more intuitively, we present some visualization results in Figures 9 and 10. We compared our methods with RetinaNet. RetinaNet can only detect large objects close to the camera and misses small objects far away. Compared with RetinaNet, we proposed that SEPNet could detect not only large objects in the image but also small objects far from the camera. This indicates that our SEPNet can capture objects of different scales more accurately while paying more attention to the small object region rather than the surrounding background. It can be seen from the visualization results that SEPNet can solve the problem of missed detection of small objects well. It can also be seen that SEPNet can adapt well to object scale changes and improve detection accuracy.

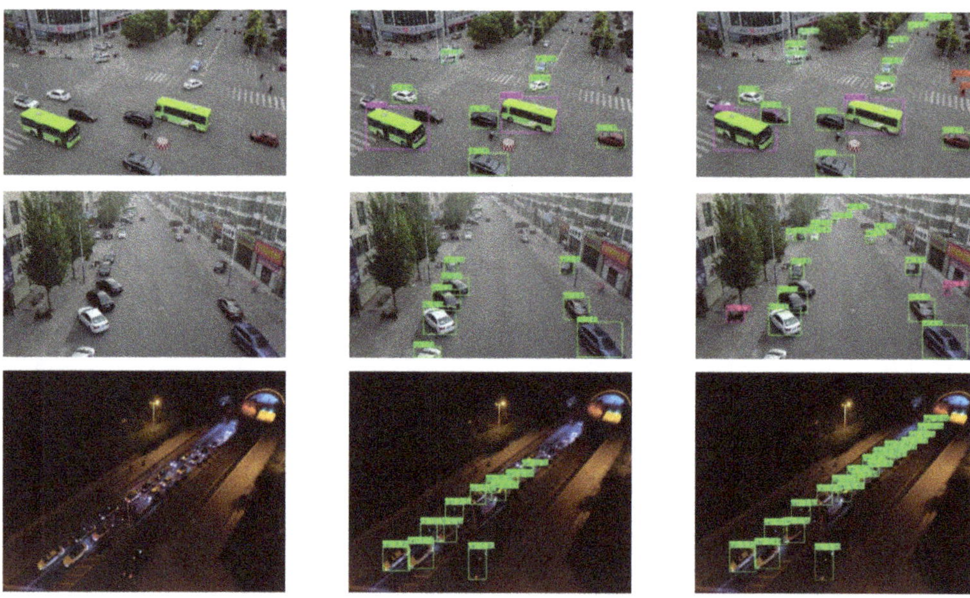

(**a**) Original image (**b**) RetinaNet (**c**) Ours

Figure 9. Visualization of detection results on VisDrone. Our SEPNet predicts more refined boundaries and learns more detailed information.

(**a**) Original image (**b**) RetinaNet (**c**) Ours

Figure 10. Continuation. Other visualization examples of detection results on PASCAL VOC.

5. Conclusions

This paper proposes a one-stage scale enhancement pyramid network (SEPNet) to solve small object and extreme scale variation problems in UAV images. The proposed method consists of two main core components: CEM maintains deep features with rich contextual information, avoiding the loss of small target information and FAM addresses the lack of effective communication between adjacent features. Our results show that the proposed components offer significant improvements. Furthermore, our SEPNet exhibits good generalization in different datasets. In future work, we will focus on designing lightweight structures for models to be deployed into embedded devices.

Author Contributions: Conceptualization, J.S.; methodology, J.S.; software, J.S.; validation, J.S.; formal analysis, H.G.; investigation, J.S.; resources, J.Y.; data curation, J.Y. and X.W.; writing—original draft preparation, J.S.; writing—review and editing, X.W. and J.Y.; visualization, X.W.; supervision, H.G. and J.Y.; project administration, H.G.; funding acquisition, H.G. All authors have read and agreed to the published version of the manuscript.

Funding: This work was funded by the Liaoning Province Higher Education Innovative Talents Program, grant number LR2019058, Liaoning Province Joint Open Fund for Key Scientific and Technological Innovation Bases, grant number 2021-KF-12-05, and the China Postdoctoral Science Foundation, grant number 2022M712756.

Data Availability Statement: The data presented in this study are available on request from the corresponding author.

Acknowledgments: The authors would like to acknowledge support from the following projects: Liaoning Province Higher Education Innovative Talents Program Support Project (Grant No. LR2019058), Liaoning Province Joint Open Fund for Key Scientific and Technological Innovation Bases (Grant No.2021-KF-12-05), and China Postdoctoral Science Foundation (Grant No. 2022M712756).

Conflicts of Interest: The authors declare no conflict of interest.

References

1. Yu, J.; Gao, H.; Sun, J.; Zhou, D.; Ju, Z. Spatial Cognition-driven Deep Learning for Car Detection in Unmanned Aerial Vehicle Imagery. *IEEE Trans. Cogn. Dev. Syst.* **2021**, 1. [CrossRef]
2. Koyun, O.C.; Keser, R.K.; Akkaya, I.B.; Töreyin, B.U. Focus-and-Detect: A small object detection framework for aerial images. *Signal Process. Image Commun.* **2022**, *104*, 116675. [CrossRef]
3. Vieira-E-Silva, A.L.B.; Felix, H.D.C.; Chaves, T.D.M.; Simoes, F.P.M.; Teichrieb, V.; dos Santos, M.M.; Santiago, H.D.C.; Sgotti, V.A.C.; Neto, H.B.D.T.L. STN PLAD: A Dataset for Multi-Size Power Line Assets Detection in High-Resolution UAV Images. In Proceedings of the IEEE Conference on SIBGRAPI Conference on Graphics, Patterns and Images, Gramado, Rio Grande do Sul, Brazil, 18–22 October 2021. [CrossRef]
4. Butte, S.; Vakanski, A.; Duellman, K.; Wang, H.; Mirkouei, A. Potato crop stress identification in aerial images using deep learning-based object detection. *Agron. J.* **2021**, *113*, 3991–4002. [CrossRef]
5. Dewangan, A.; Pande, Y.; Braun, H.-W.; Vernon, F.; Perez, I.; Altintas, I.; Cottrell, G.W.; Nguyen, M.H. FIgLib & SmokeyNet: Dataset and deep learning model for real-time wildland fire smoke detection. *Remote Sens.* **2022**, *14*, 1007. [CrossRef]
6. Zhang, R.; Li, H.; Duan, K.; You, S.; Liu, K.; Wang, F.; Hu, Y. Automatic detection of earthquake-damaged buildings by integrating UAV oblique photography and infrared thermal imaging. *Remote Sens.* **2020**, *12*, 2621. [CrossRef]
7. He, K.; Zhang, X.; Ren, S.; Sun, J. Deep residual learning for image recognition. In Proceedings of the 2016 IEEE Conference on Computer Vision and Pattern Recognition, Las Vegas, NV, USA, 27–30 June 2016; pp. 770–778. [CrossRef]
8. Huang, G.; Liu, Z.; Van Der Maaten, L.; Weinberger, K.Q. Densely connected convolutional networks. In Proceedings of the 2017 IEEE Conference on Computer Vision and Pattern Recognition, Honolulu, HI, USA, 21–26 July 2017; pp. 2261–2269. [CrossRef]
9. Liu, Z.; Mao, H.; Wu, C.-Y.; Feichtenhofer, C.; Darrell, T.; Xie, S. A ConvNet for the 2020s. In Proceedings of the 2022 IEEE/CVF Conference on Computer Vision and Pattern Recognition (CVPR), New Orleans, LA, USA, 18–24 June 2022; pp. 11966–11976. [CrossRef]
10. Yu, J.; Gao, H.; Zhou, D.; Liu, J.; Gao, Q.; Ju, Z. Deep Temporal Model-Based Identity-Aware Hand Detection for Space Human-Robot Interaction. *IEEE Trans. Cybern.* **2021**, *52*, 13738–13751. [CrossRef]
11. Chen, M.; Zheng, Z.; Yang, Y.; Chua, T.-S. PiPa: Pixel-and Patch-wise Self-supervised Learning for Domain Adaptative Semantic Segmentation. *arXiv* **2022**, arXiv:2211.07609.
12. Sun, G.; Liu, Y.; Ding, H.; Probst, T.; Van Gool, L. Coarse-to-fine feature mining for video semantic segmentation. In Proceedings of the 2022 IEEE/CVF Conference on Computer Vision and Pattern Recognition (CVPR), New Orleans, LA, USA, 18–24 June 2022. [CrossRef]
13. Yu, J.; Gao, H.; Chen, Y.; Zhou, D.; Liu, J.; Ju, Z. Deep Object Detector with Attentional Spatiotemporal LSTM for Space Human–Robot Interaction. *IEEE Trans. Human-Machine Syst.* **2022**, *52*, 784–793. [CrossRef]
14. Bochkovskiy, A.; Wang, C.; Liao, H. YOLOv4: Optimal speed and accuracy of object detection. *arXiv* **2020**, arXiv:2004.10934.
15. Lin, T.Y.; Goyal, P.; Girshick, R.; He, K.; Dollar, P. Focal loss for dense object detection. In Proceedings of the 2017 IEEE International Conference on Computer Vision, Venice, Italy, 22–29 October 2017.
16. Everingham, M.; Van Gool, L.; Williams, C.K.I.; Winn, J.; Zisserman, A. The pascal visual object classes (voc) challenge. *Int. J. Comput. Vis.* **2010**, *88*, 303–338. [CrossRef]
17. Zhu, P.; Du, D.; Wen, L.; Bian, X.; Ling, H.; Hu, Q.; Peng, T.; Zheng, J.; Wang, X.; Zhang, Y.; et al. VisDrone-vid2019: The vision meets drone object detection in video challenge results. In Proceedings of the IEEE/CVF International Conference on Computer Vision Workshops, Seoul, Republic of Korea, 27–28 October 2019. [CrossRef]
18. Li, C.; Yang, T.; Zhu, S.; Chen, C.; Guan, S. Density map guided object detection in Aerial images. In Proceedings of the IEEE/CVF Conference on Computer Vision and Pattern Recognition Workshops, Seattle, WA, USA, 14–19 June 2020. [CrossRef]
19. Hong, S.; Kang, S.; Cho, D. Patch-Level Augmentation for Object Detection in Aerial Images. In Proceedings of the IEEE/CVF International Conference on Computer Vision Workshops, Seoul, Republic of Korea, 27–28 October 2019. [CrossRef]
20. Fan, J.; Bocus, M.J.; Hosking, B.; Wu, R.; Liu, Y.; Vityazev, S.; Fan, R. Multi-Scale Feature Fusion: Learning Better Semantic Segmentation for Road Pothole Detection. In Proceedings of the IEEE International Conference on Autonomous Systems (ICAS), Montreal, QC, Canada, 11–13 August 2021. [CrossRef]
21. Luo, Y.; Cao, X.; Zhang, J.; Guo, J.; Shen, H.; Wang, T.; Feng, Q. CE-FPN: Enhancing channel information for object detection. *Multimedia Tools Appl.* **2022**, *81*, 30685–30704. [CrossRef]
22. Chen, L.; Papandreou, G.; Schroff, F.; Hartwig, A. Rethinking Atrous Convolution for Semantic Image Segmentation. *arXiv* **2017**, arXiv:1706.05587.
23. Cao, J.X.; Chen, Q.; Guo, J.; Shi, R. Attention-guided context feature pyramid network for object detection. *arXiv* **2020**, arXiv:2005.11475.
24. Xiao, J.S.; Zhao, T.; Yao, Y.T.; Yu, Q.Z.; Chen, Y.H. Context augmentation and feature refinement network for tiny object detection. *ICRL* **2022**. submitted.
25. Lian, J.; Yin, Y.; Li, L.; Wang, Z.; Zhou, Y. Small object detection in traffic scenes based on attention feature fusion. *Sensors* **2021**, *21*, 3031. [CrossRef]
26. Law, H.; Deng, J. CornerNet: Detecting Objects as Paired Keypoints. *arXiv* **2018**, arXiv:1808.01244. [CrossRef]
27. Ren, S.; He, K.; Girshick, R.; Sun, J. Faster R-CNN: Towards real-time object detection with region proposal networks. *IEEE Trans. Pattern Anal. Mach. Intell.* **2015**, *39*, 1137–1149. [CrossRef] [PubMed]

28. Liu, W.; Anguelov, D.; Erhan, D.; Szegedy, C.; Reed, S.; Fu, C.Y.; Berg, A.C. SSD: Single shot multibox detector. In Proceedings of the European Conference on Computer Vision, Amsterdam, The Netherlands, 11–14 October 2016. [CrossRef]
29. Zhou, X.Y.; Wang, D.Q.; Krähenbühl, P. Objects as points. *arXiv* **2019**, arXiv:1904.07850.
30. Tian, Z.; Shen, C.; Chen, H.; He, T. FCOS: Fully convolutional one-stage object detection. In Proceedings of the 2019 IEEE/CVF international conference on computer vision, Seoul, Republic of Korea, 27 October–2 November 2019. [CrossRef]
31. Zhu, C.; He, Y.; Savvides, M. Feature selective anchor-free module for single-shot object detection. In Proceedings of the 2019 IEEE/CVF Conference on Computer Vision and Pattern Recognition, Long Beach, CA, USA, 15–20 June 2019. [CrossRef]
32. Kong, T.; Sun, F.; Liu, H.; Jiang, Y.; Li, L.; Shi, J. FoveaBox: Beyond anchor-based object detector. *IEEE Trans. Image Process.* **2020**, *29*, 7389–7398. [CrossRef]
33. Tong, K.; Wu, Y. Deep learning-based detection from the perspective of small or tiny objects: A survey. *Image Vis. Comput.* **2022**, *104471*. [CrossRef]
34. Min, K.; Lee, G.-H.; Lee, S.-W. Attentional feature pyramid network for small object detection. *Neural Netw.* **2022**, *155*, 439–450. [CrossRef]
35. Lin, T.Y.; Dollár, P.; Girshick, R.; He, K.; Hariharan, B.; Belongie, S. Feature pyramid networks for object detection. In Proceedings of the IEEE Conference on Computer Vision and Pattern Recognition, Honolulu, HI, USA, 21–26 July 2017. [CrossRef]
36. Liu, S.; Qi, L.; Qin, H.; Shi, J.; Jia, J. Path aggregation network for instance segmentation. In Proceedings of the 2018 IEEE/CVF Conference on Computer Vision and Pattern Recognition, Salt Lake City, UT, USA, 18–23 June 2018. [CrossRef]
37. Tan, M.; Pang, R.; Le, Q.V. EfficientDet: Scalable and efficient object detection. In Proceedings of the IEEE/CVF Conference on Computer Vision and Pattern Recognition (CVPR), Seattle, WA, USA, 13–19 June 2020. [CrossRef]
38. Jiang, Y.Q.; Tan, Z.Y.; Wang, J.Y.; Sun, X.y.; Lin, M.; Li, H. GiraffeDet: A heavy-neck paradigm for object detection. *arXiv* **2022**, arXiv:2202.04256.
39. Hong, M.; Li, S.; Yang, Y.; Zhu, F.; Zhao, Q.; Lu, L. SSPNet: Scale selection pyramid network for tiny person detection from UAV images. *IEEE Geosci. Remote Sens. Lett.* **2021**, *19*, 1–5. [CrossRef]
40. Li, Y.; Chen, Y.; Wang, N.; Zhang, Z.-X. Scale-aware trident networks for object detection. In Proceedings of the IEEE/CVF International Conference on Computer Vision, Seoul, Republic of Korea, 27 October–2 November 2019. [CrossRef]
41. Huang, S.; Lu, Z.; Cheng, R.; He, C. FaPN: Feature-aligned pyramid network for dense image prediction. In Proceedings of the IEEE/CVF International Conference on Computer Vision, Montreal, QC, Canada, 10–17 October 2021. [CrossRef]
42. Vaswani, A.; Shazeer, N.; Parmar, N.; Uszkoreit, J.; Jones, L.; Gomez, A.N.; Kaiser, Ł.; Polosukhin, I. Attention is all you need. In Proceedings of the 31st International Conference on Neural Information Processing Systems, Long Beach, CA, USA, 4–9 December 2017; pp. 6000–6010. [CrossRef]
43. Liu, Z.; Lin, Y.; Cao, Y.; Hu, H.; Wei, Y.; Zhang, Z.; Lin, S.; Guo, B. Swin transformer: Hierarchical vision transformer using shifted windows. In Proceedings of the IEEE/CVF International Conference on Computer Vision, Montreal, QC, Canada, 10–17 October 2021. [CrossRef]
44. Grainger, R.; Paniagua, T.; Song, X.; Wu, T. Learning patch-to-cluster attention in vision transformer. *arXiv* **2022**, arXiv:2203.11987.
45. Hu, J.; Shen, L.; Sun, G. Squeeze-and-excitation networks. In Proceedings of the IEEE Conference on Computer Vision and Pattern Recognition, Glasgow, UK, 1 August 2020.
46. Wang, Q.; Wu, B.; Zhu, P.; Li, P.; Zuo, W.; Hu, Q. ECA-Net: Efficient channel attention for deep convolutional neural networks. In Proceedings of the 2020 IEEE/CVF Conference on Computer Vision and Pattern Recognition, Seattle, WA, USA, 14–19 June 2020. [CrossRef]
47. Li, X.; Wang, W.; Hu, X.; Yang, J. Selective kernel networks. In Proceedings of the 2019 IEEE/CVF Conference on Computer Vision and Pattern Recognition, Long Beach, CA, USA, 15–20 June 2019. [CrossRef]
48. Wang, X.; Girshick, R.; Gupta, A.; He, K. Non local neural networks. In Proceedings of the 2018 IEEE Conference on Computer Vision and Pattern Recognition, Salt Lake City, UT, USA, 18–23 June 2018. [CrossRef]
49. Woo, S.; Park, J.; Lee, J.Y.; Kweon, I.S. CBAM: Convolutional block attention module. In Proceedings of the European Conference on Computer Vision, Munich, Germany, 8–14 September 2018. [CrossRef]
50. Park, J.; Woo, S.; Lee, J.-Y.; Kweon, I.S. Bam: Bottleneck attention module. *arXiv* **2018**, arXiv:1807.06514.
51. Zhang, Q.-L.; Yang, Y.-B. Sa-net: Shuffle attention for deep convolutional neural networks. In Proceedings of the IEEE International Conference on Acoustics, Speech and Signal Processing (ICASSP), Toronto, ON, Canada, 6–11 June 2021. [CrossRef]
52. Chen, L.-C.; Zhu, Y.; Papandreou, G.; Schroff, F.; Adam, H. Encoder-decoder with atrous separable convolution for semantic image segmentation. In Proceedings of the European Conference on Computer Vision, Munich, Germany, 8–14 September 2018. [CrossRef]
53. Duta, I.C.; Liu, L.; Zhu, F.; Shao, L. Pyramidal convolution: Rethinking convolutional neural networks for visual recognition. *arXiv* **2020**, arXiv:2006.11538.
54. Dai, J.; Qi, H.; Xiong, Y.; Li, Y.; Zhang, G.; Hu, H.; Wei, Y. Deformable convolutional networks. In Proceedings of the 2017 IEEE International Conference on Computer Vision, Venice, Italy, 22–29 October 2017. [CrossRef]
55. Zhu, X.; Hu, H.; Lin, S.; Dai, J. Deformable convnets v2: More deformable, better results. In Proceedings of the 2019 IEEE/CVF Conference on Computer Vision and Pattern Recognition, Long Beach, CA, USA, 16–20 June 2019. [CrossRef]
56. Lin, T.Y.; Maire, M.; Belongie, S.; Bourdev, L.; Girshick, R.; Hays, J.; Perona, P.; Ramanan, D.; Zitnick, C.L.; Dollar, P. Microsoft COCO: Common Objects in Context. *arXiv* **2014**, arXiv:1405.0312. [CrossRef]

57. Yu, J.H.; Jiang, Y.N.; Wang, Z.Y.; Cao, Z.; Huang, T. UnitBox: An Advanced Object Detection Network. *arXiv* **2016**, arXiv:1608.01471. [CrossRef]
58. Zheng, Z.H.; Wang, P.; Ren, D.W.; Liu, W.; Ye, R.; Hu, Q.; Zuo, W. Enhancing Geometric Factors in Model Learning and Inference for Object Detection and Instance Segmentation. *arXiv* **2020**, arXiv:2005.03572.
59. Girshick, R. Fast r-cnn. In Proceedings of the 2015 IEEE International Conference on Computer Vision, Santiago, Chile, 7–13 December 2015.
60. Zhang, S.; Wen, L.; Bian, X.; Lei, Z.; Li, S.Z. Single-shot refinement neural network for object detection. In Proceedings of the IEEE conference on computer vision and pattern recognition, Salt Lake City, UT, USA, 18–23 June 2018.
61. Cai, Z.; Vasconcelos, N. Cascade r-cnn: Delving into high quality object detection. In Proceedings of the IEEE conference on computer vision and pattern recognition, Salt Lake City, UT, USA, 18–23 June 2018.
62. Li, Z.; Peng, C.; Yu, G.; Zhang, X.Y.; Deng, Y.D.; Sun, J. Light-head r-cnn: In defense of two-stage object detector. *arXiv* **2017**, arXiv:1711.07264.
63. Liu, S.; Huang, D.; Wang, Y. Receptive field block net for accurate and fast object detection. In Proceedings of the European conference on computer vision (ECCV), Munich, Germany, 8–14 September 2018. [CrossRef]
64. Fu, C.Y.; Liu, W.; Ranga, A.; Tyagi, A.; Berg, A.C. Dssd: Deconvolutional single shot detector. *arXiv* **2017**, arXiv:1701.06659.
65. Redmon, J.; Farhadi, A. Yolov3: An incremental improvement. *arXiv* **2018**, arXiv:1804.02767.
66. Dai, J.; Li, Y.; He, K.; Sun, J. R-FCN: Object detection via region-based fully convolutional networks. *Adv. Neural Inf. Process. Syst.* **2016**, *29*.

MDPI
St. Alban-Anlage 66
4052 Basel
Switzerland
www.mdpi.com

Entropy Editorial Office
E-mail: entropy@mdpi.com
www.mdpi.com/journal/entropy

www.ingramcontent.com/pod-product-compliance
Lightning Source LLC
Chambersburg PA
CBHW040824250926
51850CB00007BA/1257
9783725802920